Structure and Fabric

Part 1

MITCHELL'S BUILDING SERIES

Structure and Fabric
Part 1

Jack Stroud Foster *FRIBA*

B T Batsford Limited London

Longman Scientific & Technical,
Longman Group UK Limited,
Longman House, Burnt Mill, Harlow,
Essex, CM20 2JE, England
and Associated Companies throughout the world.

First published in Great Britain by B T Batsford Ltd 1973
Reprinted 1975
Revised edition 1979
Reprinted 1981
Revised edition 1983
Reprinted 1985
Reprinted 1986
Reprinted 1987
Reprinted 1989
Revised edition 1991
Reprinted by Longman Scientific & Technical 1993

ISBN 0 582 21882 9

British Library Cataloguing in Publication Data
A CIP record for this book is available from the British Library

Printed in Malaysia by TCP

Contents

Acknowledgment

I am grateful to the following for permission to reproduce tables or to use drawings as a basis for illustrations in this volume: *Acier 6*, June 1963, Centre Belgo–Luxembourgeois d'Information de l'Acier; *Building Elements*, R Llewelyn Davies and A Petty, Architectural Press; *Correct Installation of Domestic Solid Fuel Appliances*, W C Moss, Solid Fuel Advisory Service; The British Precast Concrete Federation; The Cement and Concrete Association; The Timber Research and Development Association; Messrs Bawtry Timber Company Limited; Finlock Gutters Limited; Rainham Timber Engineering Company Limited; Messrs R K Harrington and A F M Mendoza.

With the permission of the Controller of H M Stationery Office, I have drawn freely on *Principles of Modern Building*, Volumes 1 and 2 and on *Building Research Establishment Digests* and *Current Papers;* I have based Table 2 on tables in *National Building Studies Special Reports Nos 18* and *21* and have quoted from *The Building Regulations*, 1985. I have also drawn on *British Standard Specifications* and *Codes of Practice* with the permission of the British Standards Institution, from whom official copies may be obtained. I also owe much in Chapters 1 and 2 to my reading of P A Stone's excellent book, *Building Economy*, and in Chapter 3 to my reading of *The Elements of Structure* by W Morgan.

I am grateful to the librarian of the Chartered Institute of Building for providing me with statistics for up-dating material relating to the building industry.

I am also grateful to those who have prepared the illustrations, especially to Jean Marshall, John Green and George Dilks, the latter in particular being responsible for the greater part of the work.

I must also express my appreciation to Thelma M Nye, of the publishers, for her help and patience in seeing the work through to press.

1991 JSF

Preface

The period between the last and present editions of this volume has been one of considerable change and development in many areas related to the subject matter of the book. The revised Building Regulations 1985 and the 1990 Amendments have been published and the London Building By-laws have been abolished to be replaced by the Building Regulations and the Inner London Building Regulations. During this period, too, much research and development has been carried out in the field of constructional techniques and many British Standards and Codes of Practice have been revised and new ones published.

This has necessitated the revision and correction of the text at various points and recasting and adding to it at others. A number of new illustrations have been added to clarify the text and where necessary existing illustrations have been amended.

1991 JSF

1 The nature of buildings and building

Buildings exist to meet a primary physical human need — that of shelter. Shelter for man, his goods, his animals and all the mechanical and electrical equipment he requires for his present-day existence. To this need the whole development of building technology and building techniques is related. In addition to meeting this physical need, buildings and well related groups of buildings may also satisfy man's desire for mental and spiritual satisfaction from his environment. To achieve this buildings must be well designed as well as efficiently constructed.

1.1 THE FUNCTION OF A BUILDING

A shelter is basically a protection from the elements and the function of a building is to enclose space so that a satisfactory internal environment may be created relative to the purpose of the particular building. That is to say, the space within the building must provide conditions appropriate to the activities to take place within it and satisfactory for the comfort and safety of any occupants. Thus the space will be designed in terms of size and shape and in terms of environmental factors such as weather and noise exclusion, and the provision of adequate heat, light and air. The fabric of the building must be designed to ensure that any standards in respect to these are attained.

1.1.1 Functional requirements

The building fabric can be seen, therefore, as the means by which the natural or external environment may be modified to produce a satisfactory internal environment and for this reason it has been called the *environmental envelope*. In fulfilling this function the building and its parts must satisfy certain requirements related to the environmental factors on which the design of the spaces within it is based.[1] These functional requirements are the provision of adequate weather resistance, thermal insulation, sound insulation, light and air. In addition, adequate strength and stability must be provided together with adequate fire protection for the occupants, contents and fabric of the building (see figure 1.1). The importance of any of these will vary with the particular part of the building and with its primary function, but some indication of their relation to the various parts is given in table 1.1. More detailed discussion of this will be found in the following chapters.

1.2 THE NATURE OF BUILDING

Building is concerned with providing in physical form the 'envelopes' to the spaces within buildings and it has been a primary activity of man throughout history. It is now, to a large extent, an erection process in which the products of other industries are assembled; a complex process, more so than for most other products, both organisationally and technically, involving on most jobs many trades and many different operations, the majority of which are carried out on site and subject, therefore, to the hazards of weather.

1.2.1 Building as an organisational process

Organisationally the building process is concerned with the rational and economic use of the resources for building activity — men, materials, machines, money, in order to produce buildings in the quickest and most economic manner. Practically the building process involves two broad and related activities — design and production. The design process is concerned with the size, shape and disposition of the spaces within the building and defined by its fabric and with the nature and form of the building fabric and its services. The production process is concerned with the nature and sequence of the operations which are involved in the erection of the building fabric and through which the resources for building are deployed.

The design of the building largely determines the nature and sequence of the building operations. These in turn will determine the methods which can be adopted in carrying out the operations, and the opera-

[1] These environmental factors are discussed fully in *Mitchell's Building Series: Environment and Services* Peter Burberry and standards relating to them which are normally required to be met, are given.

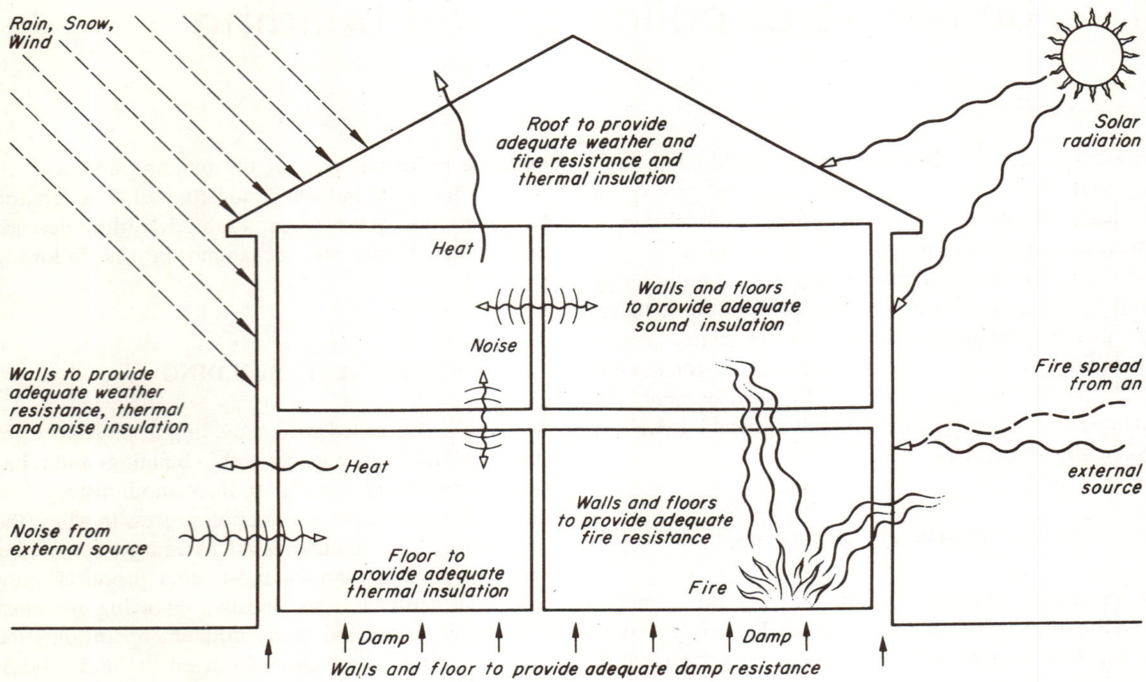

1.1 *Functional requirements of the building fabric*

Element	*Strength and stability*	*Weather resistance*	*Fire resistance*	*Thermal insulation*	*Sound insulation*
External walls					
Loadbearing	★	★	☆	☆	×
Non-loadbearing	×	★	☆	☆	×
Internal walls					
Loadbearing	★	×	☆	×	★
Non-loadbearing	×	×	☆	×	★
Frame	★	×	☆	×	×
Floor					
Ground	★	×	×	☆	×
Upper	★	×	☆	×	☆
Stairs	★	×	★	×	×
Roof	★	★	☆	☆	×

★ usually a critical factor ☆ usually an important factor × not usually an important factor

Table 1.1 *Functional requirements of elements of construction*

tional methods will determine the manner in which the building resources can be deployed. Thus there is a significant relationship between the design of the building and the use of the building resources. The possibility of the rational and economic use of these resources is, therefore, latent in the building design and the implications of every design decision in respect of this must be exposed at the design stage to ensure that such a rational use of resources can be made at the production stage. Such an exposure is often difficult because of the separation existing in general between designer and constructor. The former, being divorced from actual production activity, is not sufficiently aware of the operational significance of many of his decisions; the latter, being divorced from the design process, is not always able to relate his production knowledge and skill to design decisions at a sufficiently early stage. This weakness in the industry has been recognised and attempts to overcome it have been made in building education and in various ways in practice.[1]

1.2.2 Building as a technology

In the past a limited number of available materials resulted in a limited number of structural forms and methods of construction which, after a long period, became fully developed and standardised in practice. These could be, and were, then used on an empirical basis established on their proved performance in use.

This is no longer possible nor, indeed, has it been for a long time past. The introduction of new materials, which is now a continuing and expanding process, with properties and characteristics differing from those of the traditional materials, requires the rapid development of new building techniques and new forms of structure appropriate to the nature of these materials. At the same time it is necessary to develop a better understanding of the older materials so that they may be used more efficiently and effectively. Demands on the building industry require an increase in the productivity of the building process which, amongst other things, may necessitate the development of new techniques. Traditional building materials are bulky and heavy and, therefore, relatively difficult to handle on site and expensive to transport. This has encouraged the search for new, lighter materials which will fulfil the same, or even a greater range of functional requirements than the old. Such problems and many others such as these cannot be solved with the aid of empirical knowledge but require a scientific approach as a basis of investigation and development. For this reason building, of necessity, is being transformed from a craft-based industry into a modern technology with its repository of knowledge based on scientific principles applied to the problems of building, and using scientific methods of investigation and research.

That part of building technology dealing primarily with the design of the fabric of buildings and the manner in which it is put together, is known as *building construction* and draws, in particular, on the sciences of materials and structure, on the environmental sciences and on building economics. In the past this subject was concerned exclusively with the traditional forms of known and proved performance which could continue to be used in precisely the same way with the same materials to provide the same performance. For reasons already given this is no longer a reasonable approach. New materials with new properties, new performance standards required to be met by the fabric and the need for greater productivity and economy in building all make it essential that the subject be dealt with as a technology and be considered as a part of the whole field of building technology.

The environmental requirements of the internal spaces set the performance standards of the building fabric and the attainment of these standards sets the practical problems in fabric design. The task of solving these problems is largely that of selecting materials, components and structures which will meet these performance standards in the most economical way. The designer must know the limits within which his choices must be made in terms of the properties of his materials, of structural principles and of the economics of the end result, and these he will derive from building technology.

The architect, however, in trying to meet performance standards also seeks architectural significance for his buildings. This he must do through the fabric, for it is this which defines and gives character and form to the spaces within it. The building form develops from the functional requirements of the building as a shelter, the materials of which it is built, the type of construction used for the fabric and the methods used in its produc-

[1] Negotiated contracts which bring in the contractor at an early stage go some way in solving this problem, although they have not always been entirely successful in this respect. 'Design and build' contracts have been developed in which the contractor offers also the design service and more recently the designer himself has become involved directly in both the development and erection processes (See *Architects Journal*, 22 February 1978 and 14 October 1987).

tion. The architect, therefore, makes choices in these spheres not only in the light of the required performance standards but also in the light of the architectural end he seeks.

The choice of materials for the building fabric and the manner in which they are used depends to a large extent upon their properties relative to the environmental requirements of the building and upon their strength properties. The strength which the fabric of the building must possess in order to function as an 'environmental envelope' is derived from materials of appropriate strength used in accordance with known structural principles. Thus, an appreciation of building construction and the ability to devise new forms of construction is developed from a knowledge of materials and an understanding of structural principles and the overall behaviour of structures under load.

Building is no longer limited to a number of standardised techniques based on the use of a few well known materials, but involves an understanding of the properties and characteristics of an increasing number of materials, of structural principles and of building economics so that existing techniques may be used more efficiently and new forms of construction may be developed for the solution of environmental and structural problems. For purposes of current practice, and as illustrations of the ways in which performance standards are met by the component parts of buildings produced by current techniques, it is necessary to study current methods of construction such as those discussed in this series.[1]

1.3 STRUCTURAL CONCEPTS

The building fabric, having been broadly conceived in terms of an environmental envelope, must be of such a nature that it can withstand safely all the forces to which the building will be subjected in use. In other words it must be developed as a *structure*, a fabrication which for practical purposes does not move in any appreciable manner under its loads. Buildings vary widely in form and appearance but throughout history they have all developed from three basic concepts of structure. These are known as skeletal, solid and surface structures.

1.3.1 Skeletal structure

As the term implies this consists essentially of a skeleton or framework which supports all the loads and resists all the forces acting on the building and through which all loads are transferred to the soil on which the building rests. Simple examples are the North American Indian and the mid-European wigwams in which a framework of poles or branches supports a skin or treebark enclosing membrane (figure 1.2). This elementary form has developed throughout history into frameworks which consist essentially of pairs of uprights supporting some form of spanning member as shown in figure 1.2. These are spaced apart and tied together by longitudinal members to form the volume of the building. In these frames the vertical supports are in compression (see page 35). Skeletal structures in which the floors are suspended from the top of the building by vertical supports in tension are generally called *suspended* or *suspension* structures. Other forms of the skeletal structure are the frameworks or lattices of interconnected members known as *grid structures*, an example of which is shown in figure 1.2.

By its nature the skeleton frame cannot enclose[2] the space within it as an environmental envelope and other, enclosing, elements must be associated with it. The significance of this clear distinction between the supporting element and the enclosing element is that the latter can be made relatively light and thin and is not fixed in its position relative to the skeleton frame — it may be placed outside or inside the frame or may fit into the panels of the frame as may be seen in examples of contemporary steel or concrete frame structures. The practical implications of this distinction are discussed in chapter 6. Skeletal structures are suitable for high and low rise, and for long and short span buildings.

1.3.2 Solid structure

In this form of structure the wall acts as both the enclosing[2] and supporting element. It falls, therefore, within the category of loadbearing wall structures, an inclusive term implying a structure in which all loads are transferred to the soil through the walls. The characteristic of this particular form is a wall of substantial thickness due to the nature of the walling materials and the manner in which they are used, such as in masonry and mass concrete work. The Eskimo igloo is an interesting example of this type of construction (figure 1.2) although for technical and economic

[1] The types and nature of materials used for building work are covered fully in *MBS: Materials* Alan Everett.

[2] *Enclose* here implies also the *division* of the internal space.

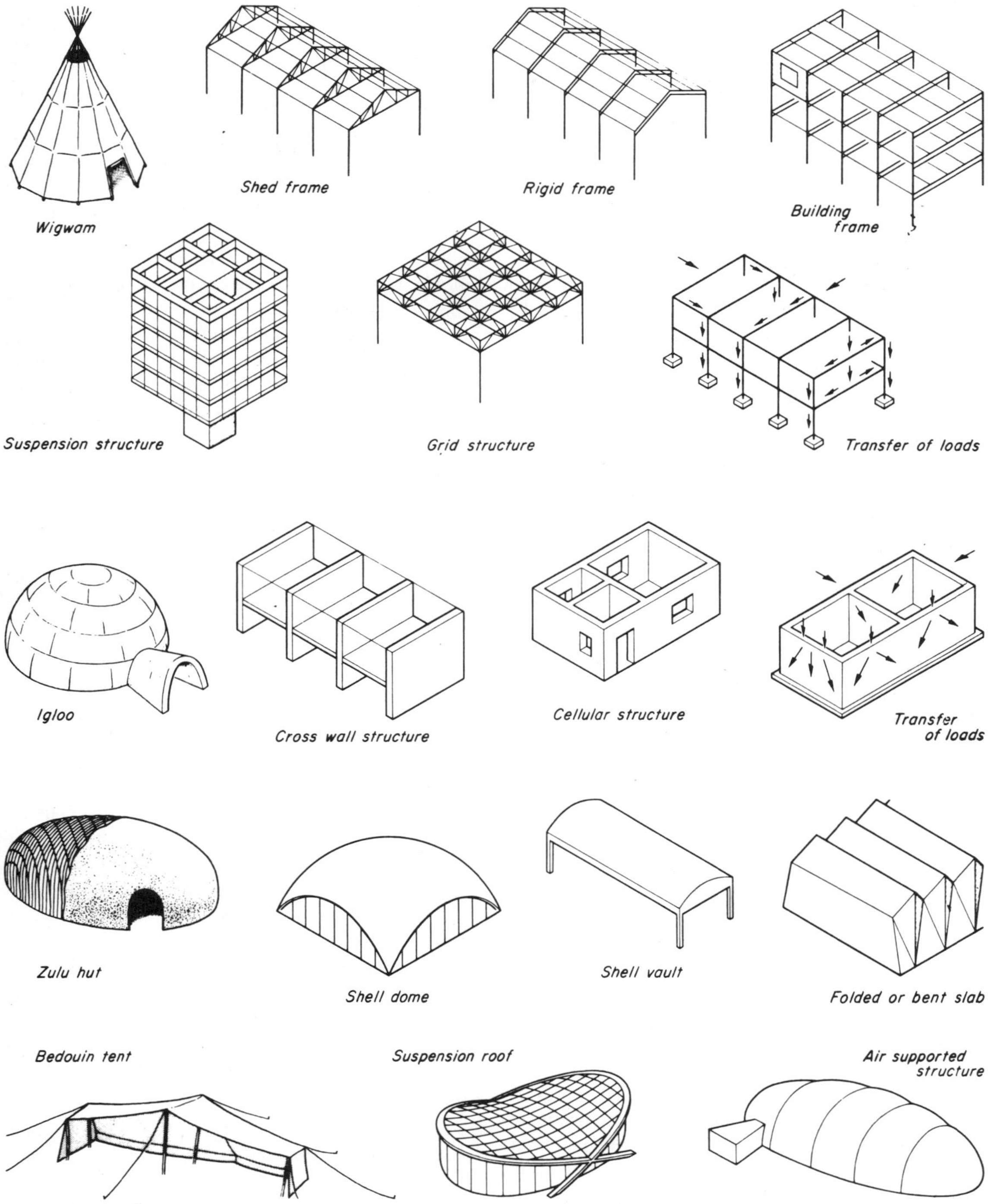

1.2 *Structural concepts*

reasons circular plan forms have been less used than rectangular forms for buildings constructed in this way. Solid construction in the form of brick and stone wall buildings has been used over the centuries and, in certain circumstances, in its various modern forms it is still a valid and economic type of construction for both high- and low-rise buildings if these are of limited span permitting types of floor structure which impose an even distribution of loading on the wall (figure 1.2).

Roof structures are not vaulted over in solid construction, even over limited spans such as that of the Eskimo igloo, because of the problems of construction and the existence of cheaper, lighter and more quickly erected alternatives.

1.3.3 Surface structure

Surface structures fall into two broad groups (i) those in which the elements are made of thin plates of solid material which are given necessary stiffness by being curved or bent, and (ii) those in which the elements consist of very thin flexible sheet membranes suspended or stretched in tension over supporting members. A Zulu woven branch and mud hut (figure 1.2) and modern reinforced concrete shell and folded slab structures are typical of the first. In this form also the wall, and the roof, may act as both the enclosing and supporting structure but the manner in which particular materials are used results in quite thin wall and roof elements. Those in the second group are used for roofs and are known as tension structures. One form is typified by the traditional Bedouin tent (figure 1.2) of which delightful modern applications were first made by Frei Otto[1] for roofing temporary exhibition buildings. Utilising suitably developed membranes this form can now be used for roofing permanent structures. Another form in this group, using compressed air as the supporting medium for similar types of membrane, dispenses with compression members over which, in the tent form, the membrane is stretched. In this the membrane is fixed and sealed at ground level and is tensioned into shape and supported by air pumped into the interior and maintained under slight pressure (figure 1.2). Alternatively, inflated tubes may be incorporated which form supporting ribs to the membrane stretched between them. These are called *air-stabilised* or *pneumatic structures*. In a third form in this group the membrane consists of steel cables suspended from supports and carrying a thin applied cladding and weatherproof covering (figure 1.2).

Surface roof structures are particularly economic over wide spans and for a more detailed consideration of them reference should be made to Part 2.

1.4 FORMS OF CONSTRUCTION

Ways of constructing the building fabric, that is to say the manner in which it is formed of different materials, vary with (i) the structural concept on which it is based, (ii) the nature of the materials used and (iii) the manner in which the materials are combined. For example, if solid loadbearing wall construction is adopted this may be constructed of masonry units or of concrete: the type of masonry units can vary and may be combined in different ways and the concrete may be formed into walls by in situ casting or by precasting.[2] The form of construction will also vary with the functional requirements of particular parts of the fabric since these may be satisfied by various materials in varying combinations. As explained in chapter 5, for example, adequate weather resistance can be provided in external walls by using either solid masonry of considerable thickness or by using a thinner wall incorporating a cavity which prevents the passage of moisture from the external, wet face to the interior face of the wall.

Different forms of construction are, fundamentally, organisational devices used for economic reasons. They vary with the availability and the relative costs of building resources, especially of labour and materials and develop for reasons of economy of time and labour and of materials.

Over the course of history building materials have been, and very largely still are, heavy and bulky and the earliest buildings were constructed of local materials in the absence of cheap and easy means of transport. As the supply of these materials became locally depleted and the need to import from other areas arose the economic use of materials became increasingly important and new forms of construction were introduced, developing from a better understanding of these materials, in which they were used more economically, thus requiring less labour in obtaining them and less transport.

[1] See *Frei Otto—Structures* by Conrad Roland, Longmans 1972.

[2] 'Cast in situ' means cast in the actual position to be occupied in the completed structure. 'Precast' means cast in a mould in a position other than that which will be occupied in the structure and requiring to be placed in position.

Economy of labour in actual building also exercises considerable influence on forms of construction, either because of rising costs of labour or scarcity of labour. Thus, as building passed from the 'self-build' stage of the early building days into the 'contractor-built' era, involving paid labour, forms of construction developed in which less labour was required, for example, the development of brick construction to supersede labour-intensive forms such as traditional rammed earth construction for walling.

As well as the actual cost of labour the relative costs of labour and materials can have a significant effect upon forms of construction. Where the cost of labour is considerably higher than that of materials methods tend to develop in which the labour content of the building operations is reduced at the expense of an increase in the amount of material used, as in the American 'plank and beam' form of timber floor and roof construction which uses large, widely spaced joists or beams spanned by thick boarding involving a greater timber content than forms of floor construction shown in chapter 8 but involving less labour in fabrication.

Scarcity of labour, particularly skilled labour, has a similar effect to that of rising costs of labour. Both bring about forms of construction which are economic of labour, such as the use of concrete blocks instead of bricks for masonry work because these are quicker to lay (see page 100) or the use of modern trussed rafters (see page 175) for small span timber roof construction which greatly reduces the labour content of site fabrication compared with that of traditional methods. Because of the scarcity of plasterers in the past the use of plasterboard in place of lath and plaster has become an accepted technique.

Forms of construction which reduce labour content often decrease the time required for the operations, that is they result in increased productivity. When scarcity of labour and the need to increase productivity are current problems of the building industry the development of new forms of construction requiring less labour is an important exercise. But production can also be increased by good organisation as well as by changes in construction and this has led to a general re-appraisal of the whole process of production in the field of building. This is considered in the next chapter under the section *The industrialisation of building*, page 11.

2 The production of buildings

2.1 METHODS OF BUILDING

The component parts of the building fabric, whatever the form of construction, must be fabricated and then assembled or erected on the site to produce the completed building.[1] These processes must be organised and the manner of organisation differs from country to country and from time to time in any particular country. There are a number of ways of building current in Great Britain and these are briefly described below. The names given to them, apart from the first, are arbitrary but they will serve to define them for use later in this book.

2.1.1 Traditional building

As known at the present time this has developed from the use of forms of construction evolved by the traditional building crafts, particularly those of bricklaying, carpentry, plastering and tiling and slating. The proportion of skilled labour required is, therefore, fairly high: about two-thirds of the work in a traditionally built house is skilled craft work.

The product of traditional building, apart from speculative housing, is the 'one-off' building designed and constructed for specific requirements and a specific site. Indeed, this holds for building generally where based on methods developed from the traditional way of building. The processes of design and construction of this type of building are normally carried out by separate groups described in the previous chapter, the materials for the work being obtained from a third group — the materials and component suppliers.

A considerable amount of fabrication as well as the assembly of parts takes place on site and, indeed, in situ,[2] since traditional forms of construction involve the combining of many small units. The amount of on-site fabrication, however, has for many years now been reduced by the introduction of prefabricated factory produced components, especially in the field of joinery and carpentry in the form of windows, doors, cupboard and other internal fitments and of prefabricated structural components such as roof components in the form of trussed rafters, for example. Such changes, of course, do not fundamentally alter the nature of this way of building since a change in craft techniques is not involved. When a change does involve the development of new techniques, even though allied to the older crafts, it is less readily acceptable. For example, plasterboard has been used for many years; this involves craft work, but development in the use of sprayed or projection plastering, requiring quite a new technique, has been very slow.

In traditional building the craftsmen are not only familiar with the content and order of operations in their own trade but, because of the limited range of materials and forms of construction, are also aware of the relation of these operations to those in other trades and of the order in which they follow each other. The organisation of the work and its nature is, therefore, so well known by all concerned that the work may be carried out with a minimum of detailed information.

The traditional craft-based building method is very flexible and is able to meet variations in the demand of the market or on the work of the craftsmen much more readily and inexpensively than methods based on highly mechanised factory production. This is because production is by craftsmen and therefore, little fixed assets in the form of plant, especially mechanical plant, and buildings are necessary. A builder operating on this basis can with little loss expand or contract the size of his organisation according to the fluctuations of demand because his capital investment is small, and he can readily transfer his craftsmen from one type of operation to another with far less loss in productivity than would occur if similar changes were made in factory production (see page 19). This method of building is, therefore, of necessity adopted by small building firms with little capital to sink into their undertakings or to enable them to carry over slack periods in demand. It is, for this reason, the method used in the construction of the major part of small-scale building carried out at present, especially in the field of housing (see page 12).

[1] In the context of building, *fabrication* means the making of the component parts of the building from smaller units, such as framing up a window, and *assembly* means the erection or putting together of these components to form the total building.

[2] In situ: for definition see footnote on page 6.

2.1.2 Post-traditional or conventional building

To some extent traditional building has always been in a state of change due to the introduction, from time to time, of new materials and developing techniques but the most significant changes probably occurred with the development of Portland cement and of mild steel. This brought new materials into the field of building and with them large and complex buildings made feasible by new forms of construction arising from their use. Together with this arose the need for the efficient organisation of the construction processes related to these forms of construction. Portland cement concrete was a new material but the technique of casting mass concrete in situ in formwork is basically the same as that of traditional cob and pisé wall construction. The development of reinforced concrete, however, introduced new techniques and forms of construction. Steel which, by its nature, is produced as a pre-formed material, lends itself to off-site fabrication and has resulted in forms of construction for skeleton frames involving the prefabrication of the parts and only their assembly on site by operatives with specialised skill.

This extension of the available forms of construction and their increasing complexity has been accompanied by an increasing complexity in building services and this has been accompanied by the need for specialised knowledge distinct from the old crafts and for detailed information for carrying out the work.

With increasing size of buildings there has been an increasing use of mechanical plant, but its application to a construction process which is one of assembly is limited. It is most readily applied to excavating and earthmoving operations and to operations related to the structure of the building, particularly where this is large, such as the mixing and transporting of concrete and the handling of materials and component parts. With increasing size of buildings the process of construction has increasingly become one of moving earth, placing concrete and fixing steel but in which traditional craft work although declining in importance has not been eliminated.

This post-traditional method of building is, in fact, a mixture of traditional and new forms of construction involving both the old craft and the newly-developed techniques based on new materials. It varies from traditional building not so much in radical differences but mainly in the scale of the work carried out and, as a consequence, in the use of expensive mechanical plant for many operations. This is usually accompanied by a greater attention paid to planning and organisation of the work necessitated by the scale of the work and the use of plant.

Apart from 'specialised' work in reinforced concrete and steelwork the remainder of the carcassing and the finishing work tends to be carried out on craft lines. This makes much use of prefabricated components but these are not designed and used relative to each other to produce a rationalised system related to the whole construction process for a particular building. In fact, although this method of building is largely concerned with the organisation of the systematic supply and assembly of materials to produce a specific building, the design of the building is not necessarily evolved to facilitate this.

It is a somewhat less flexible method than the traditional method because, although still labour-intensive, the labour for many operations is closely tied to the operations of the mechanical plant being used (see Part 2 chapter 2), and because the scale of the work necessitates a greater investment of capital in fixed assets.

2.1.3 Rationalised building

In the present context this term refers to a method of building in which organisational techniques used in the manufacturing industries are applied to the erection process without involving a radical change in forms of construction or necessarily in techniques of production in current use.[1]

Increases in the complexity and size of buildings resulting in a more complex construction process, the demand for more buildings of all types and the need to economise in labour and reduce costs have turned the attention of the building industry to methods used in the manufacturing industries, where productivity is high and the products are relatively cheap, and the principles on which these methods are based have been applied to the construction process through proper planning and organisation. This seeks to achieve a properly integrated system of design and production leading to continuity in all the production operations. Work is planned to ensure that all operations fit into a continuous time sequence so that construction proceeds as a continuous operation and this necessitates thorough organisation of the whole construction process to ensure a proper flow of labour and materials so

[1] In a wider context the term 'rationalised building' is related to the industrialisation of the whole building process.

that these are always available when and where required. Standardisation and prefabrication of components as far as possible and the introduction of mechanical plant where this will achieve continuity and reduce labour are used as means to these ends. Continuity is further ensured by the separation of fabrication operations from those of assembly and by designing so that the work in different trades is separated. It will thus be seen that it is essential that the design of the building and the production operations be considered together at design stage as already emphasised on page 1.

Rationalised building takes further what has already happened to some degree in post-traditional building but it can as well be applied to construction carried out entirely by craft processes with traditional materials provided the design and organisation are developed with a view to continuity of operations and economy of labour, and suitable aids to craftwork are introduced to reduce the number of separate operations and to save time.[1] Rationalised building is discussed more fully later in this chapter and some of the practical means by which the rationalisation process may be accomplished are described.

2.1.4 System building

This term refers to a method of building based on forms of construction in which the component parts of the building fabric are wholly factory produced and site assembled.

The components relate to each other only as parts of a single integrated system of construction, usually related to a specific building type such as houses or schools or to a restricted range of types. They will not normally fit with components of other systems and this method, therefore, is commonly referred to as *closed system building*.

The term *industrialised building* is often applied to this method but this is far too narrow an application of this word as explained in the section on *The Industrialisation of Building* (page 11) and it is for this reason not used in this context.

Factory production removes fabrication from the site leaving only assembly operations to be performed, thus reducing the amount of skilled site labour required and reducing the time spent on site operations. Since this method incurs the higher overhead expenses of factory production and the charges for transport from factory to site, it is essential that the savings in site time and labour are sufficient to offset these costs. For this reason there must be thorough co-ordination of design and the production-assembly processes, with close integration of factory production and site work to ensure continuity of operations throughout the assembly period. The economic success of this method depends probably more on efficient organisation than on the method of fabrication of the parts.

Systems are based on skeletal structures in steel, concrete or timber or on loadbearing wall construction built up either from relatively heavy precast concrete panels ranging in size up to room height and width, or from light panels in timber construction which might be up to two storeys in height. Some systems incorporate room-size units in the form of kitchen/bathroom 'heart' units while others consist entirely of factory produced dwelling units of one or more rooms.

Economically system building is appropriate only to large-scale production and necessitates a large market which it can supply. Systems are expensive to develop and most are proprietary and are marketed by the developer although a number have been developed by consortia, mainly of local authorities, co-operating with a manufacturer. By this latter means demands are combined and economies of scale are obtained.

This method of building is more fully discussed later in this chapter.

2.1.5 Component building

Like the previous method this is a method of building in which the component parts of the building fabric are factory produced and site assembled. These components may, however, be used freely in conjunction with parts of the fabric constructed on traditional lines, such as brickwork, blockwork and roof tiling.

The method differs from system building in that (i) the production of the components is not limited to one manufacturer or developer and (ii) each component is interchangeable with those produced by any other manufacturer. It has for this reason often been called *open system building*. It is thus a method of building which uses interrelated factory produced components from a variety of sources and in a relatively wide range of materials, so that the economic advantages of mass production may be combined with the greatest possible freedom to design to meet user and site requirements more precisely and over a wider range of building types than is possible with system building.

[1] Traditional small-house building rationalised in this way has been referred to by an ugly abbreviation — 'rat-trad' building.

In order to keep variety within acceptable limits for mass production (see page 19) such a system must operate within a framework of co-ordinated dimensions[1] by means of which component may be related to component and component to structure, thus permitting a standardisation of each manufacturer's products such that all components fit together with little, if any, adjustment and no waste. This, of course, necessitates a standardisation of main controlling dimensions such as floor to floor heights and floor and roof spans, within which the components will fit, and standardised designs for junctions of various types with agreed tolerances[1] on the size of components and on their positioning in the building.

Components are produced for both frame and loadbearing wall construction of different materials and all can be used separately or together to produce economic solutions to a wide range of problems. In Great Britain consortia of local authorities have used the component building method with some considerable success.

2.2 THE INDUSTRIALISATION OF BUILDING

Increasingly heavy demands for the products of the building industry were made in the years following the end of the second world war and it was appreciated that a considerable increase in productivity would be essential in order to meet the building needs of Great Britain in the then foreseeable future. This demand on the industry was made at a time when a shortage of labour existed, especially in skilled labour, and which, it was anticipated, was unlikely to improve to any significant extent.

In view of this there was a search for new methods to reduce the amount of site labour involved in building operations and to increase the productivity of the industry generally. Such methods should produce buildings at no greater cost than by traditional methods — if possible, at less cost. As already indicated, in this search attention was given to the methods of organisation and production which are characteristic of other, highly mechanised, industries in which productivity is high and products are relatively cheap.

In these industries production is carried out in factories and this, taken together with the difficulties of building on open sites in all weathers, has resulted in much emphasis being laid on the idea of all building parts or even large parts of the whole building being produced in a factory with only the assembly of the parts taking place on site. But the 'industrialised' methods used in other industries and, therefore, the industrialisation of building involve more than production in a factory. For the building industry industrialisation involves the rationalisation of the whole process of building (which includes the process of design, the forms of construction used and the methods of building adopted), in order to achieve an integration of design, supply of materials, fabrication and assembly so that building work is carried out more quickly and with less labour on site and, if possible, at less cost.

Industrialisation is essentially an organisational process and has been defined as — 'continuity of production implying a steady flow of demand; standardisation; integration of the different stages of the whole production process; a high degree of organisation of work; mechanisation to replace manual labour wherever possible; research and organised experimentation integrated with production.'[2] From this can be derived certain characteristics of industrialised production:

1 continuous, 'flow-line' production
2 standardised production
3 planned production
4 mechanised production

none of which implies that industrialised production is necessarily concomitant with factory production. As an organisational process industrialisation may be applied to any method of building and whether applied to traditional methods or factory methods will introduce these four characteristics, a mechanised and continuous fabrication and assembly process to speed up production and reduce labour requirements, a standardisation of components to reduce costs and facilitate continuous production and a properly integrated system of design, fabrication and assembly to speed up the whole building process, with feedback from the fabricating and assembly processes to the designer so that changes and developments may be made leading to reduced costs and greater productivity.

The principles of industrialised production summed up in these four characteristics have been applied to building in a number of ways. Firstly (but not necessarily so chronologically), in the rationalisation of the erection process through proper planning and organisation; secondly, in the mechanisation of the erection process in those operations which are most suited to the application of the machine (see pages 9 and 14);

[1] See page 21.

[2] *Government Policies and the Cost of Building* United Nations, Geneva 1959.

thirdly, in the use of standardised prefabricated component parts. The increasing use of these in the form of relatively small units, as explained on page 8, has developed over the years but very much larger components, such as storey-height wall panels have also been prefabricated for use on 'one-off' jobs, developing into methods of building based entirely on the assembly of large components forming part of a single integrated system of construction.

None of these alone, however, is wholly sufficient. They must be part of a rationalisation of the whole building process in which all the component parts are related at the design stage taking account of all the production processes involved from fabrication to erection or assembly on site. In this context industrialisation may be based on the rationalisation of building methods as dissimilar as the traditional craft method and the system building method using completely factory-produced components, both of which have already been described. Some further aspects of this rationalisation process are discussed in the following pages and this will link with and expand on what has already been discussed under *Methods of Building*.

2.2.1 Rationalised building

The term 'rationalised building' used here has a limited, but usually well understood, meaning as defined on page 9 and is applied to both traditional and post-traditional methods of building. In both cases the rationalising of the organisation of the work by proper planning is paramount. Techniques of construction and production vary with scale and complexity of building.

Traditional building, for reasons already given, is mainly used for small-scale work carried out by the smaller building firms and there are very good reasons for rationalising this sector of the industry.

The building industry throughout the world is made up largely of quite small firms employing few operatives and having little capital resources,. The industry in Great Britain is no exception and in this country these smaller firms, employing less than eight people, account for about twenty five per cent of the total output of the industry and are responsible, in particular, for a very large proportion of the housing output. This type of firm, using mainly traditional materials and methods, plays an important part in the industry and will do so for a long time to come so that within their limited resources they must be fully utilised. These firms have insufficient capital to sink into factory production and for this reason and others given on page 8, they continue to operate by traditional methods. It has been shown, however, that traditional building is susceptible to the organisational process of industrialisation which, where seriously applied, results in increased productivity and reduction in site labour, even using methods which have been common for many years, by improving working methods and site organisation.

The use of factory-produced components and the introduction of mechanical plant to handle materials on the site are not recent features of traditional building but the fundamental nature and organisation of the process has not necessarily been changed thereby. As indicated earlier industrialisation of building 'is concerned, with systems of construction which can be organised in an integrated way, the whole process of construction being organised as a single interlocking process in which materials and labour are available in an organised flow, so that construction can proceed without hindrance as a continuous operation. Industrialisation does not necessarily imply a completely new system of construction . . . (but) may be achieved partly by the use of mechanical plant, partly by replacing in situ work by prefabricated units and partly by rationalising the organisation.' The last phrase of this quotation from P A Stone[1] should be noted for it is the rationalising of the organisation of materials, components, labour and equipment which is the most important factor of all. Thus, the industrialisation of traditional building requires the introduction of prefabricated components and the use of mechanical plant to be accompanied by the thorough planning and rationalisation of the organisation of the whole job.

The organisation of the construction process should have as its aim the introduction of the characteristics of industrialised production given on page 11. In traditional building this can largely be achieved by organising the following:

(1) an efficient layout of the site
(2) a practical, orderly sequence of work
(3) similar work to be done in series
(4) the rational use of prefabricated, standardised components
(5) the rational use of mechanical plant.

(1), (2) and (3) promote continuous, flow-line type production, (4) can reduce site labour and time and is significantly related to the previous item. (5) will speed up operations and reduce labour requirements.

[1] P A Stone *Building Economy* Pergamon Press 1966.

(1) An efficient layout of the site

The layout of the site must be studied in advance of operations so that the flow of materials and work will be orderly and will involve the minimum movement of operatives and materials and the least double-handling of materials during the progress of the work. A satisfactory flow-line relationship between materials, plant and work must be achieved. For example, aggregates, concrete mixer and building should be so related that aggregates and cement feed directly into the mixer on one side and the concrete is discharged on the other towards the building so that the shortest route, with no obstruction by stacked materials, has to be travelled in its transport to the final position on the job. A good roadway is essential to permit easy delivery of materials to stacking areas which should be correctly positioned relative to their place of use on the job.

(2) A practical, orderly sequence of work

On a building site the work, due to its nature, does not move to the operative and plant but these move to the work. This is the reverse of the situation in a factory but by planning an orderly sequence of operations, which is facilitated by good site layout, the characteristics of normal manufacturing flow-line production can be developed on a building site. Continuity of work is an essential aim. The lack of continuity results in delays and unproductive time for men and machines, increasing overheads and costs for labour and plant and decreasing production. An orderly sequence of operation ensures the necessary continuity of work.

Delays can be caused either during the course of work at one particular work-place or in commencing operations at a new work-place. The first type of delay is due to interruption of the work of one trade at a particular position while work in another is carried out in the same position, such as the delay in bricklaying caused by the casting of a concrete lintel over a window opening in a brick wall. This form of delay can be avoided by separating operations especially those of different trades and those of fabrication and assembly. The second type of delay occurs when some operations at a work position take longer than others, thus holding up the commencement of following operations. This can be overcome by planning the size of the gangs for different operations so that the time required by each for its own particular operation is the same.[1] Each can then proceed to its different work-places and commence work without being held up while a preceding gang completes its work at that place.

(3) Similar work to be done in series

This means the repeated performance of the same operations by the same operative or gang of operatives in order to reduce production time. It has been shown that the repetition of the same operation results in decreasing completion times due to the increased skill gained through the experience provided by repetition. This effect of repetition is referred to as *routine effect*. In order to achieve this it is necessary to break the whole job down as far as possible into sections each of which involves only one type of operation, or a group of similar operations, for its completion, and to plan the job so that the same gang moves from work-place to work-place to carry out the same operation.

(4) The rational use of prefabricated standardised components

Used in a rational way, with due consideration of the effects of their introduction on the work as a whole, prefabricated components can reduce the requirements for skilled labour, simplify construction by reducing the number of separate operations to be carried out and can facilitate continuity in the remaining operations. However, in order to achieve these results careful consideration must be given to the possible effects of prefabricated components upon related parts of the work both in respect of labour and plant involved. The use of these components may not necessarily result in quicker or more economic building. For example, the use of precast concrete wall panels requiring lifting plant could be uneconomic if the plant could be used for no other purposes on the job and studies on the introduction of large prefabricated factory-made panels into buildings of otherwise traditional construction[2] have shown that unless these replace the whole of the corresponding trade operations, the number of operations necessary to complete what is left of the traditional work is likely to be increased, each change

[1] This is known as *the balancing of trade gangs* and is discussed in Part 2 chapter 1.

[2] *Study of Alternative Methods of House Construction*: National Building Study Special Report No 30, HMSO 1959.

of operation resulting in unproductive time so that little economy in overall cost or time is achieved.

(5) The rational use of mechanical plant

A large proportion of work in traditional building is skilled craftsmen's work (see page 8) which is difficult to mechanise. In this work, however, some types of powered tools are useful and also the use of non-mechanical aids such as setting-out devices and jigs. The mechanisation of small jobs depends very much on the size and nature of the job. Various pieces of plant designed primarily for the small builder are available such as powered barrows, loader-shovels, scaffold cranes and small concrete and mortar mixers. In large post-traditional jobs much plant is now considered essential in those operations which can most easily be mechanised, that is in the handling of materials, in excavation and the movement of soil and in concrete mixing.

Careful consideration must be given to the relevance and economics of plant use on any particular job. If a piece of plant can be used for only one or two operations it may be cheaper not to adopt it and even if reasonably continuous use is anticipated it may be found, particularly on very small jobs, that the traditional methods of manhandling materials and components is the most economical because of the greater flexibility of labour they permit and because of the greater amount of double handling often involved when mechanical plant is used.

To a very large extent the rationalising process is significantly affected by the design of the building and, therefore, the consideration of the design and building methods together at the design stage is essential, so that at all points the design is so related to the production process that the latter can be rationalised in a satisfactory manner.[1] Ways in which design decisions facilitate better job organisation are described in Part 2 chapters 1 and 2, where a more detailed examination of these five organisational areas will also be found.

In addition to and related to these considerations, attempts are made to transfer some of the advantages of the factory to the site (i) by transferring in situ fabricating operations from elevated heights to bench or ground level where productivity can be much greater, for example, by prefabricating formwork, making-up reinforcement 'cages' and using mechanical equipment to lift these into position, or even by casting in large areas all the floor and roof slabs for a multi-storey building on top of each other at ground level and hoisting to position by means of hydraulic jacks on top of the structural columns (see *Lift-Slab* Part 2 chapter 6); (ii) by the application of factory techniques, such as steel formwork units combining wall and floor shutters and incorporating heating facilities, to in situ concrete work, thereby simplifying erection and stripping of the formwork and reducing the curing time of the concrete; (iii) by site casting of precast concrete work either on open casting beds or in a temporary site 'factory'; (iv) by enclosing the complete building operation in a translucent plastic sheeting and scaffold shelter or in an air-supported shelter, within which the operatives are protected from the weather and the temperature may be raised during cold periods. Other techniques for winter working, widely used in Europe, are now adopted in Great Britain and assist considerably in maintaining production during cold weather.

2.2.2 System building

This term, as defined earlier, refers to a method of building based on the use of factory-produced components which relate to each other only as part of a single integrated system of construction. Since the components of one system will not usually fit with components of another these are often referred to as 'closed systems' in contrast to an 'open system' using components developed on the basis of co-ordinated dimensions and the standardisation of jointing techniques referred to on page 10.

The transfer of certain building operations from site to factory is seen as a means of increasing productivity and reducing the demand for skilled site labour. Greater productivity results from the greater speed of modern factory production and from carrying out a large proportion of the necessary work in an environment free of the delays caused by the vagaries of weather. A number of important factors, however, must be taken into account in making this transfer.

Philip O. Reece once wrote — 'the mere transference of a simple operation from the site to the factory will not of itself reduce costs — on the contrary it may increase them' and illustrated this by indicating that it could cost five times as much to drive a nail into a piece of wood in a factory as to do it on site. Economy requires new techniques based on the latent possibilities of the machine, not simply a change of place in which

[1] See Current Paper 48/77 *The rationalisation of housebuilding*.

an operation is performed.[1] Factory overhead expenses are high compared with those on sites and transport charges to the site must be covered. Savings in site time and labour to offset these are therefore essential. For this reason the prefabrication of large rather than small components has developed to reduce the number of site assembly operations and thus the labour requirements and assembly time. Linked with this is the endeavour to eliminate where possible on-site finishing work. In the field of precast concrete panels, for example, the use of steel forms produces smooth self-finished surfaces which do not require plastering.

In order to obtain increased overall productivity site operations from the outset must be closely co-ordinated with factory production in terms of the site erection programme, mechanical plant to be used and storage areas available, so that the rates of factory production and site assembly coincide and continuous production is maintained. It is often assumed that system building is necessarily a wholly rationalised building method. This is not so. The component parts may be rationalised in design and factory production but unless these are closely related to the assembly operations by good planning and organisation the total building process itself will not be rationalised.

Factory production of components requires a large capital investment for plant and premises and necessitates a steady flow of demand for the products in order to maintain the plant at its maximum level of production and to avoid expensive plant with its continuing overhead costs lying idle.

Compared with traditional means of production factory production is relatively inflexible. Large runs of any one component are essential for economic production, since changes from one type to another involve changes to the plant causing delay, loss of time and falling off of productivity, and an increase in costs. This form of production, therefore, is usually economic only when it is applied to individual buildings large enough to permit considerable repetition of the same type of components or to a sufficiently large number of small buildings of the same type for each of which the same set or range of components is required. It is for this reason that it needs to be applied to building types in which the functional requirements are, to a great extent, standardised.

In order to obtain full advantage from factory production thinking in terms of this method of production must go right back to the design stage and right forward to the organisation and planning on site. The designer must have a grasp of the economics of factory production and must understand the processes involved and design for them. The contractor must understand the implications of a system which provides him with a set of components already manufactured which he must assemble into a building rather than himself constructing the building from the ground up. One simple but important aspect, for example, is that extreme accuracy in setting-out at ground level is essential for system building since the nature of the components and the principles of the system are such that mistakes cannot be corrected during the assembly process.

Reference should be made to *Mitchell's Building Series: Components* where other practical aspects of system building are discussed.

2.3 MATERIALS AND THE CONSTRUCTION OF BUILDING ELEMENTS

The term *element* related to buildings refers to a constituent part of a building which has its own functional identity and these are, therefore, correctly called *functional elements*. They are the wall, partition, floor, roof and structural frame, by means of which the main functional requirements of the building described on page 1 are satisfied. To these are added the foundation, as a separate element at the base of a wall or frame, and the stair.

These elements are made up from smaller parts. The raw materials for building are rarely used in their basic form but are treated or processed in some way to suit them to building purposes. These materials, such as timber, stone, sand, brick earth, aggregate, lime, gypsum, are relatively cheap to win from their natural state and are also relatively cheaply converted into forms and products suitable for building use and assembly into building elements, so that they increase in value only a few times over their initial cost in the raw state. This is in contrast to the manufacturing industries in which the materials used, such as steel and plastics, are highly processed and may increase in value by as much as twenty times.

The structure and fabric of most buildings involves considerable use of these relatively cheap materials which are, however, bulky and heavy, with the result that buildings are heavy but cheap relative to their

[1] As, for example, the use of *toothed plate connectors* instead of nails for trussed rafter fabrication (chapter 7).

weight. For this reason the more expensive and often lighter materials have, generally, been used mainly for finishes and fittings. The cost-weight ratio is, however, only one amongst many factors to be considered in the design process.

The ultimate basis of cost comparison for the substitution of lighter, more expensive materials is not, of course, the cost of the material or product itself, but its cost fixed in place in the fabric of the building. Thus the use of more expensive products may effect an overall economy because it results in consequent economies in other parts of the fabric, such as reduced loads on foundations and structure deriving from a reduction in the overall weight of the building, or in the operations involved in the construction process due to a reduction in the number of separate parts required, a reduction in the number of assembly operations or a reduction in labour and in construction time. The use of a product initially costing more than a comparable product of different materials and design could, therefore, result in economies in the fabric as a whole which outweigh the extra cost of the products themselves. It must be borne in mind, however, that the cost of the materials comprises the greater part of the cost of a building even when the majority of them are traditional cheap 'heavy' materials and this increases with a decreasing use of these materials. Thus, when more expensive materials and a large amount of prefabrication are introduced these economies usually need to be substantial in order to realise a satisfactory balance of costs.

The nature and properties of a material determine the methods used in processing it and the forms into which it can be processed. For example, timber consists of longitudinal fibres and can usefully be cut into linear pieces as long as the length of the tree trunk. Its nature also permits it to be peeled off in thin sheets forming veneers which may be built up to form plywood boards. Moist clays may be moulded to shape and then burnt to form bricks. These must be limited in size, however, in order to minimise distortion during burning, a restriction which applies to all burnt clay products. Because the clay is moist and plastic before burning it can also be extruded through a die to brick cross dimensions but unlike aluminium, for example, which may also be extruded, the extruded clay must be cut into short lengths. Plastics, reinforced with glass fibres, may be formed relatively simply into thin flat sheet units or moulded into curved or other geometrical shapes, quite large in size.

The products from such processes form the *component* parts of the elements of a building and are referred to by the general terms of sections and units (table 2.1).[1]

A *section* is formed to a definite cross-section but is of unspecified length. Sections are usually produced by a continuous process such as rolling, extruding or drawing, for example steel joists and tubes.

A *unit* is formed as a simple article with all three dimensions specified. It is complete in itself but is

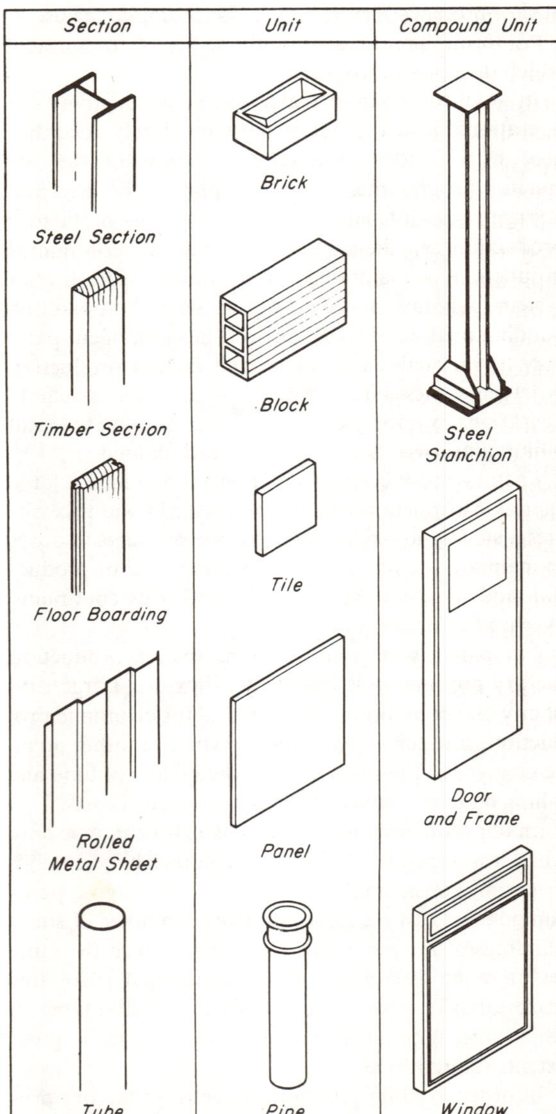

Table 2.1 *Component types*

[1] See BS 6100: Section 1: 1984.

intended to be part of a larger whole, for example a brick, block or sheet of glass.

Compound units are a combination of sections and units formed as a complex article with all three dimensions specified. It is complete in itself but is intended to be part of a complete building, such as a door and frame or a roof truss. Similar types of compound units may be formed with sections or units made from different materials but they will be different in nature and form (figure 2.1).

For example, bricks which traditionally have always been assembled in situ may, if necessary, be prefabricated into wall panels. These are heavy, but fulfil the enclosing and supporting functions by the use of one type of unit. Wall panels may also be fabricated from timber sections in the form of a built-up frame. This will fulfil the supporting function but other sections in the form of weather-boarding, for example, must be applied to the frame to form an enclosure. The timber panel, however, if not too large in size would be light enough to be manhandled but the brick panel, even of limited size, would require the use of mechanical plant to lift it into position. As a further example, timber may be used in the fabrication of small-span trussed rafter roof components built up from timber sections. These span in one direction only and must be placed short distances apart to fulfil the supporting function and must carry separate enclosing sheet units such as woodwool slabs and some form of weatherproof covering. But roof components can be constructed from glass fibre reinforced plastics large enough to require only a few to cover a given area, very light in weight and very quickly assembled. These, unlike the trussed rafters, fulfil both the supporting and enclosing functions by one or two units and at the same time provide the necessary weather resistant surface to the roof.[1] (figure 2.1)

An *assembly* is a combination of any of the above to form part of, or the whole of, any element of a building.

2.4 PRODUCTION OF COMPONENTS

Traditionally most elements were constructed in situ and were built up or fabricated in this way from units and sections, such as the brick or stone wall and the timber frame. In current traditional and conventional methods of building brick and blockwork walls are still built in situ and concrete walls and frames are still cast in situ, but for reasons given earlier in this chapter there has been an increasing use of very large prefabricated unit components, such as storey-height precast concrete slabs and compound units forming complex components, from which the elements of a building may be assembled. These and other prefabricated components permit the separation of fabricating and assembly operations thus facilitating good organisation of the erection work on site in the manner already outlined. Such components may be produced either on-site or off-site.

2.4.1 On-site fabrication

On-site fabrication requires sufficient space adjacent to the actual construction area on which to carry out the work, in the case of precast concrete panels, for example, sufficient for the casting beds or a temporary site 'factory' if justified by the size of the job. These panels may be cast in horizontal or vertical moulds

[1] The nature of the different building materials and the methods used in producing various types of units and sections from them are discussed in *MBS: Materials*.

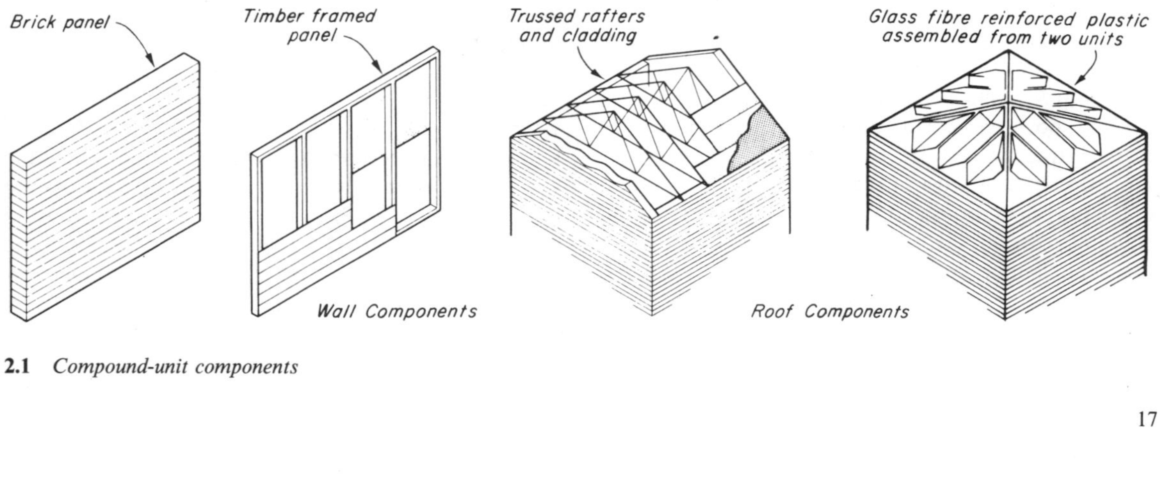

2.1 *Compound-unit components*

depending upon the shape and surface finishes required.[1] Similarly, in the case of frame construction in timber, space is required for fabricating the wall panels on the ground before erecting them in their final positions. Techniques such as lift-slab construction (see page 14) permit on-site as distinct from in situ fabrication of reinforced concrete floors without the need for extra space for casting. On-site fabrication, as already suggested, serves to transfer some of the advantages of the factory to the site and in some circumstances proves to be economically and organisationally the most satisfactory method. For example, it is shown below under the section on *Factory Production* that for economic off-site factory production large runs of any one component are usually essential. Where the size and nature of the job is such that this condition cannot be met then on-site fabrication may be the better method although the nature of the component and its method of manufacture can qualify this: it is, for example, considered to be uneconomic to produce panels for timber frame walls on site, even for a very few buildings, unless suitable workshop facilities are over twenty miles from the site. The form of the component is also significant. Plain concrete components of simple shape can be quite satisfactorily cast on site thus saving transport costs, but decorated and intricately shaped components are usually more successfully precast off-site in a factory.

2.4.2 Off-site fabrication

Many components such as bricks, blocks, pipes, doors and windows have traditionally been manufactured by other industries. Their nature has permitted a high degree of standardisation and this, together with the demand for them, has made mass production possible. Experience gained in production techniques in these fields has been applied to components of increasing size and complexity: for example, curtain walling based on metal window production and concrete wall slabs incorporating windows and, perhaps, services, developed from simple slab production.

It has, however, already been pointed out on page 14 that the transfer of work to a factory does not necessarily decrease the cost of the component nor the cost of building operations and that in order to keep overall building costs within acceptable limits savings in site time and labour are essential in order to offset the higher costs of factory production. Where such site savings are made by satisfactory co-ordination of factory and site operations and by good organisation of the erection process these are often not great enough to produce an overall cost much less than that resulting from conventional methods of construction.

There are two main reasons why the transfer of building operations from site to factory may often show comparatively small economic benefits. Firstly, the nature of the main materials and the type of finishes and equipment have not changed to any great extent so that the building process still involves the assembly of heavy components and materials (which together account for the greater part of the cost of a building) and, secondly, the traditional and conventional sectors of building have undergone a gradual process of rationalisation resulting in appreciable improvements in productivity in these fields. It is, therefore, necessary not only to effect savings on site but also to keep the costs of factory production to a minimum. This task involves the designer of the components as well as the manufacturer since the nature of and the extent of variety in the components to be produced have a significant effect upon production costs.

Factory production

Systems of production for building components are generally devised to produce a variety of models differing in design and size. To permit this to be done satisfactorily a production programme is necessary by means of which each model may be introduced into the production process in an organised manner, the nature of the production programme depending upon the method of production used. There are two basic methods used for the manufacture of building components: (a) *batch production*, in which the material or components pass in batches from one specialised work position (either manual or machine) to another at which each process involved is carried out without being related to the speed at which any succeeding or preceding operation may be done, and (b) *flow line production*, in which each process on each component is carried out in sequence at a different work position and is completed in the same time as preceding and succeeding operations. Broadly speaking, batch production is used where different parts and models require different sequences of operations and line production where the various models may be processed by the same sequence of operations and where the machines and operators may be adjusted to the requirements of the models. These methods are more fully discussed in chapter 1 of *MBS: Components* to which reference should be made.

[1] See Part 2 for details of casting methods.

In the manufacturing process a *run* means a number of identical or similar units produced in a continuous sequence of operations. Such units may not be final products but may be parts requiring further assembly to form a finished product; a *model* means (i) an individual item of production identified by a distinct catalogue reference or (ii) every product for which an adjustment in the production line has to be made; a *type* means a group of models with some common characteristics but which are not necessarily produced in the same run, for example, flush doors of the same dimensions and core construction but with different surface finishes, each constituting a different model.[1]

Cost reductions in the factory are broadly achieved by increasing the scale of output of the factory and by utilising the productive equipment to its full capacity. The former depends in one respect upon the demand for the products and in another upon the latter, that is the full utilisation of the available equipment which in turn depends upon the proper organisation of production and upon the nature and variety of the products. The costs of production may be divided into those costs which are proportionate to the number of units produced and those which are independent of the scale of output, known as *fixed costs*. With increasing scale of production the larger will be the number of units among which these fixed costs are distributed and thus the smaller the proportion attached to each.

In general, the smaller the variety of product and, therefore, the greater the number of identical or similar models produced in a run, the lower will be the production cost per unit. The economy of large, or long, runs is due to the fact that the fixed costs of setting up the machines for particular operations are spread over a large number of units and that less frequent stops in production for adjustment and set-up results in a more efficient use of plant and greater output. In addition to these factors several others combine to reduce unit costs when the number of models is decreased, in particular the decreased overheads due to design, the reduced need for different moulds and tools for different models and decreased costs in holding stocks of finished products.

Reduction in the variety of models can also result in advantages on site in the simplification of work and in the increase in the number of identical operations to be performed which produces the benefits of the routine effect referred to on page 13.

However, the extent of cost reductions by means of longer production runs resulting from the reduction in the number of models varies with circumstances. They may be considerable when, for example, plant capacity is thereby more efficiently used or they are likely to be small when the costs of the materials in the product represent more than half the total cost of production. The greater the proportion of materials costs the less will the total cost be affected by an increase in output.

Experience also indicates that the *rate* of reduction in unit costs begins to fall with continued increase in the length of run and, indeed, with very long runs an actual rise in unit cost may occur due to increased storage costs, capital tied up in stock and high maintenance costs of plant running near to its maximum capacity. This is indicated in figure 2.2. At point *A* there is very little further improvement in cost with increase in run size, the proportion of the cost of setting up having become relatively insignificant in the unit cost. The

[1] These definitions are taken from *Cost, Repetition, Maintenance — related aspects of building prices*, UN Economic Commission for Europe, Geneva 1963, from which much material in this chapter has been drawn.

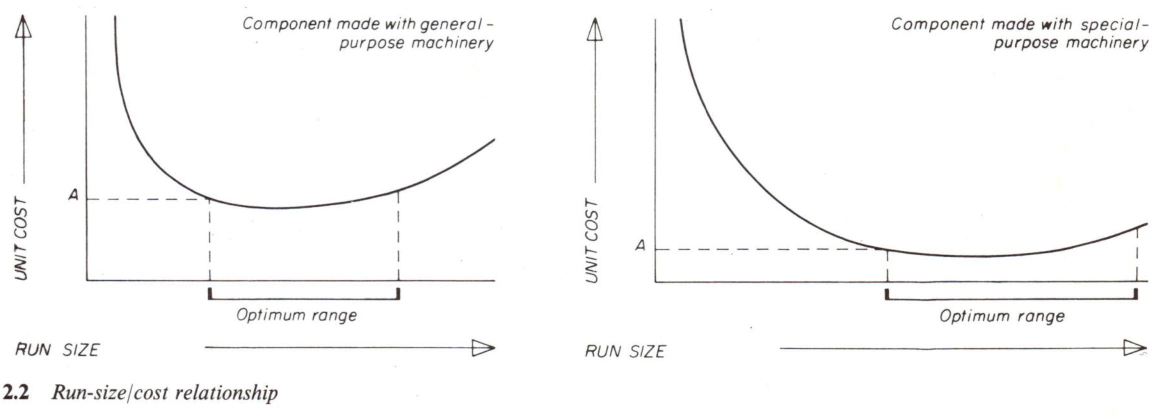

2.2 *Run-size/cost relationship*

approximately horizontal portion of the curve is the optimum range — the shape of the curve and the position of the optimum range will depend upon many factors, the most important being the production method and type of machinery used as indicated by the two curves. In most cases the curve begins to rise after the optimum range for the reasons already given.

The optimum run size also varies with the nature of the component and the market for which it is produced and can be used to classify components in relation to design/production problems. Such a classification is shown in table 2.2 in which the figures in brackets represent the mid-range of run sizes common to each group. Group I components, produced in long runs, are those manufactured as standard products based on the standards of the manufacturer or on those of the British Standards Institution or other organisations. They are manufactured by mass production methods with a high degree of dimensional standardisation and are used by the industry at large being selected from a catalogue in the design process. Group II components are those produced for a particular system of building. In this context variety is often required, for example, in external wall panels, and here the aim in design is to achieve the greatest variety within a given component type with a minimum of individual parts (see page 114 relative to flexibility in design with timber frame construction). Group III components are those produced for a particular project and include special purpose items and non-standard items such as windows of different sizes. Design of the latter should take account of the fact that standardisation of the frame sections and joints aids production by permitting the use of production line methods even though production runs may be short (see *Batch Production MBS: Components*).

Reference has already been made to the relatively small economic benefits accruing from factory production of many forms of building components because traditional heavy materials and, to some extent, traditional forms of construction are still used. The achievement of a significant reduction in costs will often necessitate completely new conceptions involving radical changes in design, new methods of fabrication and the introduction of new materials into the field of building. The modern flush door illustrates this very well. It can be produced very cheaply in long runs by mass production methods because it is fundamentally different in design to the traditional framed door, incorporating as it does different materials and using different forms of construction. The modern trussed rafter further illustrates how changes in design — in this case the overall design of the roof and the nature of the connections — not only facilitate mechanised factory production in long runs but lead to savings in site operations in roof construction (see page 175).

Group	*Run-size*	*Application*
I	High (10000)	Industry generally
II	Medium (500)	Closed system
III	Low (50)	Single project

Table 2.2 *Component classification*

The use of new materials and their effect upon production generally can be seen in the glass reinforced plastic roof components referred to on page 17. These use a relatively new material which can easily and cheaply be moulded repetitively, is light in weight, thus reducing fixing, handling and transport costs, and which results in a component which simplifies erection of the roof as a whole.

2.5 ASSEMBLY OF COMPONENTS

The process of site erection is, in fact, the assembly of a great number of components which may be large or small in size and the process itself is significantly affected by variations in shape and dimensions of the components, by the degree of accuracy in setting out the building and by the nature of the joints between the components. For reasons given below these factors gain in significance with the transition of much building to a system involving factory production of the parts.

The problem of assembling prefabricated components is primarily that of making things fit in an acceptable manner. The traditional method of building ensures satisfactory fits by dealing with variations in sizes of components and in setting out the building by well-known techniques. These, however, are not appropriate when components increase in size and adjustability in placing them and allowance for variations must be provided for in fewer joints. The nature of the joints, therefore, must be such that the assembly process and the attainment of fit is easy. To this end a minimum of joint shapes and loose accessories should be utilised in their design.

In some circumstances certain joints, the lap for example, determine the order in which the components must be positioned — from left to right or right to left as the case may be, or vertically from bottom to top and this may present difficulties in positioning the last component when they fit into a predetermined space.

Difficulties arising in assembly can be rated according to the degree of dimensional control required which may be termed 'degrees of restraint' (table 2.3). No restraint on dimensions is required with lapped joints assuming the overall height is not critical but when a panel component must be inserted into a gap it is necessary for the dimensions of both panel and gap to be controlled and produces a condition of one restraint. Additional problems arise when, for example, a panel or a window must fit into a preformed aperture, producing two degrees of restraint on dimensions. The most difficult problem arises in fitting a three-dimensional block into a hole, say a prefabricated staircase into a stairwell, where a fit all round is required as well as a fit at upper and lower floor levels. Such a situation of three degrees of restraint should be avoided in practice wherever possible.[1]

Problems of fit of this nature, which are unknown in the traditional building method, arise with the use of interchangeable prefabricated components. If they are to be solved economically they require forms of detailing in the design of the building fabric which avoid demanding of the manufacturer excessive accuracy in components with the consequent rise in costs.

2.5.1 Variations in components

Variations in the dimensions and shape of building components relative to each other and to specified dimensions are inevitable whatever may be the method of their production. This is due to a number of reasons, such as inaccuracy in workmanship during manufacture, as in the setting out and making of moulds, or to the nature of the materials, such as clay which twists and shrinks in burning or concrete which shrinks on setting and drying. These variations present difficulties when components are assembled on site to form elements of specified dimensions. They produce problems of 'fit'.

In traditional building these difficulties are satisfactorily overcome because of its nature. In this method of building the craftsman deals with the problem of variation in components by accommodating them in the joints between a large number of small components, as in masonry work in which the mortar acts as a filler, by cutting the components to fit or by making each component lie within a space between profiles or as defined by work already built, the component being fabricated after measurement of the opening into which it is to fit. Site adjustments are accepted as normal and the recognised sequence of work permits following trades to make good awkward junctions, inaccuracies and errors in earlier work as, for example, in the use of architraves and cover pieces to cover junctions and in the application of plaster finish to brickwork.

This method breaks down, however, when the building is constructed wholly or largely of prefabricated components, particularly when they are large in size, and the building process involves site assembly of completed components rather than site construction with basic materials. In order to ensure that these components will fit together on site it is essential (i) to relate their dimensions in some rational way, (ii) to define the spaces in the building in which they are located and (iii) to control any variations in their size by setting predetermined limits to them. The first and second requirements can be met by working within a system of *co-ordinated dimensions* by means of which the sizes of individual components may be related and on which may be based a reference system for locating components within the building by means of reference grids.

A system of co-ordinated dimensions also provides a framework within which manufacturing sizes may be fixed in such a way that variety in component size may be reduced while at the same time ensuring that any range of components provides the designer with sufficient flexibility.

Dimensional co-ordination based on the use of a common dimensional unit or 'module' is known as *modular co-ordination.* In Great Britain the accepted module is 100 mm as it is in metric countries. The use of a module of this size permits adequate standardisation of component dimensions but is small enough to allow adequate flexibility in the design use of the components.

The third requirement for ensuring fit on site is met by a system of *tolerances* by means of which variations in component size are limited and cutting to fit, even if feasible, involving waste of material and time during site assembly is eliminated.

A tolerance is the difference between the limits within which the size of a component must lie and the system

[1] See BRS Current Paper CP 31/68: *Metrology and the Module* where this concept is propounded and its implications are considered in relation to the building industry. Table 2.3 is based on a diagram in this paper. See also CP 7/72 *Factors influencing the choice of tolerances.*

TYPES OF ASSEMBLY OPERATION	Degrees of Restraint	Nº of critical dimensions	Problems
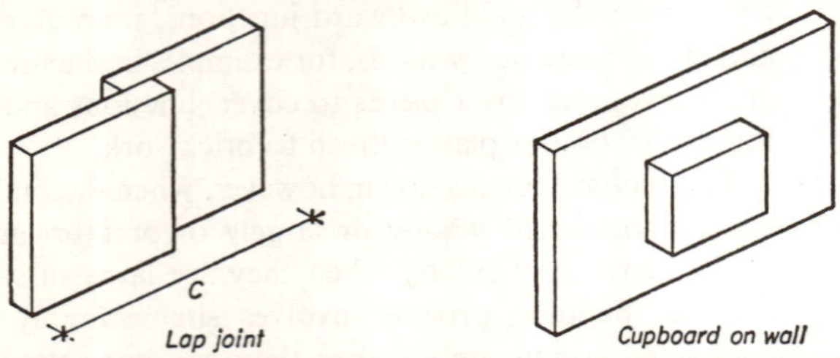	0	0	Nil Use wherever possible
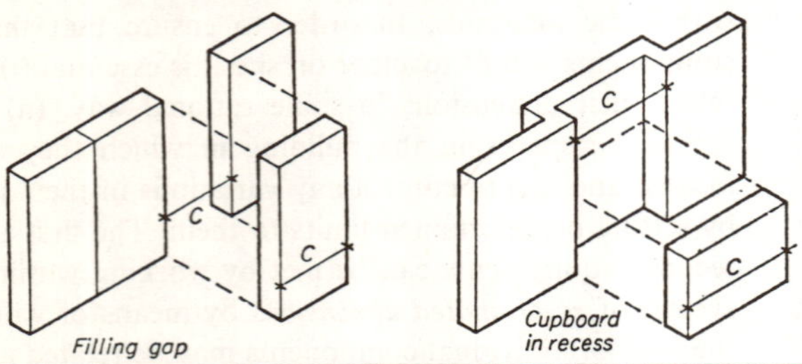	1	2	Dimensional control required
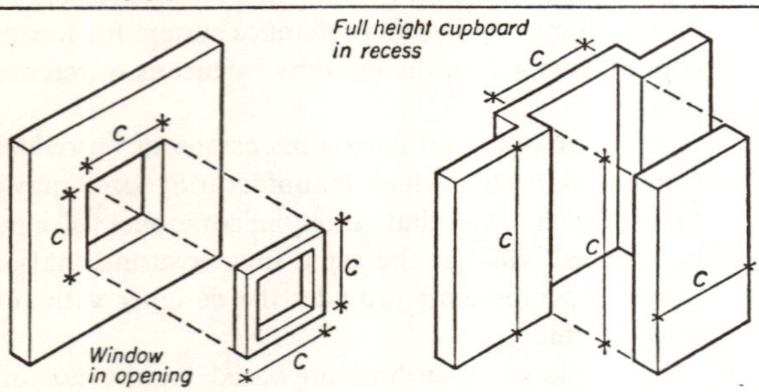	2	4	Difficult Avoid if possible
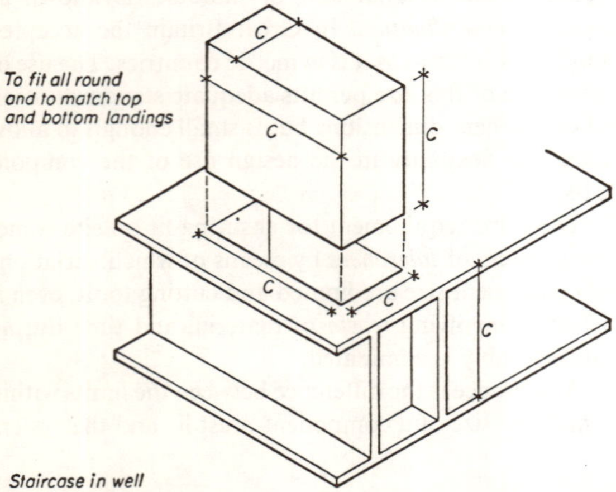	3	6	Very difficult Avoid

Table 2.3 *Assembly operations and dimensional control*

of tolerances provides a standard method of determining the limits of the manufactured sizes of a component relative to its modular size, that is the size of the space within which it should fit. The actual size must always be less than the modular size, the upper limit being determined by the minimum width of joint required and an allowance for positional variation in site assembly. The lower limit is determined by the allowance for variations inherent in manufacture such as twisting, bending, shrinkage. The limits may be indicated either by specifying the permitted maximum and minimum dimensions or by specifying the mean of these with the appropriate plus and minus variations.

The subjects of dimensional co-ordination, reference grids and tolerances are discussed at length in chapter 1 of *MBS: Components*, to which reference should be made.

2.5.2 Setting out and positioning

Problems of fit also arise from inaccuracies in the building process: in setting out the building and in positioning components. Dimensional variations are greater than is usually assumed and a range of error in setting out of up to 75 mm is not uncommon in many types of building, and substantial errors can occur in positioning components in relation to reference lines due to them being out of line, out of level or out of plumb. In small-scale steel frames column spacings, alignment and plumbs can vary up to 18 mm, requiring adjustments in fixings of ±18 mm for single-storey buildings and more for two-storey frames.[1] Such standards of accuracy while tolerable in traditional building are inadequate for component building of any sort.

Inaccuracies in setting out arise from variations in the accuracy of the instruments used and in the skill and care with which they are used.[2] In an endeavour to achieve greater accuracy and economy in time spent in setting-out experiments in the use of jigs have been made in Great Britain and abroad. Where considerable repetition is involved much time can be saved and where, as in building with prefabricated components, the assembly process is one of some precision jigs may ultimately be used to achieve the accuracy in setting out which is so essential in this method of building.

The weight of components and the ease with which they can be handled and the design of the fixings and supports are significant factors in positioning components. The subject of fixings is referred to in the section on *Claddings* in Part 2.

Over recent years information on the degree of accuracy achieved in setting out on site and in manufacture has gradually been accumulated from observations and measurements made in this country and abroad.[3] It is essential that design details, especially of joints, take this information into account so that they will function satisfactorily in practice.

2.5.3 Joints

The erection process, as indicated earlier, is essentially the assembly of parts put together in different ways to form the total building fabric. The parts meet at joints or connections. In the field of building *joint* refers to the space between components whether or not they are in contact; a *connection* has the added implication that the components are held together structurally.

For practical reasons joints are necessary between the elements of the building fabric, say between wall and floor, and between components making up the elements, in order to accommodate changes in material or because of limitations in size due to the nature of the materials used, as in bricks, or, conversely, in order to keep the parts to a manageable size. Other joints may be incorporated in order to control cracking due to thermal, moisture or structural movement of the fabric or to permit breaks in the construction process, as in casting in situ concrete.

It has been suggested that 'fundamentally there is only one need for a joint in building construction, and that is where there is a change in constructional material ... all other types of joint may be avoided by attention to design and to the method of construction'. It is pointed out that with the existence of cranes which can handle large and heavy units those joints required to reduce components to a manageable size could coincide with the essential joints occurring at changes of material. This could lead to the proposition that 'areas of one material ought to be jointless'.[4]

However, unless all components are fabricated on site transport restrictions limit their sizes and, although

[1] See BRS Current Paper, Construction Series 23: *Accuracy in Building*, and Current Papers CP 8/68, 18/69, 22/72 and 99/75 on dimensional variations in steel and in situ concrete frames.

[2] See BRE Digest 234: *Accuracy in setting-out.*

[3] See footnote[2] page 26.

[4] See BRS Current Paper CP 1/71: *Joints in the context of an assembly process.*

the crane has increased the manageable size of components far beyond the limits set by manhandling, there is an upper limit to the size of components which can be handled, even though they be wholly prefabricated sections of a whole building, so that some joints between the parts will be essential.

Functional requirements of joints

The purpose of a joint is to relate the adjacent components in such a manner that, while allowing satisfactory assembly, the functions of the components joined are preserved across the joint so that the functional integrity of the whole element is preserved. Thus the functional requirements of the joint include those of the components such as the provision of adequate weather and fire resistance, thermal and sound insulation and, if it is a structural connection, the ability to transfer forces. In addition, other requirements arising simply from the existence of the joint must be fulfilled, such as accommodating thermal and moisture movements and variations in size and in positioning of the components in order to avoid cutting and fitting on site, and permitting, in some cases, dismantling of the components for possible replacement.[1]

Types of joint

Joints may be classified according to the shape of the component edges, the positional relationship of the edges and whether or not some jointing member is required additional to the component edges. In this context there are *integral* joints in which the component edges are so shaped that they form the complete joint and *accessory* joints in which additional parts are used to form the joint. Each of these can take different forms as shown in table 2.4.

The *butt joint* is the simplest shape but requires, as an essential part, some filler or seal to enable it to function satisfactorily. The *lap joint* produces a less direct path between the components. Both butt and lap joints may present some difficulty in practice in aligning the adjacent components.

Integral joints The *partial lap joint* may be formed with the adjacent edges similarly or differently shaped as shown. This type of joint is commonly used between external vertical components. The *mated joint* is an interlocking joint in which the adjacent edges are differently shaped. It aligns the components and restricts their relative movement — normal to component face in 'edge-mating' and parallel to the face in 'face-mating'. The interlock makes penetration through the joint more difficult than through a lap joint, but dismantling may prove difficult.

Accessory joints The *spline joint* incorporates a loose accessory in the form of a spline which links the adjacent components, the edges of which may be grooved to take an 'internal spline' or tongued to accommodate an 'external spline'. The spline must either be positioned as the components are offered up or be subsequently inserted into the joint from one end. This joint possesses the advantages of the mated joint with the added advantage of identical edge shapes to the components. Dismantling could be difficult. The accessory in the *cover joint* is a cover piece applied to one or both faces making dismantling easy while preserving the advantages of the mated and spline joints. The *frame and bead joint* consists of a framing member into which the components fit and to which they are secured by beads or stops fixed to the frame. It has the advantages of the cover joint and permits adjacent components to be different in thickness.

In the frame and bead joint the frame and in the external spline joint the spline, which are in contact with both the internal and external air, may form a 'cold-bridge' through which heat loss can take place resulting in condensation on the inner surfaces. This can be prevented and methods of doing so are illustrated in Part 2 under the section on curtain walling.

A butt joint between external vertical components may be made water-resistant by the inclusion of an accessory in the form of a *zed* flashing as shown in table 2.4.

Filled and open joints Resistance to water and wind penetration is a major functional requirement of the enclosure of a building and, therefore, of any joints between the components which form it. Two diametrically opposed forms of joint are used in this context — the filled joint and the open joint. In a *filled joint* the space between the components is filled by some form of weather-proof seal upon which the joint depends for its efficiency. As indicated above provision must usually be made for movement of the adjacent components

[1] See BS 6093: 1981 *Code of practice for design of joints and jointing in building construction* for list of joint functions.

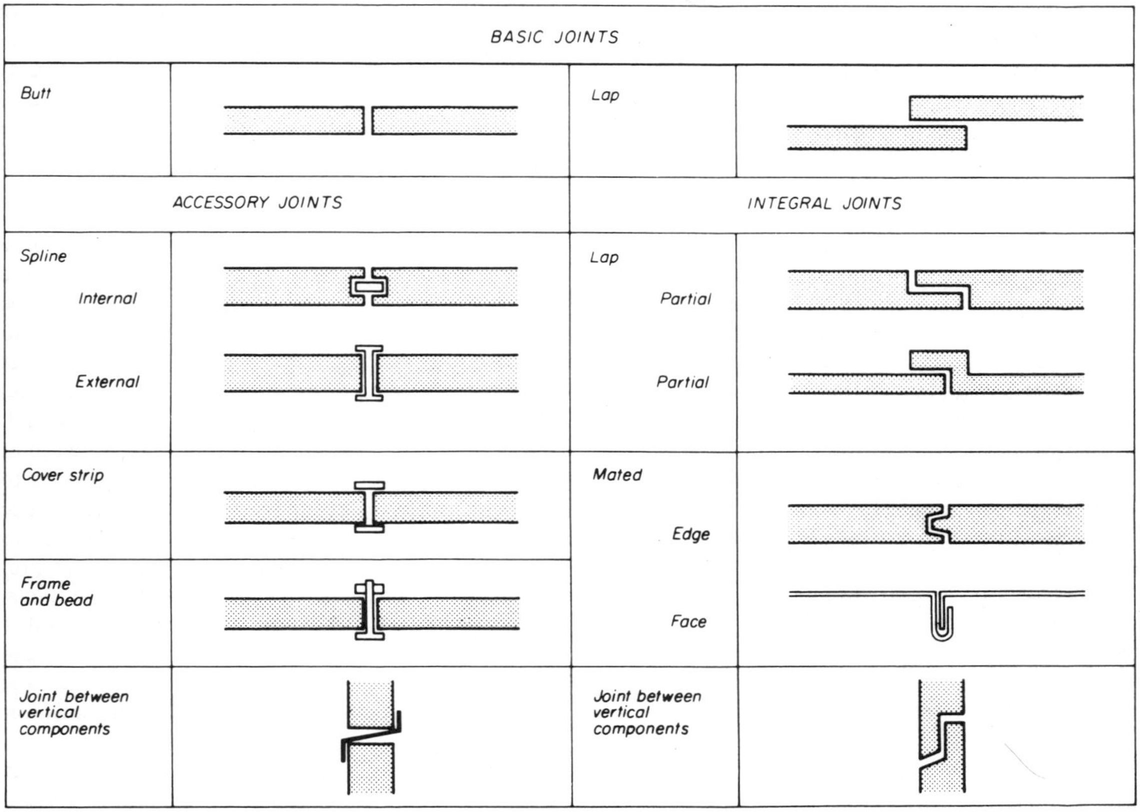

Table 2.4 *Classification of joints*

implying flexibility at some point, either in the joint itself or in the components, and this constitutes a major design factor when joints of this form are adopted. The *open joint*, in contrast, permits water to enter it but is so designed, either as an 'open-drained' joint or in conjunction with a ventilated air space within the building enclosure, that the penetration of water into the joint is minimised and the passage of any water which might enter is controlled and directed to some means of discharge to the outer face of the wall. The advantages of this joint are the ease with which movements and dimensional tolerances can be accommodated, the protection from exposure of any sealing material required as a wind barrier and the relative ease of erection.

Apart from the full-lap joint all types of joint, whether filled or open, involve edge-butting. Filled butt joints require a fairly high degree of dimensional control since the effectiveness of the seal depends upon the width of the joint relative to expected movements. The open joint demands less accuracy in the overall dimensions of the components (although the extent of the variations must be known for design purposes) but in the form of an open-drained joint requires somewhat complex edge details which complicate the production of the components.

The full-lap joint, widely used in mechanical engineering, avoids the need for close dimensional control of the components and the openings to which they are related, thus avoiding the costly demands on the manufacturer for great accuracy in components to which reference was made on page 21. It has the added advantage of tolerating greater thermal and moisture movements in the components. Problems of fit are thus reduced and assembly is simplified. The possibilities inherent in the wider use of this type of joint are discussed in Building Research Establishment Current Paper CP 1/71: *Joints in the context of an assembly process*, to which reference should be made.

Joint seals

As already indicated a seal may be incorporated in a joint to provide a water and/or wind barrier. There are two types — rigid and non-rigid.

A *rigid seal* may be formed by gluing or welding the components together and its stiffness requires movement to be accommodated in the components themselves or in special movement joints provided at intervals in the building fabric.[1] Similar requirements apply to a face-mated joint in respect of movement parallel to the component face which it restricts. A *non-rigid seal* is one which permits movement to be accommodated within the joint itself by virtue of its inherent flexibility. It can take the form of (i) a 'soft seal' using a mastic or sealant which depends for its sealing property upon adhesion to the component edges and which allows movement by its ability freely to compress or stretch or (ii) a 'compressive seal' formed with a gasket, which is a preformed resilient strip. This depends for its sealing property upon sufficient pressure being imposed on it during assembly that full contact with the component edges is maintained as the components move.

Joint design

The design of any joint requires the careful consideration of the functions it must perform and an assessment of the conditions under which it will function. It must also have regard to the dimensional accuracy of the components likely to be achieved in manufacture and in construction, and to the nature and properties of the materials to be used. On the basis of this a choice must be made of the type, shape and width of joint appropriate to particular circumstances.

Width of joint The width of the joint must allow for manufacturing tolerances and for inaccuracies in the setting out of the building and in positioning of components. It must also accommodate the changes in size of the components due to changes in temperature and moisture content and in some circumstances structural movement must also be considered.

The extent of all these variations must be determined. Manufacturing tolerances can be agreed with the component manufacturer who must, of course, employ dimensional quality control to ensure production within the agreed limits. Information on what variations in accuracy occur in practice in setting out and in the assembly of components has been gathered by the Building Research Station and other bodies and has been published[2] and information on the extent of thermal and moisture movements can be obtained from a number of sources.[3] The movements in joints which occur in practice are complex and actual movements cannot be predetermined with great accuracy. However, the Building Research Station has investigated these and some findings have been published.[4]

When the limits of these variations have been determined they must be combined to give the total effect on the joint. A method of doing so is given in BS 6954: Parts 1, 2, 3: 1988 *Tolerances for building* which introduce design procedures based on a statistical approach to the sizing of components and the design of joints.

In determining the width of joints incorporating sealants account must be taken of the properties of these materials. The ability to tolerate movement (ie deform) varies with the type of sealant and the degree of deformation with the width of the joint. The allowable movement in a sealed joint is expressed as a percentage of its width and the minimum joint width resulting from the variations referred to above thus determines the allowable movement in a particular joint. The width of joint, therefore, must be related to the particular sealant to be employed or, alternatively, a sealant must be selected appropriate to the anticipated deformation caused by movement of the components.

Filled joints Where a sealant is to be used in a joint the latter, as indicated above, must be sufficiently wide to ensure that the sealant is not overstressed by the movements which will occur, the movement per unit width of sealant increasing with a decreasing width of seal. The sealant must also be of sufficient depth because the width/depth ratio is an important factor in attaining maximum toleration for movement in the material (figure 2.3). Information on these and other

[1] See Part 2 *Movement control.*

[2] See BS 5606: 1978 *Code of practice for accuracy in building*; BRE Digest: 234 *Accuracy in Setting-out.* BRE Current Papers CP 8/68 *Dimensional Variations in Framed Structures*, CP 10/73 *Width variations of cladding joints* and CP 99/75 *Accuracy of in situ concrete.*

[3] eg *Principles of Modern Building* (HMSO), *MBS: Structure and Fabric*, Part 2, *MBS: Materials* and BRE Digest 75: *Cracking in Building.*

[4] See BRE Digests 227, 228, 229: *Estimation of thermal and moisture movements and stresses.*

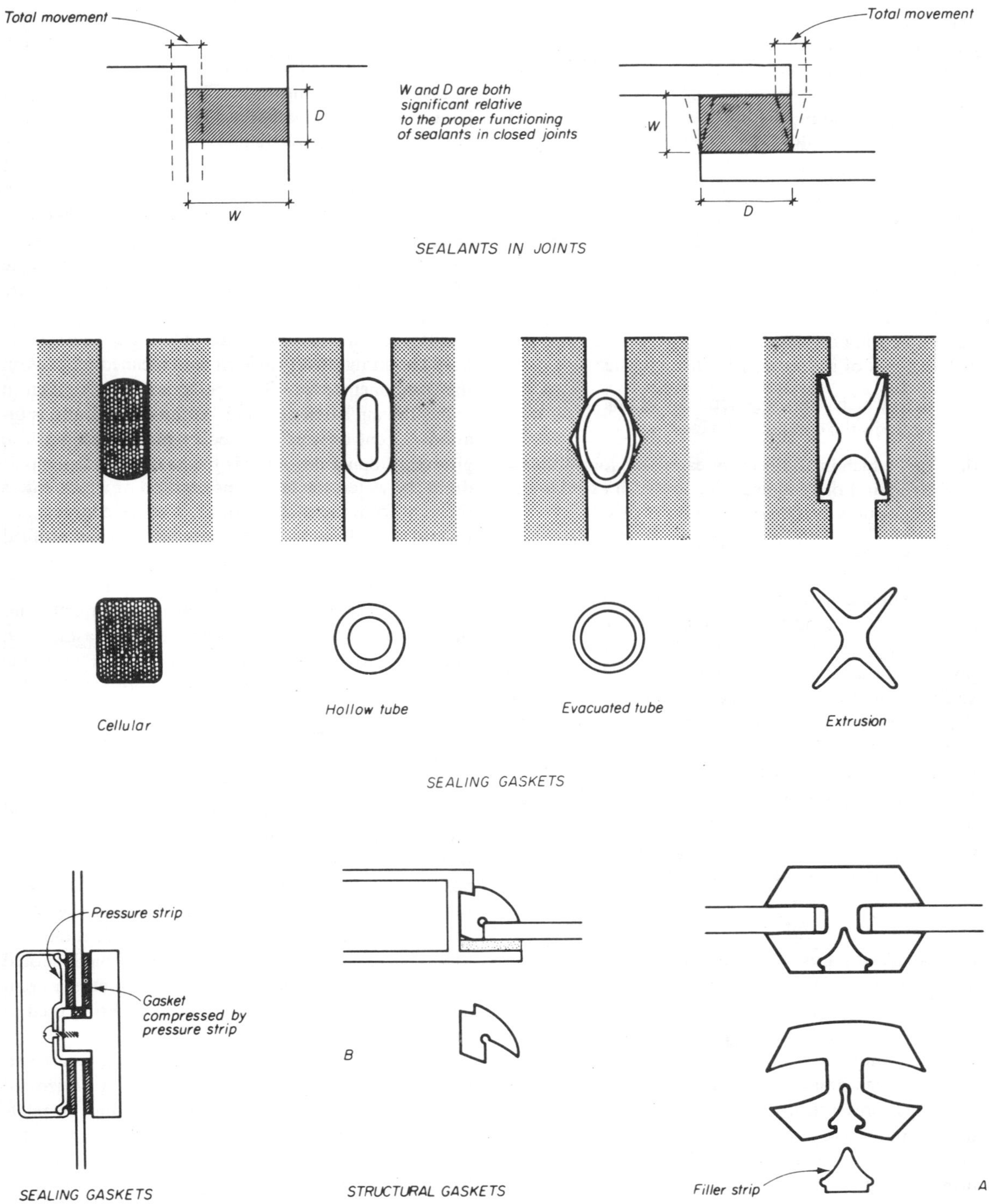

2.3 *Sealants and gaskets*

design considerations relevant to the use of sealants in joints are discussed in *MBS: Materials*, chapter 16, to which reference should be made. Good adhesion to the component edges is essential and, since exposure to light and air shortens the life of sealants, the shape of the joint should give protection to the seal while permitting access for renewal.

Where movements or joint widths are likely to be greater than sealants can accommodate, gaskets may be used since they can be sized to deal with greater variations than sealants. They can be of solid compressible material or may be hollow sections and can be shaped according to the joint requirements. They require no curing or hardening.

Two types of gasket may be distinguished:

(i) that which seals the joint by filling the space between the component edges;

(ii) that which in addition to sealing the joint fulfils a structural or retaining function such as retaining in position the components on each side.

Where gaskets of the first type can be placed in position before both components are assembled, pressure is applied to it either by bringing together the components at a butt joint or by means of applied beads or pressure strips (figure 2.3). When both components are assembled, before the gasket is applied a gasket in the form of an evacuated tube is inserted in the joint and then is expanded by readmitting air by puncturing the tube. The edges of the joint need to be smooth and its width reasonably regular.

Pressure is applied to gaskets of the second type by means of a keyed filler strip (figure 2.3 *A*) or by deforming the gasket by inserting it into a specially shaped component edge into which it keys (figure 2.3 *B*).

Open-drained joints The adoption of the open joint avoids the risk of loss of adhesion or contact of the seal, the need for extreme care in positioning the components in assembly and the need to depend on close tolerance limits.

In an open-drained joint air penetration into the building is prevented by an air-tight barrier at the back of the joint, where any sealant used is only slightly, if at all, exposed to the weather, and where it may be so disposed and sized that it can function well within its capabilities. This air barrier, by preventing an air flow through the joint, causes a build-up of air pressure at the back of the joint equal to that on the external wall face. This equalisation of external and internal air pressures minimises the penetration of wind-driven rain into the joint since there will be no pressure difference to move the water inwards. The depth of the vertical joint from the external face to the air barrier should be not less than about 100 mm to provide an outer zone in which most of the water is drained away before it penetrates to the air barrier and an inner zone into which only a very small amount of water is likely to enter and from which it can also drain away. Horizontal joints may be based on the overlapping principle of traditional tiling and weatherboarding and are simple and efficient.

These features are illustrated in figure 2.4. (*A*) shows a simple, straight vertical joint with inclined grooves on the component edges. This, provided care is taken in fixing the air barrier, would be suitable in most situations except on very exposed or high buildings. The grooves, sloping downwards to the outer face, help to drain the water, the largest proportion of which enters the joint from the external surfaces of the components, towards the front of the joints. Rough or profiled exterior wall surfaces will improve weathertightness by restricting the flow of water across the surface towards the joint. Very little driven rain will blow directly into the joint and reach the air barrier for the reason given above. A variation of (*A*) is shown at (*B*) in which the joint is joggled to provide some direct protection to the inner zone. The alternative, and preferable method, to these is the provision of a loose baffle as at (*C*) which itself provides a check against the penetration of water to the air barrier and also forms a physically separated ventilated inner zone in which the air pressure will build up to that on the exterior wall face so that little or no water will pass into it. Rough external wall surfaces are not essential with a well constructed joint of this type, but vertical profiling of the weatherside of the baffle will assist the run-off of any water which might blow on to it. The depth of the grooves holding the baffle should be such that the tolerance range of the joint is accommodated, an appropriate width of baffle being used as each joint may require.

Horizontal joints should incorporate a weathered lower edge and an overhang or overlap. A horizontal overhang can be sufficient for low buildings in relatively sheltered positions and an inclined overhang for normal exposures as indicated in figure 2.4, but an overlap is preferable varying in depth according to the conditions of exposure and ranging from 50 mm in sheltered positions to 100 mm in exposed conditions. Adequate flashings at the intersections of horizontal

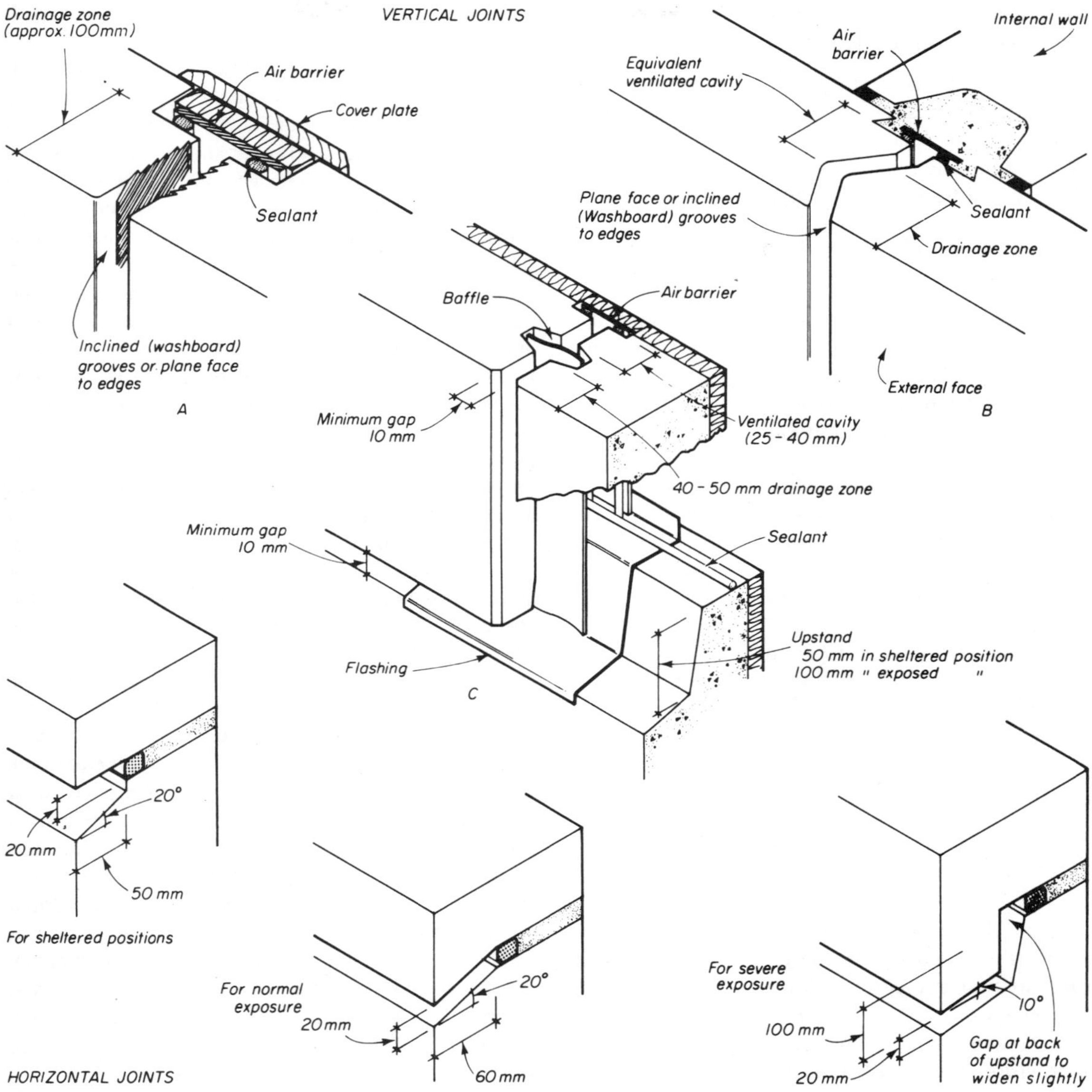

2.4 *Open-drained joints*

and vertical joints must be incorporated in order to conduct to the front any water draining down the vertical joints as at (*C*).[1]

The alternative, mentioned earlier, to the individual open-drained and ventilated *joint* just described is the open-drained and ventilated *cavity* or space between outer cladding components and the inner structure of the wall, functioning on the same principle of equalised air pressures. This method is described under rain screen construction on page 75.

[1] See BS 6093: 1981 *Code of Practice for design of joints and jointing in building construction* where the design of these joints is discussed and from which this material has been drawn.

Connections As defined on page 23 these must fulfil the following functional requirements:

(i) be able to transfer safely the imposed loads without high local stresses
(ii) not move or rotate excessively
(iii) accommodate tolerances in the components
(iv) require little temporary support, permit adjustment and be simple to make
(v) permit adequate inspection and rectification if necessary.[1]

In concrete work tolerances of necessity must be greater than in steelwork. Practical forms of connections in reinforced concrete and steel structures are described and illustrated (as are joints) in the later chapters of this volume and in Part 2.[2]

In relation to the whole topic of the assembly of components the Building Research Establishment has published a handbook *Graphical Aids for Tolerances and Fits* as a means of assistance in designing for fit. Reference should also be made to Current Paper 29/77 *Designing for fit using BRE graphical aids*, BR Digest 199 *Getting good fit* and Construction Industry Research and Information Association Technical Note 113, 1983, *A suggested design procedure for accuracy in building*.

2.6 ECONOMIC ASPECTS OF BUILDING CONSTRUCTION

Economically the design of a building is concerned with the provision of good value, in terms of standards of space, environment, construction and appearance, for the money expended. It is thus concerned with economic building rather than with cheap building, with the provision of required standards at the lowest cost. Costs must, therefore, always be considered in relation to standards and in this context the term *costs* means costs over the whole life of the building, taking into account running and maintenance costs as well as the initial costs of construction.

The resources for building are money, men, materials and machinery. Men must be employed and paid, materials must be purchased and machinery must be bought or hired. The manner in which materials are incorporated in the fabric and structure of a building at the design stage and in which materials are handled and equipment deployed on the site or in a factory all affect the degree of expenditure of money and the overall economy of a building project.

In designing the building fabric the architect may not know what equipment and methods the builder may use and will not be able to take account of these in his design. In spite of this, however, the fact that his design decisions have a direct effect upon the materials used and how they are used makes his contribution to the achievement of economy a major one. One reason is that in most cases materials account for the greater proportion of the total cost of materials and labour and this means that the economic or the excessive use of materials, which is under the control of the architect, has greater cost implications than variations in labour which is controlled by the builder. This does not mean that reduction in labour content is not important but that the relative value of efforts by the builder in this direction can easily be reduced when the design makes uneconomic use of materials. The other reason is that complicated detailing on the part of the architect can result in operational difficulties, excessive labour requirements and extra time spent in fabrication and assembly, all of which makes it difficult for the builder on his part to effect economies in the assembly process.

In addition to the choice of techniques for the construction of the building fabric, other factors, related to the planning of a building, play a significant part in the economics of the building as a whole. In the context of this book, therefore, the economics of a building may be considered under three headings — planning, design of the building fabric and production. Each affects the total cost of the building but all are inter-related and factors within each cannot be changed without affecting factors in the others. These, of course, take no account of the economic significance of the processes and patterns of building development generally which is not a subject for discussion here. Indeed, it is only possible here briefly to consider some aspects relating to these three areas.

[1] See BRS Current Paper, Design Series 17: *Large Panel Construction.*

[2] See also BRE Current Papers, Engineering Series 45 and 46; *Tests on joints between precast concrete members* and *Connections (joints) in structural concrete* respectively.

2.6.1 Planning

Site considerations On sloping sites long buildings should be sited to follow the contour lines as this involves less filling under the building and less foundation walling. Where *cut and fill* (see page 190) is adopted the ground floor level, if possible, should be so arranged that a greater volume of excavation is required than filling, because again, the higher the filling the more walling to retain it is required. When excavation into a slope is adopted it is usually more economical to continue the cutting beyond the building and slope it back until the soil is self-supporting as this avoids the expense of making the wall of the building a waterproofed retaining wall (figure 8.2 *C*).

Plan shape The plan shape of a building influences its cost due to its effect on the amount of materials and labour required and its effect on the builder's site organisation.

External loadbearing walls and cladding are high cost items so that for the same floor area, a reduction in the perimeter of a building will usually produce economies. Apart from the circle, which is expensive to construct, a square plan form requires the least perimeter and, therefore, the least enclosing area of wall for a given floor area; the longer and narrower the plan the greater the perimeter for the same enclosed area (figure 2.5 *A*). Simple plan shapes with a minimum of insets and projections are the most economical in external walling (figure 2.5 *B*) even though such shapes may necessitate an increase in internal partitions, since these are very much cheaper than weatherproof external walls. Simple shapes are also economical to build, construction time is saved and mechanical plant can be most efficiently used. (See Part 2, chapters 1 and 2.)

Irregularity in section as well as in plan results in complex building forms and higher costs because of the increased number of corners and junctions of parts, with their attendant complexity of technical problems in weatherproofing, bonding and junctions (figure 2.5 *C*).

Plan size Cost per square metre of floor area tends to decrease with increase in plan size because the perimeter length per square metre of floor area reduces as size increases (figure 2.5 *D*). In terms of cost per square metre it is usually economical to increase size even at the expense of a less efficient shape as variations in size are of greater economic significance than those of shape.

Large plan areas, however, especially in tall buildings, may reduce the ease with which the builder can get men and materials to every point and may increase construction time. They also present certain problems in relation to fire protection which are discussed in Part 2 chapter 10.

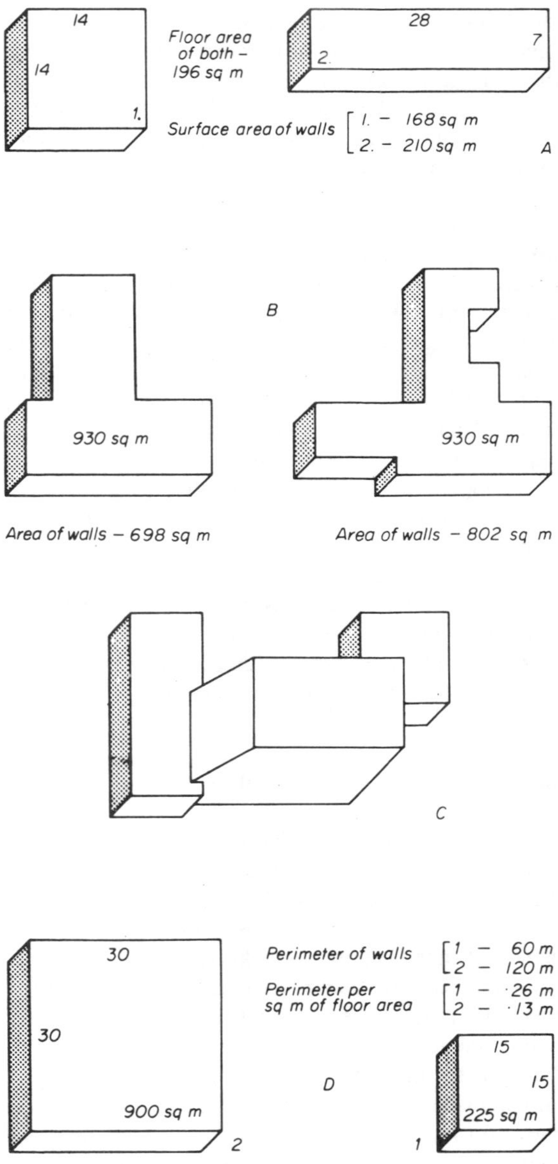

2.5 *Influence of plan shape and size*

Layout of accommodation Variations in plan layout can result in an increase or decrease in the proportion of circulation space to usable floor area. A small ratio of circulation to useful floor area is obviously economically advantageous, especially when a building is erected for letting at rents based on useful floor area. Economy in circulation space should not, however, be made at the expense of greatly increased perimeter or of an uneconomical structural form.

Height of building The average cost per square metre generally increases with the number of storeys due to (i) the increase in perimeter walling for any given total floor area, (ii) the effect of increased load on the structure, (iii) the additional hoisting of materials and the extra time taken by operatives to reach the higher storeys.

Foundation costs vary approximately in proportion with load and, thus, with height but the cost of the structure as a whole per square metre of floor area increases rapidly above four storeys because of the greater strength required in loadbearing walls or the need to introduce framed construction. The cost per square metre of a framed structure continues to increase with increase in the number of storeys due to the requirements of windbracing and because of the increasing size of columns, although the cost of these does not increase in proportion to the increase in height.

The effects on services of the various planning factors referred to above are discussed in *MBS: Environment and Services* but it may be said here that services become more costly as plan shape becomes complex and as the height of a building increases.

2.6.2 Design of the building fabric

The cost of a building is influenced greatly by the form and details of the construction adopted for the parts, the economy of which depends upon the choice and use of materials and involves the costs of materials, labour, plant and organisation. The relative costs of materials and labour at any given point in time has a significant effect upon the choice of the form of construction and is referred to on page 7. The manner of combining materials so that the number of operations in fabrication and assembly is reduced to a minimum and which allows for the repetition of similar operations can effect considerable economies in construction and is referred to on page 13.

Simplicity in detailing to take account of the operations which the craftsmen or machine must perform and detailing which allows the separation of the work of different trades to permit continuity of work results in greater speed of construction with consequent economies. This has been referred to earlier in this chapter and further reference is made in chapter 1 of Part 2.

It is essential to consider the cost of the building fabric as a whole rather than simply the costs of the isolated parts, for a more economic solution often results from the use of a more expensive component at one point in order to obtain substantial savings elsewhere. Illustrations of this are given at various points in both Parts 1 and 2: for example, the necessity of considering the interrelated actions under load of foundations, superstructure and soil in order to achieve the most economic overall design solution (chapter 4, this volume and chapter 3 Part 2) and the economies which may be derived from the use of concrete blockwork instead of brickwork even though the individual units themselves may be somewhat more expensive (chapter 5).

Some forms of structure are more economic than others for buildings of different shape, size and height; some are more economic than others for different plan forms. Different use requirements, necessitating for example heavy loading, wide spans or large storey heights, will affect the cost of the main structural elements of the building and the most appropriate forms of these, relevant to the particular requirements, must be selected to produce the most economic solution (see, for example, sections on *Choice of Structural Frames* and *Choice of Roof Structure* in Part 2).

In this section on economic aspects it has been the first, or capital, costs which have been considered. The total cost of a building is, however, made up over its useful life of the initial capital cost plus the cost of maintaining it in a useful condition, together with the running costs of heating, ventilating, lighting and cleaning of the building.

The significance of maintenance in the national economy may be gauged from the fact that a considerable part of the building labour force in Great Britain is permanently employed in work of maintenance and repair, and as much as forty per cent or more of the total expenditure on building work can be accounted for by this work.

Reduced initial expenditure often leads to increased maintenance and running costs although in some circumstances it may be more economic to use low-cost materials and to maintain them at frequent intervals as, for example, in short life buildings. In contrast it may be advantageous to spend money on a high degree of thermal insulation in order to reduce the capital cost of heating plant and the costs of heating during the life of a building. The problem of assessing what to spend initially in order to effect future savings is not an easy one but methods have been developed which take into account estimated initial, maintenance and running costs on an equivalent basis to produce a total *costs-in-use* figure for a proposed design solution and it is this figure, rather than the estimated initial cost alone, which should be used to evaluate alternative designs.[1] It should be noted that the cost of maintaining the main structure of a building is small relative to that for its fittings, finishes and services. Particular attention should, therefore, be paid to those components which account for greatest maintenance expenditure. In so doing the whole design is rationalised in relation to real cost to the owner.

2.6.3 Production

The production process is concerned with the nature and sequence of operations which are involved in the erection of a building, and through which the resources for building are deployed. It is thus an organisational process which is the responsibility of the contractor and the ease with which it can be carried through can be helped or hindered by the architect's design decisions.

This chapter has been very much concerned with the subject of the production process and at this point little remains to be said other than to refer back to some of the factors in terms of economic building.

It has been shown that proper planning and organisation of the whole process is of paramount importance in the achievement of an economical result, whatever the method of building employed. Involved in this is the need for continuity of work at all stages of fabrication and assembly and the desirability of the separation of fabrication from assembly operations on site. The ease with which this can be achieved can be greatly assisted by the architect at design stage if he is aware of the operational significance of his design decisions (see chapter 1 Part 2).

The nature of on-site and of factory fabrication have been compared and attention has been drawn to factors such as scale of output and variety reduction leading to long runs, which result in economic factory production.

In the assembly or erection of components on site mechanical handling plant has made possible the development of techniques which separate fabrication from assembly operations and which at the same time simplify the problems of fabrication by bringing the operations to ground level. Examples of this have been quoted, such as the prefabrication of large shutter panels and cages of reinforcement, taking advantage of the mechanical plant which can handle them into position at any level.

Economic aspects of contract planning, site organisation and the use of mechanical plant are discussed in the first two chapters of Part 2, to which reference should be made.

[1] See *Building Economy* (Pergamon Press) and *Building Design Evaluation: Costs-in-use* (Spon) both by P A Stone. See also *MBS: Environment and Services* chapter 7, for an application of this method to the costs of heating systems.

3 Structural behaviour

It has been suggested in the first chapter that an appreciation of building construction is developed from a knowledge of materials and an understanding of structural principles. An early consideration of these principles is, therefore, relevant to a study of building construction and it is the purpose of this chapter to introduce the broad principles underlying the behaviour of the main component parts and elements of a building under load and to discuss the reasons for the shape and form of these parts.

It is not proposed to consider here the process of structural design as such. This involves an analysis of the actual forces acting on the structure and of the forces they set up within the structure, so that these can be related to the properties of the materials to be used and the sizes of the individual members calculated. In order to carry out this process however, it is necessary to have an understanding of the general nature of the forces which act on buildings and a conception of the behaviour of a building and its parts under these loads. Such a conception is also fundamental to the production of an economic structure for from it springs the possibility of arriving at an overall economic solution at an early stage in the total design process when many of the related design factors are still under consideration.

Before commencing a study of the behaviour of buildings and their parts under load it is necessary to define certain terms which must be used in the study and to indicate the effect of forces upon the material of which a building is composed.

3.1 STRUCTURE AND FORCES

A **structure** is defined as a body, or fabrication, at rest as compared with a machine which is a fabrication of moving parts. At rest, or *static*, means in a state of equilibrium in which no movement (or, for practical purposes, no appreciable movement) occurs under the action of a number of forces. Statics is the study of forces on bodies at rest and the structural principles developed from this study are concerned with the strength and stability of structures which, in the context of this book, means building structures.

Strength is the ability of the material of a structure and its parts to resist the forces set up within them by the applied loads.

Stability is the ability of the structure to resist overall movement, for example, overturning, and of its parts to resist excessive deformation.

Force is defined as that which changes or tends to change, the state of a body whether it be of rest or of uniform motion in a straight line. Any object exerts force on account of its own weight and tends to move in a downward direction. To maintain equilibrium, that is to prevent movement, this must be resisted by an equal, opposite force. The force due to the weight of an object or structure or to any applied loads is known as the *active force* and that which resists it as the *passive force* or the *reaction* (figure 3.1).

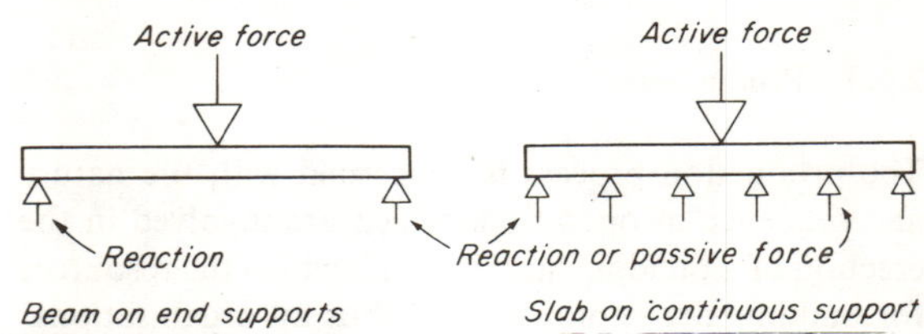

3.1 *Forces and reactions*

Loading is the term for the forces acting on a building. These are classified as (i) *dead loads*, made up of the self-weight of the building fabric, which are permanent loads; (ii) *live loads*, made up of the weight of the occupants, furniture and equipment in a building and similar applied weights which are temporary loads in the sense that they may or may not be always present. They are also referred to as superimposed or imposed loads; (iii) *wind load*, which is temporary but is considered separately. In some areas snow loads and earthquake shock must be taken into account.[1]

Loads are termed *distributed* loads if they are applied over the full area or length of a structural member and *concentrated* or *point* loads if they are concentrated at one point or over a very restricted area (figure 3.2).

[1] See BS 648 for weights of materials on which dead loads are usually based, BS 6399: Part 1 and Part 3 for superimposed loads on floors and roofs respectively and CP3: Chapter V: Part 2 for wind pressure on buildings.

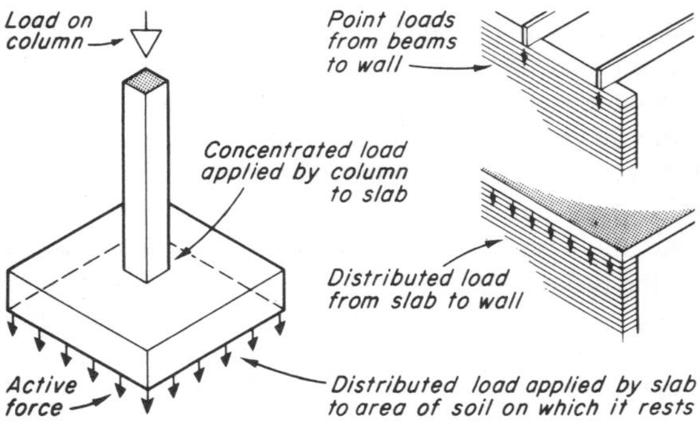

3.2 *Concentrated and distributed loads*

3.1.1 Stress

A force on a structural member may (i) stretch it, when it is termed a *tensile* force; (ii) compress it, when it is termed a *compressive* force; (iii) cause one part of the member to slide past another, when it is termed a *shear* force; and (iv) cause the member to twist, when it is termed a *torsional* force (figure 3.3).

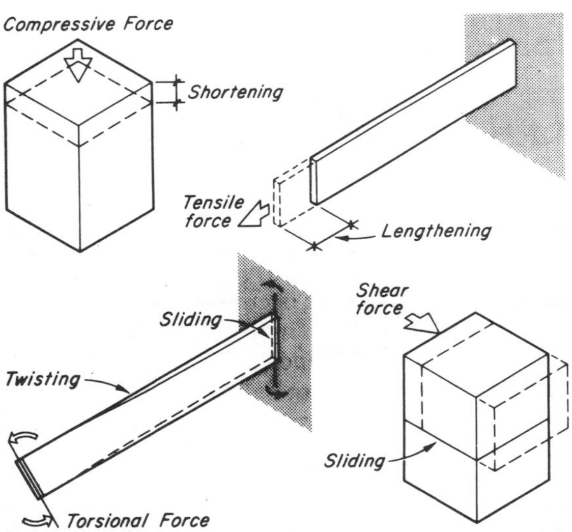

3.3 *Effect of external forces*

The effect of these various types of force is to put the material of a structural member into a state of *stress* and the material of the member is then said to be in a state either of tension, compression, shear or torsion. To resist lengthening under a tensile force the material of a member must exert an inward pull or reaction and to resist shortening under a compressive force it must exert an outward push. This is demonstrated clearly by a spring (figure 3.4). In each case as soon as the external force is removed the spring under the action of the internal forces immediately reverts to its original shorter or longer length as the case may be. In the case of shear, sliding is resisted by a force exerted by the material in a direction opposite to that of the shearing force.

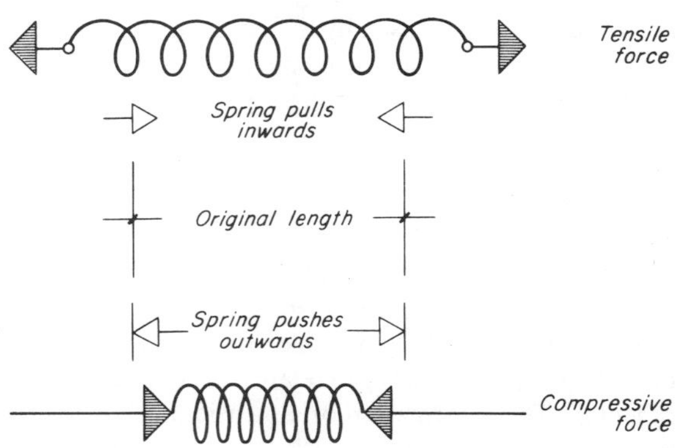

3.4 *Internal force action*

The total load or force on an element gives no indication by itself of its actual effect upon the material of the element. For example, a load of 50 kN might be applied to a column measuring 5000 mm^2 in area or to another 50 000 mm^2 in area. Both carry 50 kN but in the second ten times as much material is carrying it as in the first, so that each unit of area in the second, in fact, carries only one-tenth of the load carried by a similar unit in the first (figure 3.5). In other words the loading is less intense. The measure of intensity of loading is expressed as a load or force per unit area, known as the *intensity of stress* (in practice, simply 'stress'), and is found by dividing the applied load (W) by the cross-sectional area of the member (A) thus:

$$\text{Stress } (f) = W/A$$

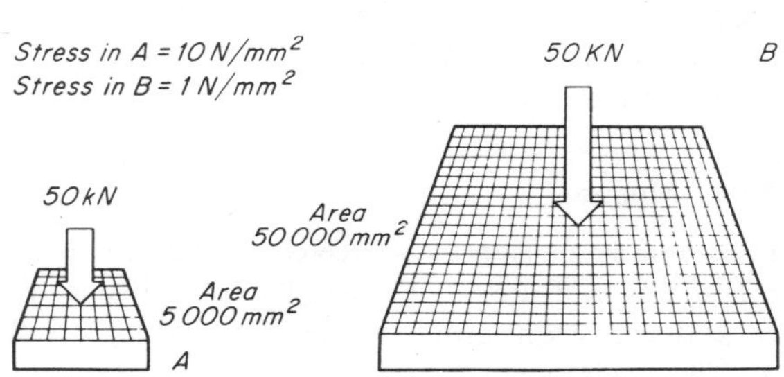

3.5 *Intensity of stress*

As with the loads causing them the stresses may be tensile, compressive or shear stresses respectively. When the stresses are caused by axial loads stretching or compressing a member in the direction of the load they are termed *direct* stresses (figure 3.3). Compression and tension at right angles to the direction of the load is caused by bending in a beam or cantilever, and these are termed *bending* stresses (figure 3.23). Shear stresses also may be caused by bending as in figure 3.25 where horizontal shear is caused by the bending action of the beam. Shear stresses are also caused by torsion or the twisting of a structural member (figure 3.3).

3.1.2 Strain

The total change in the length of a member under a given tensile or compressive stress will vary with its length and, as with loading which must be related to a unit of area, it is necessary to relate this change to a unit of length. The deformation or dimensional change in a member per unit length which occurs under load is found by dividing the change in length by the original length and is known as *strain.*

Within certain limits of loading it is assumed that stress is proportional to strain[1] so that the ratio of stress/strain is constant for any given material and is known as its *modulus of elasticity* (E).[2] This is a property of the material and is a measure of its stiffness. The higher the E value of a material the stiffer it is and the larger the stress necessary to produce a given strain. The converse is the case with materials of low E value.

Tensile and compressive stresses cause tensile and compressive strain respectively and similarly, shear stress causes shear strain, but whereas the former produce changes in length the latter produces change of shape or distortion, as the twisting in figure 3.3 or the 'parallelogramming' of a block as shown here.

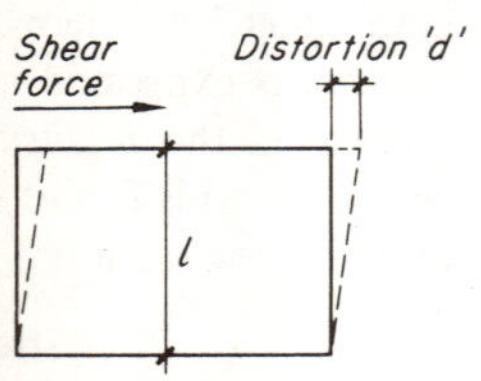

Shear strain equals d/l

3.1.3 Moments

In certain circumstances a force can cause turning or rotation as in figure 3.6 *A* where the force w tends to cause the projecting arm to rotate about its hinged end X. The tendency to rotate depends upon both the magnitude of the force and the perpendicular distance between its line of action and the point of rotation, known as the *lever arm* (L). The greater the lever arm the less the force required to rotate a given structure, and vice versa. This is well known and is the reasoning behind the use of a crowbar. Thus the same rotational effect may be obtained by different combinations of load and lever arm. The measure of this effect is given by the product of these two factors which is known as the *moment* of the force about the point, expressed in units of force and distance, for example Newton metres (Nm).

Equilibrium (no rotation) is obtained when the *clockwise* (+) moments of some forces acting on a member are balanced by the *anti-clockwise* (−) moments of others. This is illustrated in figure 3.6 *B* where rotation caused by downward load w acting at a lever arm L is prevented by the application of the upward force W acting at a lever arm l such that $Wl = wL$. If the projecting arm were not hinged at X but extended over a support or fulcrum, as in figure 3.6 *C*, the same anti-clockwise counter moment could be obtained by the force W acting downwards at the same distance l from the fulcrum.

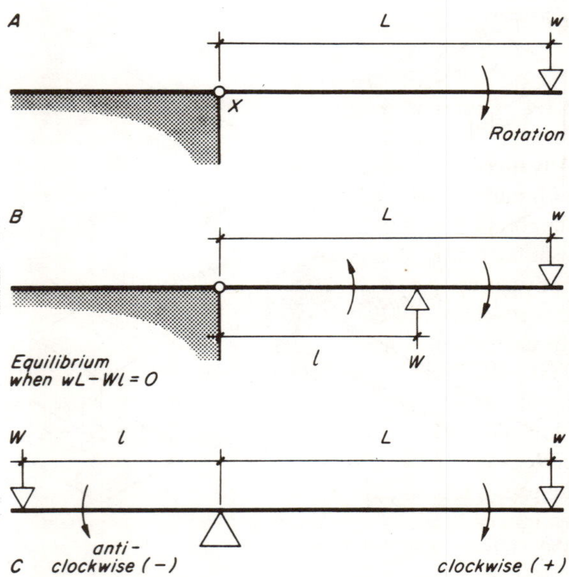

3.6 *Turning moments*

[1] The assumptions that stress is proportional to strain (known as *Hooke's Law*) only holds within the 'elastic limit' of the material, that is within the range of loading in which the material returns to its original form after removal of the load, as in the illustration of the springs on page 35.

[2] Also known as *Young's Modulus.*

When the tendency to rotate is resisted bending occurs. This can be illustrated by a door on its hinges. If the hinges function properly pressure on the door causes it to rotate and the door remains in a straight plane. If, however, the hinges are corroded rotation will not occur and under pressure the door will tend to bend (figure 3.7 *A*). Similarly the rotational effect of load *w* about the fulcrum in figure 3.6 *C* is resisted by load *W* and although the arm will not actually rotate about the fulcrum it will tend to bend along its length as in figure 3.7 *B*. A moment the rotational effect of which is resisted is called a *bending moment* (BM).

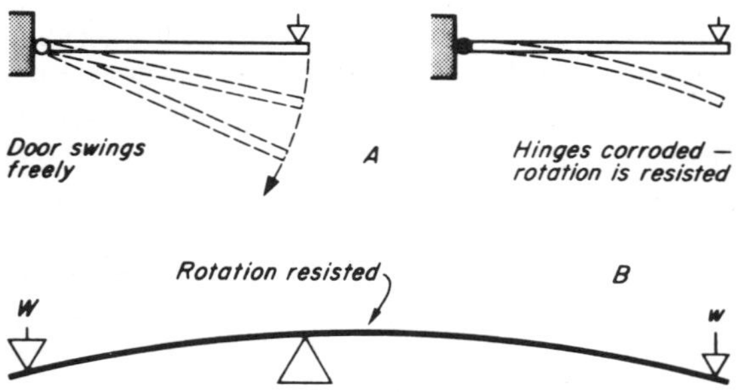

3.7 *The development of bending*

3.1.4 Centre of gravity

Any object or body can be regarded as being composed of innumerable particles all acting downwards due to the force of gravity. For practical purposes, as will be seen later, it is often useful to regard the weight of the body as acting as a single force through one point such that its effect will be the same as the total effect of all the particles of which the body is composed (figure 3.8). This point is known as the *centre of gravity* of the body. It will not always fall within the material of the body itself. The centre of gravity of a frame, for example, will be located in the space within it and for a U- or L-shaped body it will lie somewhere on the axis of symmetry according to the length of the arms (figure 3.8).

For many structural purposes the equivalent of the centre of gravity of a body is required for an area. Strictly speaking an area, which has no mass, cannot have a centre of gravity but it may be considered as a very thin plate and the same principles as for a solid body would apply. A line through the thickness of the plate, normal to its faces and passing through its centre of gravity would pierce the faces at points known as the *centroid* of the area (figure 3.8). Methods of determining the positions of the centroids of various shapes are given in the textbooks on the theory of structures.

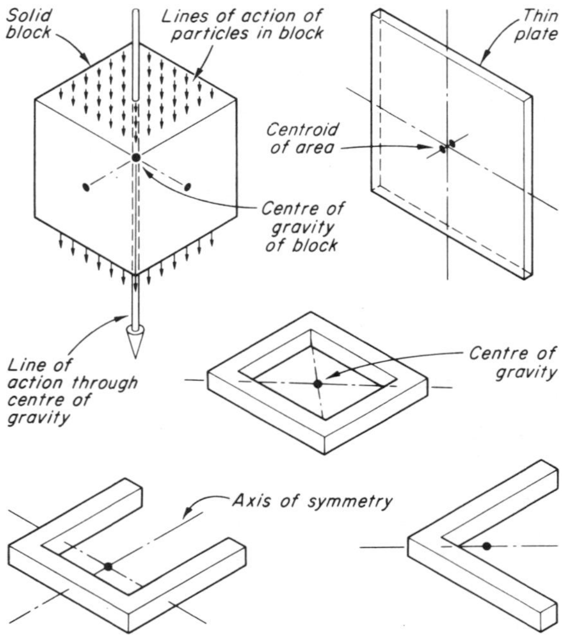

3.8 *Centre of gravity and centroid of area*

3.2 THE FORCES ON A BUILDING AND THEIR EFFECTS

The forces on a building may be considered as having an overall effect on the building or structure which tends to *move* it as a whole and local effects on its parts which tend to *deform* them but not move them out of position. The stability of the building involves the equilibrium of all the forces acting on the structure and the absence of excessive deformations.

3.2.1 Overall movement

Vertical downward forces caused by the dead weight of the building and its loads tend to force it down into the soil on which it rests or tend to force floor or roof structures, for example, down on their supports. In the first case the soil must be sufficiently strong to exert an upward force or reaction equal to the weight of the building (figure 3.9) and in the latter the supports must exert similar forces equal to that which the floor exerts on them (see figure 3.1). Vertical active forces may also be upward as in the upward suction caused by wind passing over a flat roof which tends to raise the roof and its structure (figure 3.9).

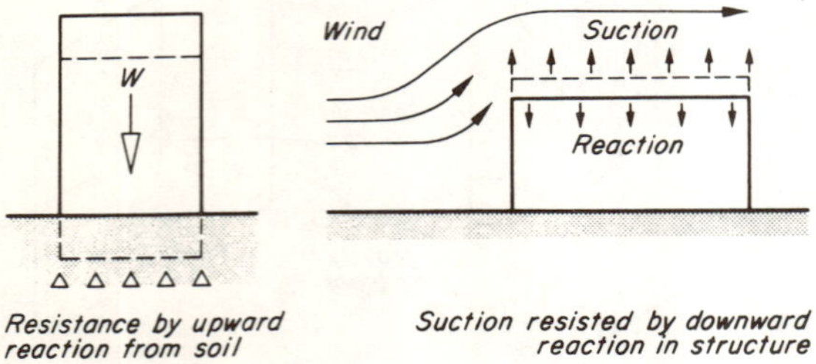

3.9 *Movement: vertical*

Horizontal forces, which may be exerted by wind or soil against the side of a wall or a building, tend to make it slide on its base or overturn. The former tendency must be resisted by the friction between the base and the soil on which the structure rests or by the passive pressure of the soil on the opposite side, the latter by the weight of the structure itself, by a strut or by a suitable tension element, any of which would cause a counter-moment (figure 3.10).

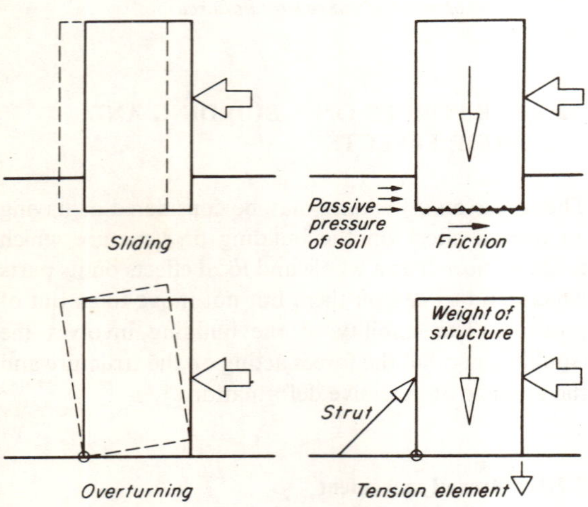

3.10 *Movement: sliding and overturning*

Oblique forces have an effect similar to that of horizontal forces. Figure 3.11 illustrates some circumstances in which oblique forces are generated and their effects. A curved roof structure springing from its foundations will exert an oblique outward thrust and the foundations will tend to slide outwards; the oblique thrusts from an arch on fairly tall supports will tend to overturn the supports. The oblique forces from an inclined roof structure will tend to cause the roof both to slide off its supporting walls and to overturn the walls. The smaller the angle of inclination the greater is the tendency for sliding and overturning to occur since the line of the force more nearly approaches the horizontal. Methods similar in principle to those for horizontal forces are adopted to maintain the equilibrium of structures under oblique loading. In the examples shown a tension member, if used, would tie the two ends of the curved or inclined roof elements causing the oblique thrusts and prevent them moving outwards. An oblique force acting downwards at any point generates both an outward horizontal force and a downward vertical force at that point, the magnitude of each of which will vary according to the inclination of the oblique force. These forces can be established graphically, with sufficient accuracy for practical purposes, by means of a parallelogram of forces as explained on the next page.

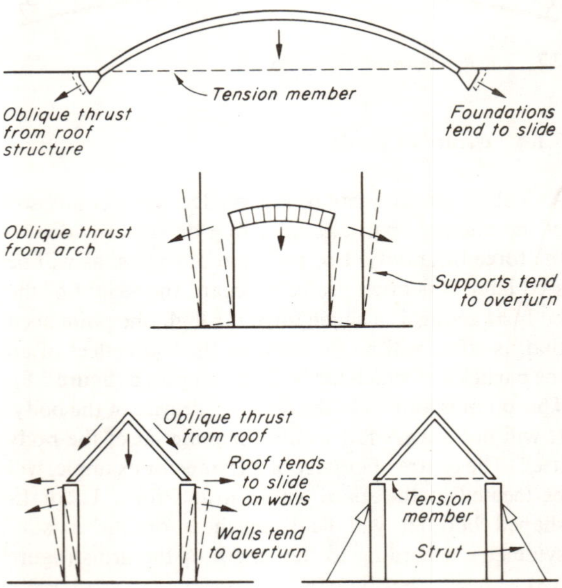

3.11 *Movement: effect of oblique forces*

3.2.2 Deformation

Vertical forces on thin walls or columns tend to make them bend in their height, in the same way as a thin stick under pressure on the top (figure 3.12). This is known as *buckling*. Similarly a load will cause vertical bending or *deflection* in a beam as in figure 3.12. Buckling or sideways bending may also occur in the top of a thin, deep beam for reasons given later (figure 3.33).

Horizontal forces acting on thin walls or columns may cause deflection in them as they act like a vertical cantilever (figure 3.12). It should be noted that this is not identical to buckling which is caused by loading in the direction of its length, not normal to its length.

Buckling and deflection are both due to bending and both are controlled by using materials and members of adequate stiffness.

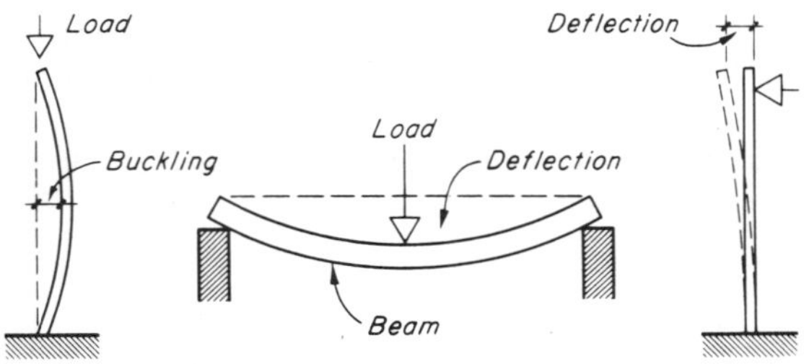

3.12 *Deformations*

3.2.3 Triangle and parallelogram of forces

In the section on oblique forces reference is made to a method of establishing graphically the magnitude of forces acting on a structure. This is possible because a force may be represented in direction and magnitude by a line drawn parallel to the direction of the force and to scale so that its length represents the magnitude of the force.

When two forces act on a body, other than being directly opposed, they tend to move the body. Thus the joint effect of the two forces *A* and *B* in figure 3.13 (1), acting in the direction shown, will be to move the body along a line somewhere between them, say along the broken line. A third force *C*, of an appropriate magnitude, acting along this line could have the same effect as the combined forces *A* and *B* and could replace them. Such a force is known as the *resultant* of the other two. A force *D* of exactly the same magnitude as *C* and directly opposed to it would produce equilibrium in the body and is known as the *equilibrant* of *C* and, consequently, of the two forces *A* and *B*.

In figure 3.13 (2)a the three forces are in equilibrium, therefore each is the equibilibrant of the other two. If three such forces are drawn as lines to scale representing their magnitudes and parallel to their directions, continuing from each other in a clockwise manner as in (2)b, they will form a closed triangle. Thus if the direction and magnitude of only two, say *A* and *B*, are known the magnitude and direction of their equilibrant can be found as the closing side of the triangle. Such a triangle is known as a *triangle of forces*.

The resultant of any pair of the forces is, of course, equal but opposite in direction to their equilibrant and can often be found more conveniently by drawing a *parallelogram of forces* as in (2)c, where the forces *A* and *B* are drawn as the adjacent sides of a parallelogram, the diagonal of which will represent their resultant in magnitude and direction.

In practice it is often necessary not only to find the resultant of a pair of forces but to *resolve* a single force into two component forces as in the case of oblique forces referred to on page 38. The known force is drawn to scale and in the correct direction as the diagonal of a parallelogram, the sides of which are drawn parallel to the directions of the required components. In figure 3.13 (3) the diagonals represent oblique forces such as in figure 3.11 and the sides of the parallelograms represent the vertical and horizontal components. It

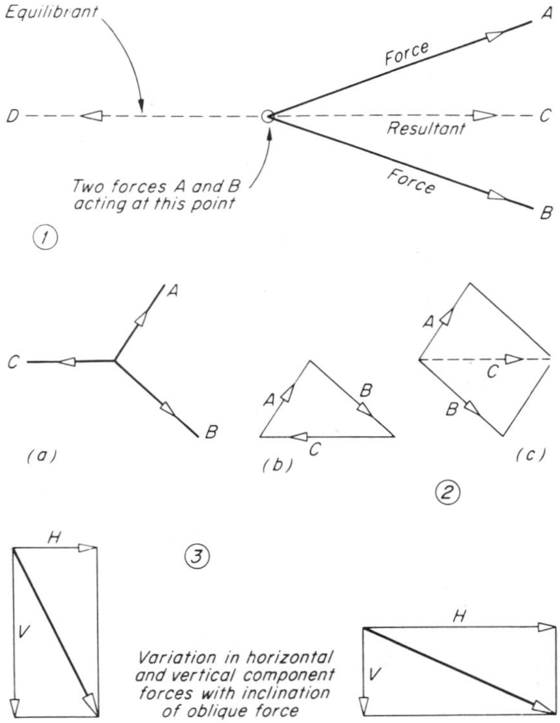

3.13 *Triangle and parallelogram of forces*

will be clear from this that the smaller the angle of inclination of an oblique force the greater will be the magnitude of its horizontal component and, therefore, the greater the tendency for the structure to slide or overturn under its action.

3.3 THE BEHAVIOUR OF BUILDING ELEMENTS UNDER LOAD

3.3.1 The wall

Under vertical loading a wall may crush, buckle or settle.

Crushing

This is caused by over-stressing the material of which the wall is constructed and is avoided by adequate thickness at all points to keep the stresses in the wall within the safe compressive strength of the materials of which it is built.

Eccentric loading, that is loading applied other than through the centre of gravity of the wall has the effect of increasing the compressive stress in the wall on the loaded side and of decreasing it on the opposite side, and tends to cause bending in the wall whatever its thickness.

The explanation for this lies in the fact that a moment is set up in the wall and to maintain equilibrium this must be resisted by an opposite moment within the wall, the forces for which must be provided by the walling material itself (figure 3.14). This causes compression on one side of the axis of the wall and tension on the other thus increasing on one side the compressive stress which would be caused by the same load W applied axially and decreasing it on the other. The result of this can be two-fold: (i) the increased compressive stress could become greater than the safe compressive strength of the walling material and (ii) if the eccentricity is too great tensile stresses will be set up in the side opposite that on which the load is applied. Figure 3.14 indicates the increasing compressive stress and the development of tensile stress with increasing eccentricity of load.

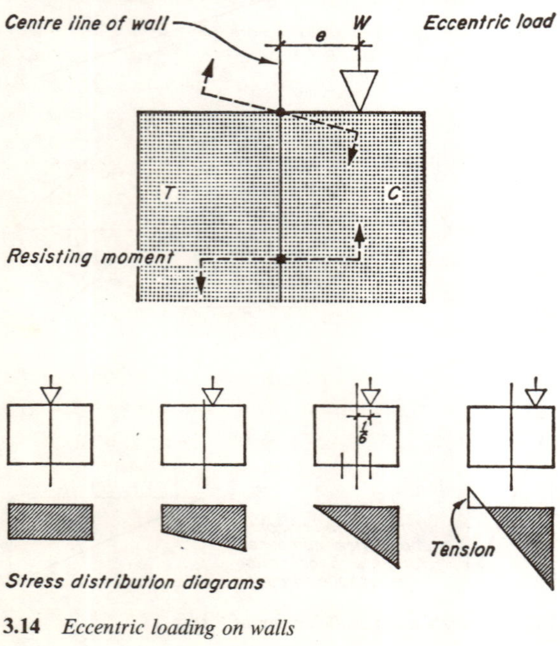

3.14 *Eccentric loading on walls*

In practice the actual stresses in the wall are determined by the formula:

$$\frac{W}{A} \pm \frac{We}{Z}$$

where W/A = stress due to the load applied axially
We = moment caused by eccentric loading
Z = a geometrical property relating to the shape and size of the cross-section of the wall[1] such that
We/Z = stress at the faces of the wall due to eccentric loading.

It can be shown that tension will occur when the eccentricity is greater than one-sixth of the wall thickness. When the stresses due to eccentric loading are too great they are reduced either by reducing the eccentricity at which the load is applied or by increasing the thickness of the wall. The last has the double effect of reducing the relative eccentricity and of increasing the value of Z.

For similar reasons an arch built of blocks such as that in figure 3.11 must be sufficiently deep in order to prevent the development of tensiles stresses. These will occur, as in a wall, if the oblique thrust within the arch due to the applied load is at an eccentricity greater than one-sixth of the depth of the arch ring. In such circumstances the arch would collapse in the manner shown in figure 3.15.

[1] Z is the symbol of the property of a section known as the *section modulus*. Methods for computing this for any given section are given in standard textbooks on the theory of structures.

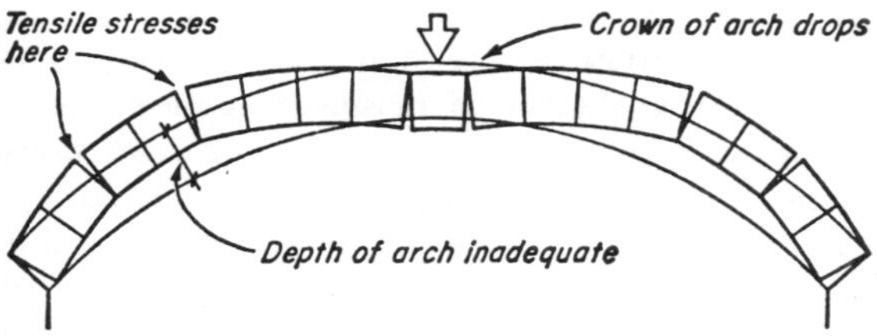

3.15 *Arch failure*

Buckling

This will occur when the thickness of the wall is small relative to its height. Short walls or piers ultimately fail by crushing, but as the height increases they tend to fail under decreasing loads by buckling. The terms 'short' and 'tall' in this context are relative to the thickness of the wall not to its actual height and are defined, broadly, in terms of the ratio of unsupported height to horizontal thickness known as the *slenderness ratio*.[1] The greater this is the greater is the tendency to buckle. It should be noted that buckling is not related to the strength of the walling material but to the stiffness of the wall. Buckling may, therefore, be controlled either by restricting height, increasing thickness, stiffening by buttresses[2] or intersecting walls or by reducing the applied load.

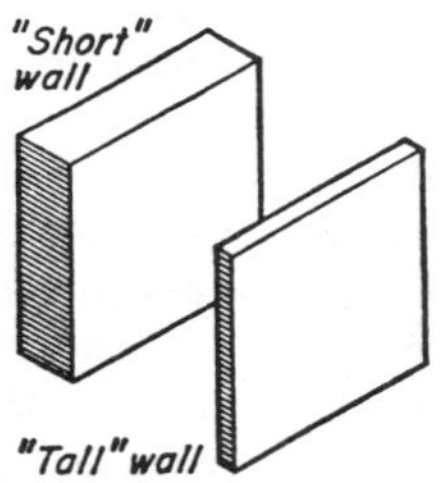

Settlement

The downward force of a wall must be resisted by an equal, upward reaction from the soil on which it rests in order to maintain equilibrium. Soils vary in strength. Some, verging on rock, are very strong, others are relatively weak. All of these consolidate under load but the former can resist very high stresses with little consolidation while the same stresses would cause excessive consolidation in the latter.

This consolidation causes a vertical downward movement of the wall which is known as *settlement* and in order to keep this within acceptable limits the stress in the soil due to the load from the wall must not exceed that which it can safely resist. This is ensured by making the base of the wall of such a size that the load is distributed over a sufficiently large area of soil. On some soils this will necessitate the provision of an extended foundation strip while on others no increase in the base of the wall may be necessary.

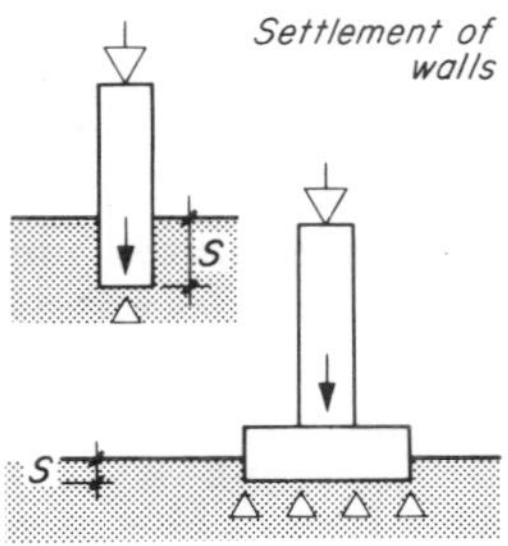

Eccentric loading of foundations has the same effect upon the stress distribution in the soil as on that in an eccentrically loaded wall and the practical implications of this are considered later in chapter 4.

Under horizontal loading a wall may tend to slide or overturn:

Sliding

This is more likely to occur in a free-standing retaining wall than in a wall forming part of a building which can provide the weight necessary to assist stability. It has already been indicated that friction and the passive pressure of the soil on which the wall rests are utilised to prevent sliding action (figure 3.10).

The amount of friction, or the frictional resistance, existing between the base of the wall and the soil depends upon the weight exerted on the soil, that is, the pressure between the two surfaces, and upon the degree of smoothness of the surfaces. For any pair of surfaces in contact the frictional resistance increases approximately in proportion to the applied weight. Thus the ratio of frictional resistance to weight is constant. This ratio is termed the *coefficent of friction* and it varies according to the types of surface in contact. In any given case, therefore, the frictional resistance is equal to the coefficient of friction for the particular surfaces in contact multiplied by the applied weight. The area of the surfaces in contact does not affect the frictional resistance.

The other force which may resist the tendency of the wall to slide in the passive pressure of the soil in which the wall bears, which is opposite in direction to the horizontal active pressure on the wall (figure 3.10).

[1] A *short* masonry wall or column is defined as one having a slenderness ratio not exceeding 12. *Tall* is applied to those having a slenderness ratio greater than this.

[2] A buttress is a thickening of a wall at intervals along its length: see figure 3.17.

Frictional resistance may be increased only by an increase in the weight of the wall, either by increasing the height or the thickness. When, in certain circumstances, insufficient frictional resistance is available further resistance must be provided by the passive soil pressure. The stresses in the soil caused by this pressure must be kept within the safe limits of the particular soil and this may necessitate taking the wall deeper into the soil so that the pressure is distributed over a greater area.

Overturning

This may be caused (i) by rotation or (ii) by settlement. Overturning by rotation occurs when the counter-moment *We″* in figure 3.16 set up by the weight of the wall acting through its centre of gravity is too small to resist the moment *Fe′* set up by the overturning force.

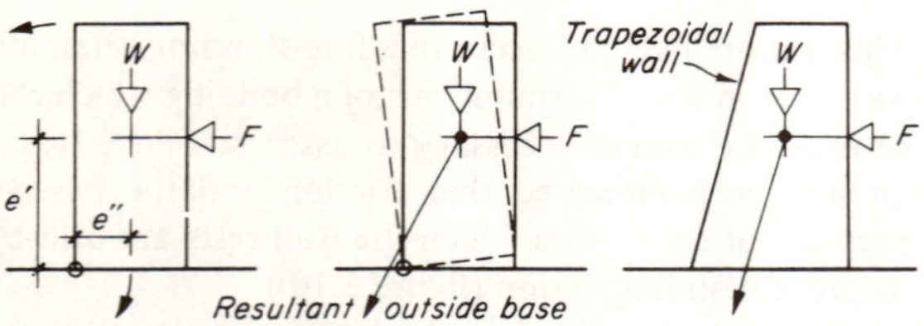

3.16 *Overturning of walls: rotation*

In these circumstances the resultant (see page 39) of the weight of the wall *W* and the overturning force *F* falls outside the base of the wall so that the base is wholly under tension and overturning occurs. To prevent this the weight of the wall can be increased either by increasing its height or thickness. The latter is most beneficial because it also increases the width of the base within which the resultant must fall. Alternatively, or in addition, the shape of the wall may be made trapezoidal to shift its centre of gravity relative to the base towards the overturning force, thus reducing the eccentricity of the resultant at the base. Another option is to use buttresses as shown in figure 3.17. The thrust on the wall would then be transmitted to the buttresses the added weight and extended width of which would bring the resultant force within their base as shown.

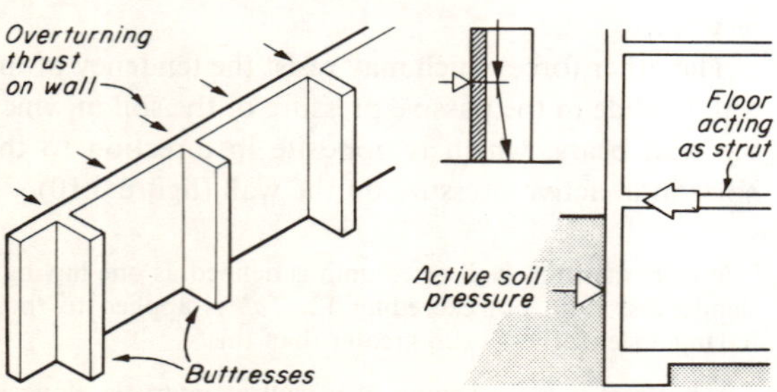

3.17 *Stabilising of walls*

These methods are adopted for walls having little tensile strength. Alternative and, in the case of tall walls, more economic methods may be adopted when materials with adequate tensile strength are used, such as reinforced concrete (see Part 2 *Retaining Walls*).

The use of a strut to prevent rotation (figure 3.10) may be adopted and where a wall undergoing a lateral force forms part of a building a floor can often be made to function as a strut for this purpose, as shown in figure 3.17.

Overturning due to settlement may occur through overstressing of the soil causing excessive consolidation under the wall. The resultant of the weight of the wall and the overturning force will always cause eccentric pressure at the base of the wall leading to similar stress distributions to those in an eccentrically loaded wall. This will result in a distribution of pressure in the soil similar to that shown in figure 3.18 with a pressure at the toe which might be considerably greater than the average pressure. If this should overstress the soil excessive consolidation might occur at this point causing overturning through unequal settlement of the wall. This problem can be overcome by reducing the eccentricity of the resultant either by increasing the thickness of the wall or the width of its foundation or by making the wall trapezoidal in shape for the reason given above.

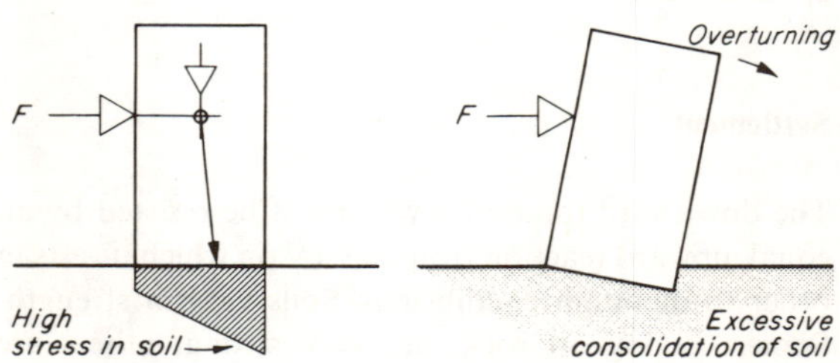

3.18 *Overturning of walls: settlement*

3.3.2 The beam

The consideration of the behaviour of a beam under load can be approached by first examining the action of a *cantilever*, which is a beam supported at one end only (figure 3.19 *A*). According to the nature of the bearing, that is the method of fixing to the support and the nature of the material at the support, the cantilever under load could break away from the support as at (*B*) if, for example, a weak glued junction gave way, or the bearing could deform as at (*C*) if, for example, it were made of rubber. In both cases the cantilever would be unstable and rotation would have occurred, but the cantilever itself would have remained straight.

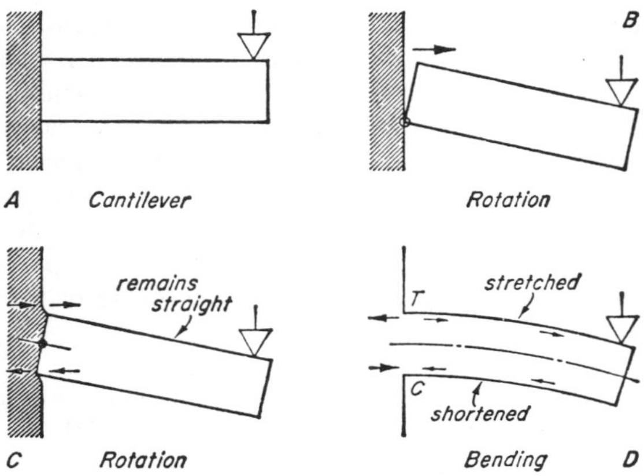

3.19 *The cantilever beam*

Bending

If, however, the cantilever were firmly fixed to a solid support so that rotation could not occur, then under load it would tend to bend as in figure 3.19 *D*. The cantilever would still tend to pull away from the support at the top and push it in at the bottom and to maintain equilibrium these forces must be opposed by equal and opposite forces in the support so that tensile stresses occur at the top and compressive stresses at the bottom as shown.

Tension and compression is also set up along the length of the cantilever because in bending the upper side will stretch and the material will be stressed in tension and the lower side will shorten and be in compression. At some point between the two the material neither stretches nor compresses and, assuming a symmetrical section, this is on the centre line where the arc is the same length as that of the straight cantilever. Since the cantilever at this point is neither stretched nor compressed there is no strain in it and there can, therefore, be no stress at this point. The line along which no strain nor stress occurs is known as the *neutral axis* (NA) and can be shown to pass through the centroid of the beam section, which in a symmetrical section is at mid-depth.

It will be apparent from this that the tensile strain at the top gradually reduces to nothing at the neutral axis and the compressive strain at the bottom reduces in a similar way. Thus the strain and, therefore, the stress varies from the outside to the neutral axis in proportion to the distance from the neutral axis, being at a maximum at the top and bottom or, as it is often termed, 'in the outer fibres' as indicated in figure 3.20.

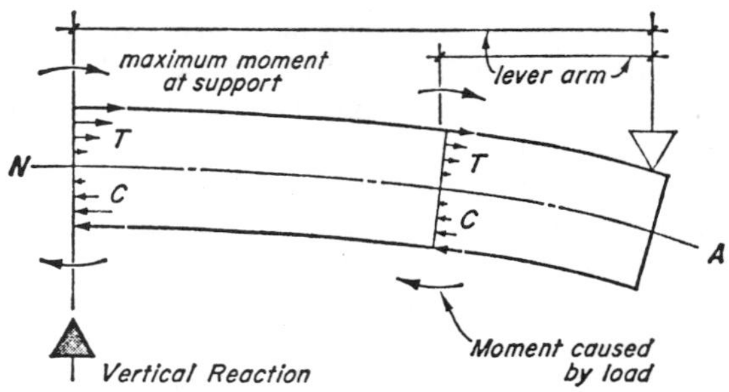

3.20 *Stress distribution in cantilever beam*

In addition to the horizontal reactive forces in the bearing there is also a vertical reaction necessary to resist the vertical downward pressure caused by the load on the cantilever.

To prevent the material of the cantilever crushing at the bottom and tearing apart at the top the cantilever beam itself must provide the necessary resistance to this. It is provided by forces in the material working about the neutral axis to produce an opposing or *counter-moment* to the bending moment *Wl* set up by the external load which tends to make the beam rotate in a clockwise direction about its bearing (figure 3.21 (*a*)). If all the material in the beam were concentrated

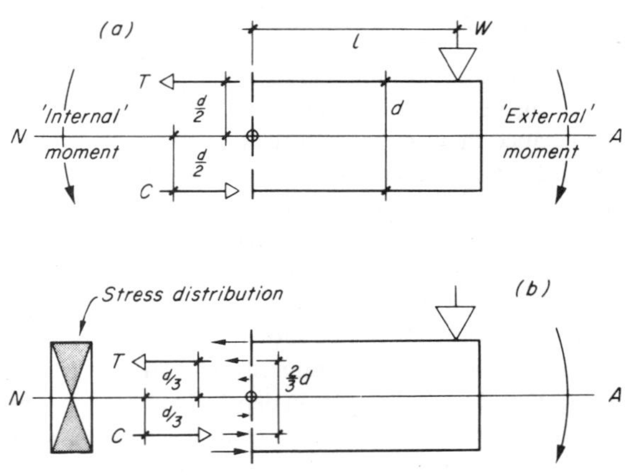

3.21 *Moment of resistance*

at the outer edges this internal counter-moment would be $(T \times d/2) + (C \times d/2)$ where T and C are the tensile and compressive resisting forces at the edges of the beam and d is the depth of the beam. Two opposite forces acting together like this to cause a moment are known as *a couple* and the moment set up in a beam by such a couple is known as the *moment of resistance* (MR) of the beam section. In a state of equilibrium MR = BM.

The magnitude of the bending moment caused by the load varies along the length of the cantilever, being a maximum at the bearing since the lever arm is at its greatest, and reducing along the length as the lever arm to the load point reduces (figure 3.20).

In a rectangular beam the material is not concentrated at the outer edges and the total resisting force is spread over the whole cross-section, only a part acting at the full lever arm $d/2$. The forces at other points are acting at smaller lever arms and as shown above, at varying intensities across the section as indicated in figure 3.21(*b*). It can be shown that the total of the moments of these forces is equivalent to that caused by the total tensile force T and total compressive force C each acting at a lever arm $d/3$.[1] Thus, because the lever arm is smaller than when all the force is assumed to be acting at the outer edges, the moment of resistance of the section will be smaller.

For this reason, therefore, greater efficiency results from concentrating the maximum amount of material as far as possible from the neutral axis because this economises in material since in a rectangular section that near the neutral axis is stressed to a lesser degree than that at the edges and is, therefore, not fully utilised. This accounts in practice for typical beam sections such as those shown in which most of the material is concentrated in the top and bottom *flanges* leaving a minimum at the centre portion or *web*.

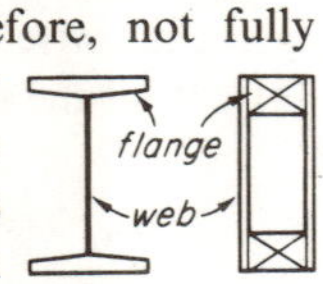

For this reason also the depth of a beam is of greater significance than its breadth relative to its strength in bending. Over the same span a given section of beam will be stronger in bending when set with its longest dimension on the vertical axis, as in figure 3.22, since the material in the cross-section is then situated at a greater distance from the neutral axis and functions at a greater lever arm. It can be shown that this increase in bending strength for the same cross-sectional area is proportional to the increase in depth. Thus section A in figure 3.22 will be twice as strong when used as in section B. It can also be shown that when strength is to be increased by an increase in area, greater efficiency results if this is obtained by an increase in depth rather than in breadth. This is because the bending strength increases only in proportion to any increase in breadth whereas it increases in proporion to the *square* of any increase in depth. Thus in figure 3.22, section C is twice the area and twice the breadth of section B and is twice as strong, whereas section D, which is twice the area but twice the depth of B is four times as strong. This being so it is clear that for any required moment of resistance less material will be required in a beam if it is disposed vertically rather than laterally in the section.

In the case of open web sections such as lattice beams (figure 3.35) an increase in depth results in almost negligible increase in area (only in the thin web members) so that the increase in bending strength tends to be directly proportional to the depth, that is in proportion to the lever arm, and not to the square of the depth.

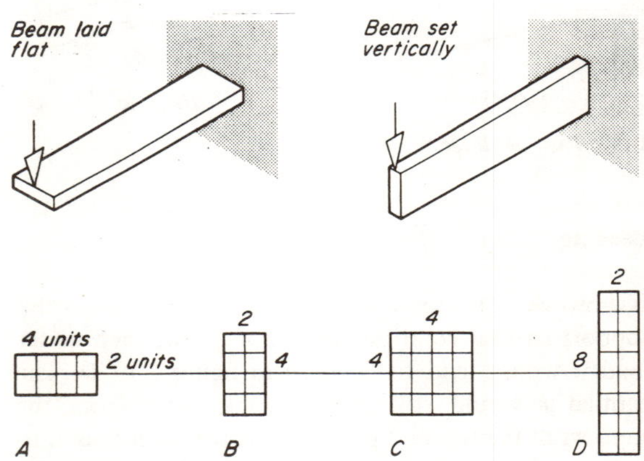

3.22 *Distribution of material in a beam*

From all the foregoing it will be seen that to produce a required moment of resistance with the least amount of material, the deepest practicable section should be used.

A *beam* as distinct from a *cantilever* is supported at both ends and in considering its behaviour under load may be viewed as a reversed double cantilever, the support reaction becoming the imposed load and the end loads becoming the support reactions (figure 3.23). It will be seen that the compressive stresses are now above the neutral axis and the tensile stresses below. The distribution of these stresses across the beam section is identical with that in a cantilever and the

[1] On the assumption that T and C act through the centroid of their stress triangles.

development of a moment of resistance is the same. For a central load as shown in figure 3.23 the maximum bending moment is at the centre reducing to zero at the supports (compare with figure 3.37).

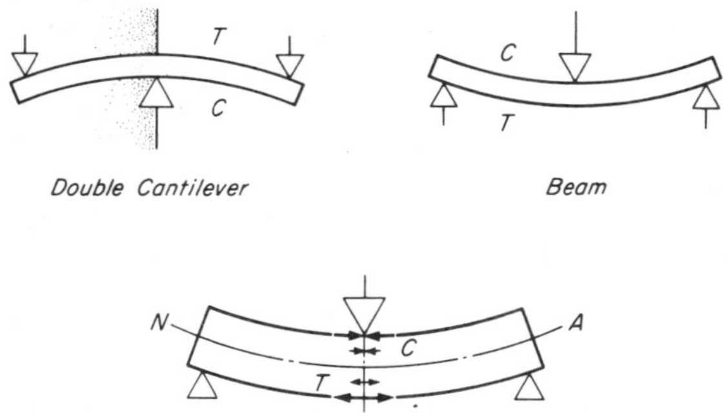

3.23 *The beam*

Deflection

The bending of a beam under load has, up to this point, been considered in terms of the forces set up within it and the mechanism by which it is able to resist them. For clarity the beams in the illustrations are shown in a state of considerable bending. In practice the actual bending is usually so small as to be invisible to the eye. Nevertheless, a beam may be strong enough to resist the bending moment set up by its load and yet may bend excessively and sag but not collapse. This sag is termed *deflection* as explained earlier and it must be kept within acceptable limits for practical reasons.

For a given set of conditions of span, load and size and shape of beam the actual deflection depends on the elasticity of the material. The greater the elasticity the greater the strain under stress and the greater the deflection since this is caused by the shortening of the beam at the top and the lengthening at the bottom. As explained on page 36 the measure of the stiffness of the material is given by its elastic modulus, E, the higher the value of this the greater the stiffness.

Similarly, for given span, load and material the deflection of a beam depends on the stiffness of the beam section, which varies with its shape and size. This is because variations in size, shape and especially depth cause variations in the stresses in the material (see page 44) and thus in the strains. The greater the depth of a beam of given area the stiffer it is because the strain is less and the less will it deflect. The measure of stiffness of a beam section is given by its *moment of inertia* (I) which is a geometrical property of the shape and size of a particular section irrespective of the material of which it is made.[1] The greater the value of this the stiffer will be the beam.

This relationship of stiffness of material and stiffness of section to deflection can be seen in the formula relating to deflection:

$$\text{Deflection} = (\text{constant}) \times \frac{Wl^3}{EI}$$

where W is the load and l is the span of the beam.[2] EI, the measure of the overall stiffness of the beam, is known as the *modulus of flexural rigidity* and it is clear that the greater this is the smaller will be the deflection. Thus when the material of a beam has a low elastic modulus, E, a beam section of high I value is necessary to keep deflection within a given limit. Conversely, when a stiff material is used the stiffness of the section can be reduced, which normally means its depth.

It can be seen from this formula also that deflection is directly proportional to the load but proportional to the cube of the span. Thus, for example, if the load is doubled the deflection is doubled, but if the span is doubled the deflection will be eight times as great. It is clear, therefore, that variations in span can be of greater significance relative to deflection than variations in loading. In wide span beams, especially when the loading is light, the critical factor in design is likely to be deflection rather than strength in bending so that deep beams with a minimum amount of material are advantageous as they produce the necessary stiffness with small self-weight, resulting in low dead/live load ratios (see page 126).

Beam action A beam will vary in the way it tends to bend and the extent to which it deflects under load according to the manner in which it is supported. A beam may be simply supported, that is to say, its ends simply rest on the supports with, theoretically, no fixing. Under the action of a load the beam may freely bend as in figure 3.24 *A*, sliding or pivoting on its bearings. Its ends may, however, be fixed rigidly to its supports (figure 3.24 *B*) as in the cantilever already

[1] The moment of inertia is not, in fact, a moment but a measure of the potential moment latent in a section for resisting deformation. Methods for computing this for any given section are given in standard textbooks on the theory of structures.

[2] The constant is a factor relating to support conditions and the nature of the loading, ie whether it is concentrated or distributed.

considered. In this case the ends cannot move freely as in (*A*) but must bend like the cantilever, the beam action occurring only at the centre. The beam, therefore, bends in two directions along its length: convexly at the ends, concavely in the middle. The usual convention is to call the former *negative bending* and the latter *positive bending*. In practical terms the effect of the negative bending moments due to the fixed supports is to reduce the positive moment and the deflection at the centre, leading to a greater load capacity or, alternatively, to a reduction in the necessary size of beam.

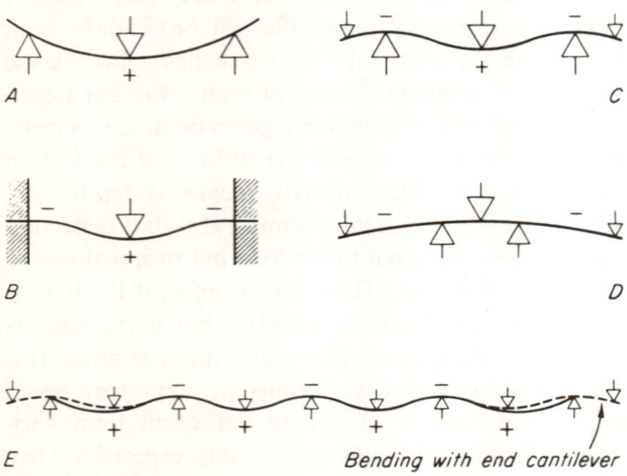

3.24 *Beam action*

A similar pattern of bending is produced if the ends of a beam project or cantilever beyond the supports as in (*C*). Although the beam is not built-in at the supports the loads on the cantilever ends set up negative bending over the supports in the same way and with the same effect upon the positive bending moment. If, however, the cantilever projections are large relative to the centre span the effect may be to eliminate the positive bending and produce negative bending along the full length as in (*D*). When this occurs it is known as 'hogging'. Continuous beams over a number of supports behave in a similar way to that in (*C*), the internal positive moments being reduced relative to those in a series of simply supported beams. The positive moments in the end bays will, however, be larger unless the end bearings are made sufficiently rigid or the beam cantilevers to produce negative moments over the end supports (*E*).

Shear

The downward pressure of the load on a beam together with upward pressure of the end reactions tends to make the centre portion slide down between the supports as in figure 3.25 *A*, causing shear stresses at the vertical planes of sliding at each end. Since, however, the upward and downward forces exist at any point along the beam these shear stresses exist at all sections of the beam, (*B*). The load also makes the beam bend and this tends to cause horizontal shear action in the beam. This can be visualised if the beam is considered to be made of thin horizontal layers as in (*C*) which will slide slightly over each other in bending. Since in a normal beam this sliding cannot occur horizontal shear stresses are generated.

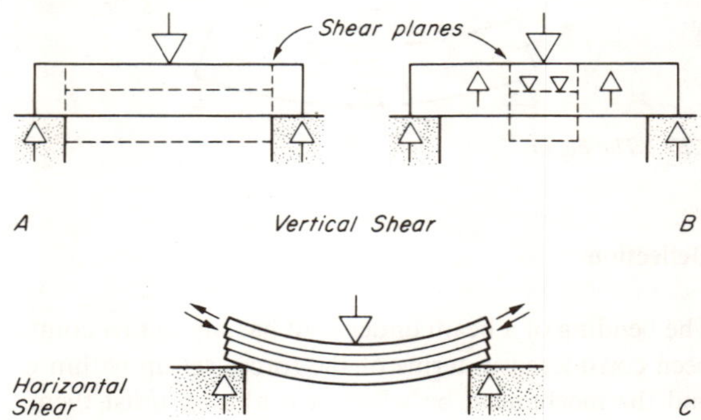

3.25 *Shear action*

These horizontal shear stresses at any point are of equal intensity to the vertical shear stress at the same point. This is essential in order to maintain equilibrium and can be understood by imagining a very small square particle within the section of the beam to be greatly enlarged (figure 3.26). The vertical shear forces on the left and right hand sides of the square constitute a couple tending to rotate it in a clockwise direction and in order to maintain equilibrium this must be resisted by an opposite couple of equal magnitude. This is provided by horizontal shear forces on the top and bottom sides of the square.

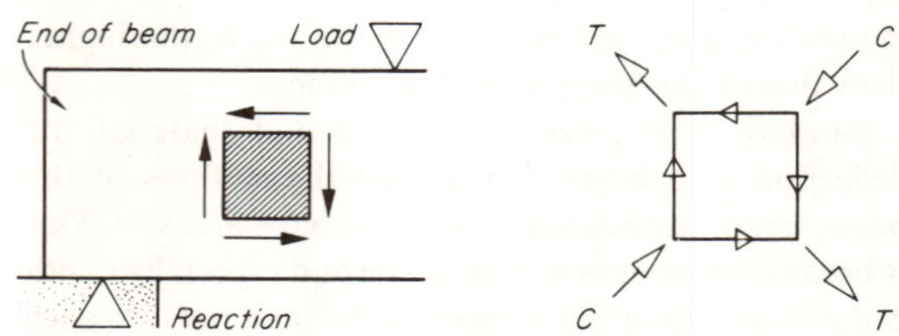

3.26 *Vertical and horizontal shear*

These two sets of equal horizontal and vertical shear forces resolve into a compressive force *C* and a tensile force *T* acting on the diagonals of the square (figure

3.26) and it can be shown that the stresses on these planes at 45 degrees to the horizontal are equal in intensity to the vertical shear stress at the same point. These diagonal forces are of particular significance towards the bearings of a beam where they are likely to be at a maximum. In deep thin-webbed beams, such as steel plate girders, the high compressive stresses may cause buckling of the thin web across the 45 degree line (figure 3.27) and local stiffening may be necessary to prevent this (see Part 2). In the case of a beam of concrete which is weak in tension it could fail under the high tensile stresses across the other 45 degree plane (figure 3.27). It is for this reason that reinforcing bars are often bent up at the ends to provide resistance to these forces.

Shear forces depend only on the load applied to the beam and are the same whether a given load be applied to a long or short beam. In some circumstances the bending moments set up over short spans may be relatively small, even with heavy loads, due to the limited length of span and a thin-webbed beam of sufficient size for the bending stresses might be too small in cross-section for the shear stresses. The critical design factor in such cases is, therefore, shear strength rather than strength in bending or deflection. This is in contrast to wide-span lightly loaded beams where deflection is likely to be the critical factor and where depth rather than cross-sectional area is significant.

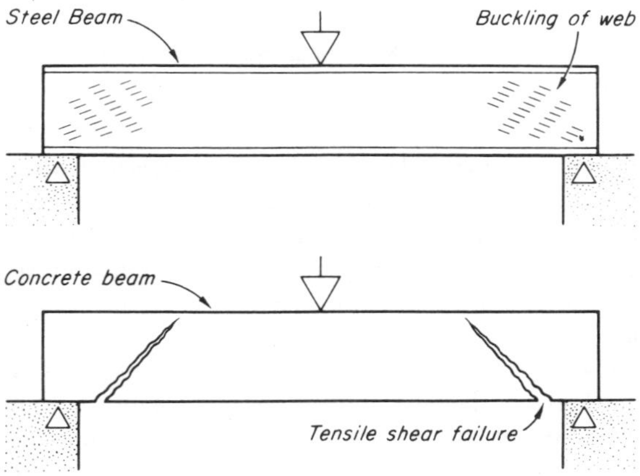

3.27 *Effect of shear forces in beams*

3.3.3 The column

Under load a column may crush or buckle in a similar manner to a wall. How it actually behaves depends primarily upon the material of which it is made, its shape and its slenderness, that is the relation of its thickness to its height.

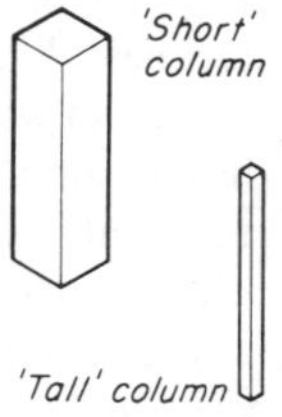

If the height of the column is small relative to its thickness (this is termed a *short* column) it will remain stable under increasing axial load until the material finally crushes. The stronger the material the greater the load it will carry before crushing.

If the height is great relative to its thickness (this is termed a *tall* or *slender* column)[1] it will become unstable by buckling at a load much smaller than that which would crush a short column of the same cross-section and material. This is called the *critical load* and it varies with the column slenderness, decreasing with an increase in slenderness. Thus the bearing capacity of a tall column depends less upon the strength of the material of which it is made than upon its stiffness and this becomes smaller as its slenderness increases.

Buckling

As with a beam relative to deflection the shape, or disposition of the material, of a column is important relative to buckling. The direction of deflection in a beam is known (figure 3.28 *A*) and the beam can be arranged to present its greatest resistance to bending in that direction as discussed on page 45. A column, however, may buckle in any direction under a vertical load and this has a significant effect upon the forms which a column may take (*B*).

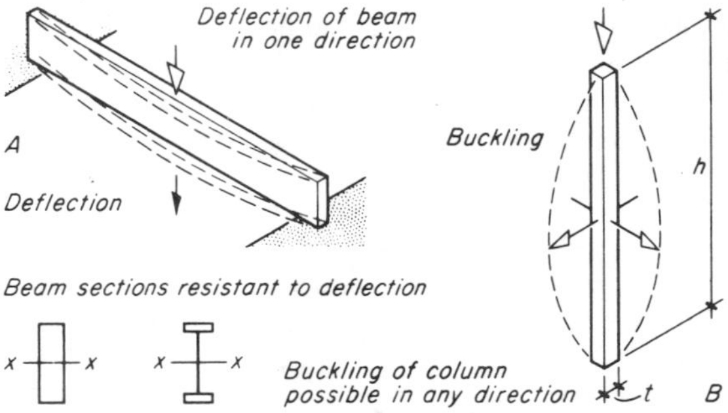

3.28 *Buckling of columns*

[1] As with walls the slenderness of a column is dictated by the relation between its thickness and its height, not solely by its physical height. A *tall* concrete column is defined as one having a slenderness ratio of more than 15 if braced against lateral forces or more than 10 if not so braced.

Since the tendency to buckle is related to the height and thickness of the column the measure of this tendency may be expressed as the ratio of column height to thickness, *h/t*, which is termed the *slenderness ratio* as in the case of walls. The greater this ratio the more slender the column and the greater its tendency to buckle. If a rectangular section, as used for a beam, is used for a column there will be two values for the slenderness ratio because of the two values for *t* and buckling will occur in the direction of the least resistance, that is in the direction of the least thickness (figure 3.29).

Since the crushing strength of a column of any given material depends upon its cross-sectional area and not on its shape, it is clear that in the case of columns which are likely to fail by buckling maximum economy of material will result from disposing it symmetrically around the centre axis of the column, so that the greatest thickness in each direction may be obtained. This can be appreciated from figure 3.29 which shows two columns each with a cross-sectional area of 10 000 square mm. Although the slenderness ratio of the rectangular column about its short axis *xx*, *h*/200, is half that of the square column, *h*/100, the former will buckle under a smaller load about its long axis *yy* since the slenderness ratio in this direction is greater, *h*/50.

For this reason the deep rolled steel sections suitable for beams are less suitable for columns and sections with broad flanges are rolled for use as columns in order to reduce the difference in slenderness ratios about the two main axes. Of the two Universal steel sections both with the same cross-sectional area, shown in figure 3.29, the 408 × 152 mm beam section has, for any given height, a slenderness ratio about its long axis *yy*, approximately five times as great as that about the *xx* axis, whereas with the broad flange column section that about the *yy* axis is only about one and three-quarter times as great as that about its *xx* axis. Furthermore, because of the greater flange width of the latter

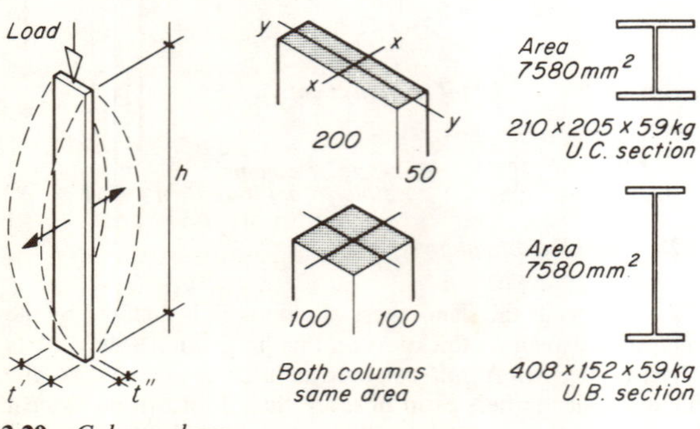

3.29 *Column shape*

the slenderness ratio about its *yy* axis is considerably smaller than that for the beam section. The column section is, therefore, stiffer in that direction.

As a deeper beam, with its material disposed further away from the neutral axis than in a shallow beam of the same cross-sectional area, offers greater resistance to deflection, so a column shaped with the material disposed at greater distances from its axes offers greater resistance to buckling. This, of course, obtains in steel beam and channel sections where most of the material is concentrated in the flanges, but the increase in resistance to buckling is obtained only about one axis as shown above. The use of hollow sections, however, permits the material to be disposed both at greater distances from and symmetrically to the axes of the column and results in equivalent stiffness about different axes. This can be seen in figure 3.30 which shows four sections of equal cross-sectional area so that the material in the hollow sections is at a greater distance from the axes than that in the solid sections and the former are, therefore, stiffer. Of the hollow sections, of the same overall size, the hollow square is the stiffest because its shape permits material to be disposed at the greatest distance from the axes.

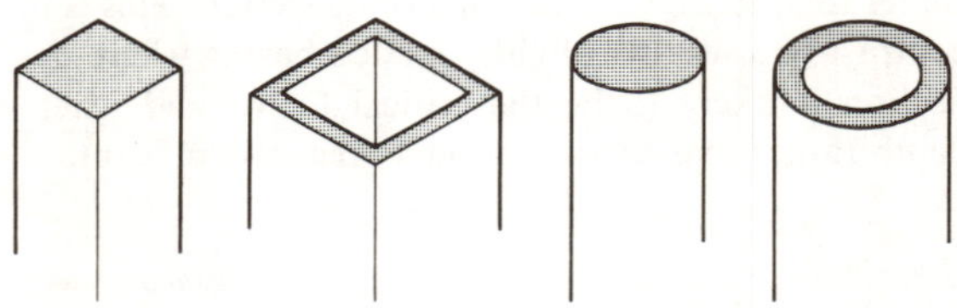

Load carrying area in all sections is the same

3.30 *Column shape*

From these considerations it will be seen that when columns are slender and the critical factor in their design is resistance to buckling rather than loadbearing capacity (that is in lightly loaded tall columns) maximum economy of material results from the use of hollow columns of symmetrical shape.

It will be clear from the sections shown in figures 3.29 and 3.30 that the outer dimensions of a column give no indication of the distribution of material within the column section and, therefore, the width or breadth is not suitable as a basis for calculating the slenderness ratio. Some factor is required which takes account of both size and distribution of material. For this purpose a concept from dynamics is used — the *radius of gyration*. This is the radius at which all the material in a particular wheel, for example, is assumed to be concentrated while still providing the same inertia against rotation as the wheel itself (figure 3.31). In statics the radius of gyration (*r*) is the distance at which all the material of a column, of whatever section, is assumed to be concentrated on either side of a particular axis, such that in terms of stability the same effect would be produced as by the actual section.

A normal rolled steel section about its axis *xx* may be likened to the section of the wheel and the radius of gyration is the distance from *xx* at which all the material above or below the axis is assumed to be concentrated into an infinitely thin flange separated by an infinitely thin web (figure 3.31). Since the material is disposed differently about the other axis *yy* a similar assumption must be made relative to that axis, giving a different, smaller radius of gyration. For other sections, whether hollow or solid, the assumption regarding concentration of material is the same and in circular sections the radius of gyration is the same about all axes.

The slenderness ratio is, therefore, based on the radius of gyration rather than on the outer column dimensions and becomes h/r which may have more than one value. The respective *r* values for the two steel sections shown in figure 3.29 are:

broad flange section: *r* (*xx*) 3·53, *r* (*yy*) 2·04
beam section: *r* (*xx*) 6·49, *r* (*yy*) 1·25

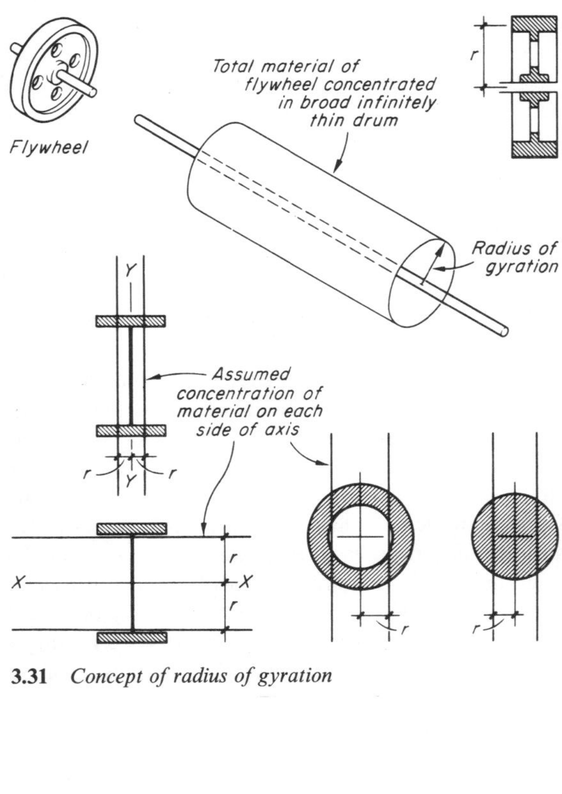

3.31 *Concept of radius of gyration*

The relative closeness of the two values for the broad flange section and the greater value about its *yy* axis when compared with those for the beam section, indicate its greater suitability for use as a column than a beam section of the same cross-sectional area.

Effective height

As beams are affected by their support conditions (page 45) so the behaviour and loadbearing capacity of columns are affected by their end conditions.

The end of a column may be so attached to the adjacent structure that although it is maintained in position under load it can nevertheless freely rotate or pivot. Such a junction is called a *hinged end.* Alternatively, the end may be fixed rigidly in position so that any tendency to pivot is restrained as in a beam with similar support conditions. This is called a *fixed end,*[1] and the greater the degree of fixity or restraint at the column ends the less will the column buckle under load.

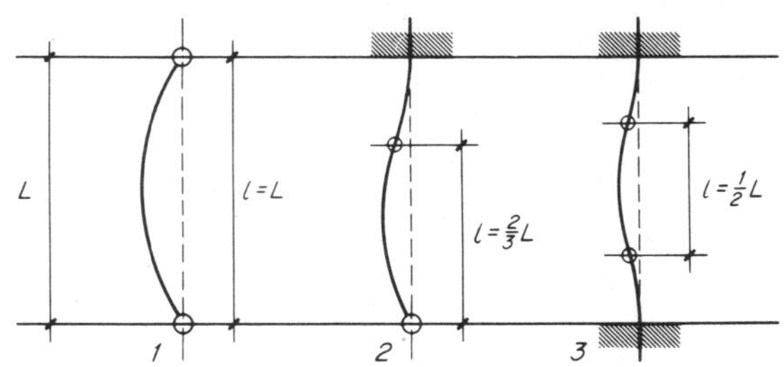

3.32 *Column action*

The reasons for this can be seen from figure 3.32. In (1) both ends are hinged so that the full length of the column bends or buckles under load. In (2) one end is hinged and one is fixed with the result that only a portion of the column length is able to buckle and buckling is reduced. The column, in fact, may be considered to behave as a shorter column with both ends hinged so that the length of curve and the tendency to buckle are less. In (3), with both ends fixed an even shorter length may be viewed as acting with both ends hinged and the tendency to buckle is further reduced. The length of column in each case which bends as if it had hinged ends is known as the *effective height or length (l)* of the column. In (1) this is obviously the full height of the column and in (2) and

[1] More fully — 'fixed in position and direction' ie the end cannot rotate to another direction.

(3), assuming complete fixity at the respective ends, the proportions of the full height shown are usually taken although in practice full fixity or restraint is not usually achieved and codes of practice lay down greater proportions of column length to be assumed for varying degrees of restraint at the ends of the column.

To take account of the effects of varying end conditions the slenderness ratio must be based on the effective length of the column which becomes less with increasing end restraint, resulting in a smaller slenderness ratio and, therefore, greater loadbearing capacity. Thus the slenderness ratio becomes l/r rather than h/r used earlier.

Buckling can also occur in the compression zone of a beam due to the 'column effect' at this point. The compression forces have the same effect as those in a column and the tendency to buckle is similarly related to the length and thickness of the beam. The deeper and thinner the beam the greater the tendency to buckle. Buckling, if it occurs, is sideways (figure 3.33) and this can be resisted by suitable lateral stiffening, either continuous or at intervals to keep the ratio of beam length to depth of beam within an acceptable limit which is laid down by codes of practice (see page 198).

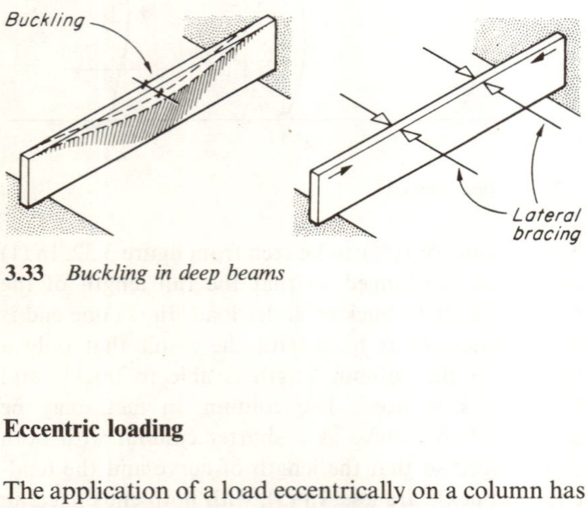

3.33 *Buckling in deep beams*

Eccentric loading

The application of a load eccentrically on a column has exactly the same effect on the column as on a wall, the overall stress W/A being reduced on one side and increased on the other due to the moment We (figure 3.34), the stresses being calculated in the same manner (see page 40). In columns, however, the eccentricity of load is often much greater than in the case of walls because beams are commonly fixed to the side of a column so that the point of application of load is at a greater distance from the column axis.

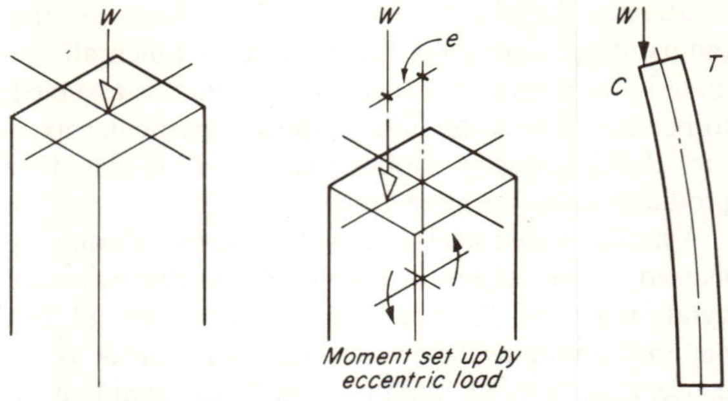

3.34 *Eccentric loading on columns*

3.3.4 Framed or trussed elements

Reference has already been made to the greater efficiency and stiffness which results from the use of deep beams. Over wide spans very deep beams or *girders* may be required and economies in material and reductions in weight can be made by replacing a solid web joining the top and bottom flanges by separate members placed on the lines of diagonal tension and compression in the web as shown in figure 3.35 (*A*). This is known as a *lattice, framed* or *trussed* girder.

If figure 3.35 is compared with figure 3.26 it will be seen that the members in heavier lines, sloping down towards the bearings, lie on the lines of the diagonal compressive shear stresses to which they provide resistance and the other members lie on the lines of the tensile shear stresses to which they also provide the necessary resistance. This particular form of framing is not commonly used as strength and stability may be obtained by using only one set of members as shown in figure 3.35 (*B*) and (*C*). The vertical and horizontal members shown in broken lines are not essential to the structural design of the girders but are often included for constructional reasons.

It will be seen that the top members or flanges are in compression and the bottom in tension while those in the web are either in compression or tension according to whether they lie on the lines of compressive or tensile shear stresses: the compression members are called *struts* and the tensile members *ties*. The struts must be stiff enough to resist buckling and will, therefore, be larger than the ties which can in principle be quite thin and flexible. In order to economise in material in the struts another form of framing may be used as shown in figure 3.35 (*D*) which has the effect of making the struts shorter in length and therefore stiffer, thus permitting a smaller cross-section. The length of the ties is of no significance in this respect because under load they stretch tight rather than buckle.

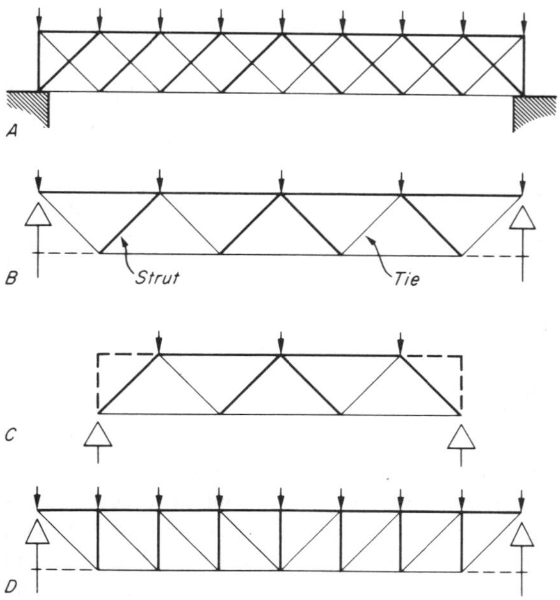

3.35 *Trussed or lattice girders*

This type of framing produces a *triangulated* structure which permits the use of pinned or hinged joints at all the connections or *nodes* while still producing a stable structure. This is because the triangle cannot be made to change shape whereas the square or rectangle with hinged nodes will easily deform into a parallelogram and collapse (figure 3.36). It is possible to prevent this by the use of rigid joints but in resisting deformation bending and bending stresses are set up in the members of the square frame as shown. In a triangulated structure with hinged joints, however, since rotation can occur at these points there is no bending in the members and all stresses are *direct* stresses.[1] This assumes that all loads are applied at the nodal points as indicated in figure 3.35 otherwise bending would be caused in the flange members if loaded between the nodes. Members which undergo direct rather than bending stresses are fully stressed at all points and the material is fully utilised. Under bending the stress intensity varies along the member and the material at certain points is often under-utilised (figure 3.37).

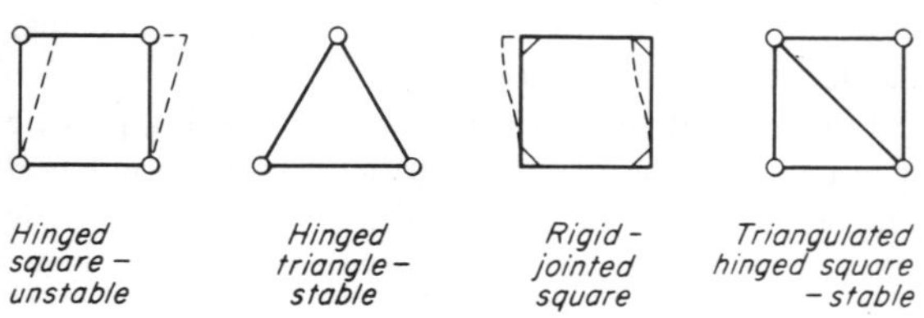

3.36 *Triangulation*

Although when nodally loaded no bending occurs in the individual members of these beams the beam as a whole will still deflect under load due to the lengthening and shortening of the tension and compression members respectively under direct stress.

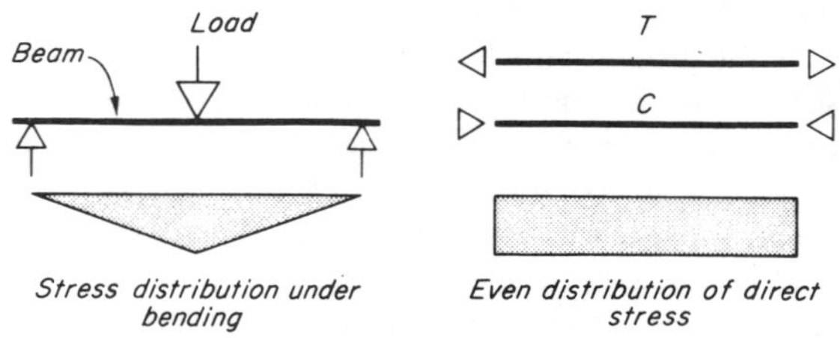

3.37 *Bending and direct stressing*

The top flanges, or booms as they are called in this type of girder, are often pitched at an angle to provide slopes to a roof surface as in figure 3.38. Those requiring a considerable pitch for small unit coverings such as roofing slates take the triangular form of the traditional *roof truss* and are framed, or trussed, in various ways to take account (i) of the need to keep compression members short, (ii) of the number of purlins so that these are positioned at nodal points and (iii) of the need to prevent excessive deflection of the bottom tie under its own weight (see also page 166). The principles of triangulation and of loading at nodal points are the same.

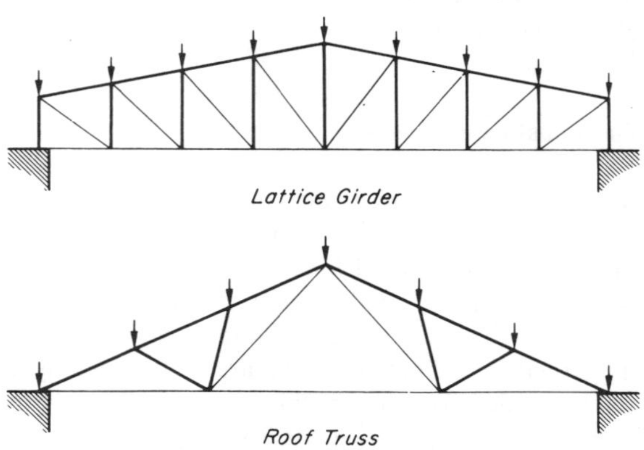

3.38 *Roof trusses*

[1] In practice a certain degree of fixity is usually present at the nodes because of the nature of normal jointing techniques. See *Roof Trusses* chapter 6.

4 Foundations

The foundation of a building is that part of it which is in direct contact with the ground and which transmits the load of the building to the ground. Apart from solid rock the ground on which a building is founded consists of soil of one type or another, all of which are compressible in varying degrees, so that under the building load foundations on such soils will, to some extent, move in a downward direction. This is known as *settlement* and is due mainly to the consolidation of the soil particles. Excessive settlement will result from overloading the soil to such an extent that the loaded area of soil shears past the surrounding soil in what is known as *plastic failure of the soil*. In addition, settlement may be caused by a reduction in the moisture content of certain soils which shrink on drying out or by general movement of the earth due to various causes.[1]

Provided that settlement is uniform over the whole area of the building and is not excessive, the movement does little damage. If, however, the amount of settlement varies at different points under the building, giving rise to what is known as *relative* or *differential settlement*, distortion of the structure will occur which, if too great, may result in damage to fabric and finishes or possible failure of the structure. Such differential movements must, therefore, be kept within limits which avoid harmful distortion. These limits will vary with the type of structure and its ability safely to withstand differential movements.

Functional requirements

It will be seen from these considerations that the function of a foundation is to transmit all the dead, superimposed and wind loads from a building to the soil on which the building rests in such a way that settlement, particularly uneven or relative settlement of the structure, is limited and failure of the underlying soil is avoided. To perform this function efficiently the foundation must provide in its design and construction adequate strength and stability.

The strength of a foundation to bear its load and to resist the stresses set up within it is ensured by the satisfactory design of the foundation itself. Its stability depends upon the behaviour under load of the soil on which it rests and this is affected partly by the design of the foundation and partly by the characteristics of the soil. It is necessary, therefore, in the design of foundations to take account not only of the nature and strength of the materials to be used for the foundations but also of the nature, strength and likely behaviour under load of the soils on which the foundation will rest.

Soils and soil characteristics

The topmost layer of soil at ground level is an unsuitable material on which to found. It has been weathered, is relatively loose and usually contains decayed vegetable matter. It is soft and excessively compressible. This is known as *topsoil* or *vegetable soil* and varies in thickness usually from about 150 mm to 230 mm. Below this lies the *subsoil* from which the topsoil has developed and which consists of solid particles of varying shape and size derived from the weathering of solid rock, the spaces between which are filled with water and to some extent air. It is to this sub-soil that the word 'soil' refers when used in relation to foundations.

Soils vary widely in nature and characteristics and some classification is essential in order to identify a particular soil and judge its likely behaviour from that of similar soils. The physical properties of a soil are those most relevant to foundation design, and as these are known to be closely linked with the size of the particles of which the soil is formed, a classification according to particle or grain size distribution is adopted. This classification defines five broad types apart from cobbles and boulders: *gravels, sands, silts, clays* and *peats*, of which the last is not a satisfactory bearing soil. The first four categories, the names of which are widely but loosely used, are given precise meaning by the allocation of each to a particular particle-size range (see table 4.2).

Fine-grained soils, because of their varying particle characteristics necessitate a further classification based on the variations in volume and consistency of these

[1] For general earth movements see chapter 3 Part 2.

soils with changes in moisture content.[1]

The gravels and sands and the silts and clays form two broad groups: the coarse-grained or granular soils and the fine-grained soils, which have quite different properties. Some of these are related in table 4.1.

GRAVELS SANDS *Coarse-grained or granular: cohesionless soils*	SILTS CLAYS *Fine-grained: cohesive soils*
Low proportion of voids between particles	High proportion of voids between particles
Slightly compressible	Highly compressible
Permeable	Almost impermeable
Compression occurs quickly	Compression occurs slowly
Negligible cohesion between particles	Considerable cohesion between particles
Little variation in volume with change in moisture content	Considerable change in volume in some clays with change in moisture content

Table 4.1 *Properties of soils*

These characteristics result in greater settlements in cohesive soils than in cohesionless because of their greater compressibility and in less risk of shear failure in cohensionless soils because their strength depends on the friction existing between the particles, which increases with an increase in applied load, whereas the fine-grained soils depend for their strength upon the cohesion between the particles, which is constant at all loads.[2]

Variation in moisture content is of particular significance in the case of fine-grained cohesive soils which change considerably in volume with changes in moisture content. The shrinkable clays found in the south-east of England are most susceptible to these changes, which may be caused either by climatic changes or by the effects of tree roots. The significant effect of normal seasonal changes does not extend to a depth of more than about 1 m except in times of long drought, so that if foundations are placed at or below this depth, the structure is unlikely to be affected by settlement due to drying out. The Building Research Establishment does, in fact, recommend a minimum depth of 0.9 m on *all* clay soils. Since the depth of foundations to small and lightly loaded structures need not be great for structural reasons,[3] this moisture movement of clay soils often dictates the foundation depth in smaller types of buildings.

Tree roots can extend radially greater than the height of the tree, depending on the type of tree, and they extract water from the soil to considerable depths — large trees can extract water to depths up to 4.5 m or more. The proximity of trees to a building site, particularly fast-growing and water-seeking trees such as poplars, elms and willows, should therefore be noted in any site investigation. Within a few years of planting the roots of these trees will extend 15 m or more and dry out clay soil below any nearby building. Movements of 25 mm to 50 mm are common and settlements as much as 100 mm have been known to occur in buildings as far as 24 m from black poplars. Buildings on foundations not deeper than 1 m should, as a rough guide, not be erected closer to single trees than their mature height or to groups or rows of trees not closer than one and a half times to twice the mature height of the trees.

Shrinkable clays which have supported trees for many years will have dried out and when the trees are felled to permit building the clay will swell over a long period as moisture returns to the soil. This swelling can be substantial and the consequent uplift can cause considerable damage to buildings erected on the site.[4] This must be avoided by the use of special foundations (see table 4.3 and page 65).

In some soils, in particular fine sands, silts and chalk, the sub-soil water may freeze due to frost penetration and form layers or lenses of ice. As a result of the expansion due to freezing the ground surface is raised in what is known as *frost heave* and this, in some circumstances, may be sufficient to lift some parts of a structure. In Great Britain during winter frost may penetrate to a depth of around 500 mm, except in extreme conditions in open ground when it may be more. At the foundation of a heated building, however, the ground is unlikely to have frozen to this depth, if at

[1] See Part 2 for a fuller discussion on these systems of classification.

[2] For a fuller discussion of these differences and of means of investigating soil characteristics, see Part 2.

[3] For practical purposes in soils other than clay a foundation depth of 450 mm is usually accepted as a minimum to allow for variations in topsoil thickness and for minor variations in ground level.

[4] See BRE Current Paper 73/75: *Long-term heave of a building on clay due to tree removal* and BRE Digest 298.

all. It is considered,[1] therefore, that a depth of 450 mm to the underside of the foundations should protect against frost heave although an increase in depth may be advisable in upland areas and in others known to be subject to long periods of frost.

Site exploration

Tall, wide-span or heavily loaded buildings exert greater pressures on the soil resulting in greater settlements, and lead to greater possibility of shear failure of the soil than do small-scale buildings. To overcome this, types of foundations may be required which affect the soil to considerable depths. In such cases the various soil characteristics take on greater significance and a closer consideration of the soil and its properties is required than is often necessary for small-scale buildings. This may require an extensive examination of the subsoils below the site involving boring to considerable depths and carrying out field and laboratory tests on the soils. Such an examination is called a *site exploration* or *subsoil survey*, the procedures for which are described in Part 2.

An extensive investigation of the soil is not usually necessary in the case of small-scale buildings[2] on soils of adequate strength where, by means of simple strip or pad foundations near the surface, the pressure on the soil can be kept well within the known safe bearing capacity of a particular soil type. Provided the foundations are placed at a depth adequate to avoid the effects of moisture movement and freezing of the soil and provided the soil to a sufficient depth is of similar type or similar strength to that on which the foundation actually bears, it will be enough to have regard only to the bearing capacity of the soil. With light loads and small soil pressures settlement will be small.

What is required is a simple method of establishing the type of soil to a sufficient depth and a means of determining its bearing capacity. The latter can be found from standard tables and the soil type can be established by exposing the soil to view for the shallow depths involved by digging holes known as *trial pits* and using simple visual and tactile means of identification.

Trial pits should not be further apart than about 30 m and not less than one per 930 m^2 of site. The depth should be at least that to which the soil will be significantly affected, based on a preliminary assessment of the required width of foundation.[3] For small-scale work this is not likely to exceed 2.75 m to 3 m deep. The soil should be inspected at all levels as soon as possible after excavation. As a check on the possible existence of a very weak layer of soil below the trial pit a probe of about 1 m may be made by means of a hand auger.[4] This can also be used when the presence of ground water makes the completion of the trial pit difficult.

The soils excavated and exposed in the trial pits can be identified within broad types by simple field tests of a visual, tactile and physical nature. These, with the bearing capacities of the soils, are indicated in table 4.2. By this means the bearing capacity of the soil immediately under the proposed foundation may be determined and provided the soil to the bottom of the pit is not less in strength the foundation can be designed in accordance with this as described later.

Should a stratum weaker than that on which the foundation is to bear be exposed in the pit, unless the foundation can be designed to accord with its lower bearing capacity, a closer investigation into the pressure likely to be exerted at the level of the weaker stratum would be necessary. Simple approximate means for doing this are described in Part 2, where the methods of more detailed soil exploration are discussed.

[1] See BS 8103: Part 1: 1986 *Structural design of low-rise buildings*.

[2] A small-scale building in this context means one which is not very tall or wide in span so that the dead loads are relatively small and of a type in which the imposed loadings are light so that the pressures on the soil are not excessive and are evenly dispersed. These conditions are usually met by houses and many types of low-rise buildings. Reference should be made to BS 8103 *Structural design of low-rise buildings*.

[3] This depth is about one and a half times the width of a pad foundation and three times the width of a strip foundation, see Part 2.

[4] See post hole auger figure 4.11.

Note on Table 4.2
Sands and gravels: In these soils the permissible bearing capacity can be increased by 12·5 kN/m^2 for each 0·30 m of depth of the loaded area below ground level. If ground water-level is likely to be less than the foundation width below the foundation base the bearing capacities given should be halved. The bearing capacities given for these soils assume a width of foundation around 1·00 m but the bearing capacities decrease with a decrease in width of foundation. For narrower foundations a reduced value should be used: the bearing capacity given in the table multiplied by the width of the foundation in metres.

Subsoil types	*Condition of subsoil*	*Means of Field Identification*	*Particle size range*	*Bearing capacity kN/m²*	*Minimum width of strip foundations in mm for total load in kN/m of loadbearing wall of not more than*					
					20	30	40	50	60	70
Gravel	Compact	Require pick for excavation. 50 mm peg hard to drive more than about 150 mm	Larger than 2 mm	>600	250	300	400	500	600	650
Sand		Clean sands break down completely when dry. Particles are visible to naked eye and gritty to fingers. Some dry strength indicates presence of clay	0.06 to 2 mm	>300						
Clay	Stiff	Require a pick or pneumatic spade for removal. Cannot be moulded with the fingers	Smaller than 0.002 mm	150—300	250	300	400	500	600	650
Sandy clay		Clays are smooth and greasy to the touch. Hold together when dry, are sticky when moist. Wet lumps immersed in water soften without disintegration	See Sand and Clay	150—300						
Clay	Firm	Can be excavated with graft or spade. Can be moulded with strong finger pressure	See above	75—150	300	350	450	600	750	850
Sandy clay			See Sand and Clay	75—150						
Gravel	Loose	Can be excavated with a spade. A 50 mm peg can be easily driven	See above	<200	400	600	For loading of more than 30·0 kN/m run on these types of soil: the necessary foundations do not fall within the provisions of Approved Document A, Part E from which these figures are taken.		Pad foundations generally and surface rafts are designed using the bearing capacities for soils given in this Table. *Note* See note on facing page regarding the use of values given in this table for bearing capacities of sands and gravels.	
Sand			See above	<100						
Silty sand			See Silt and Sand	May need to be assessed by test						
Clayey sand			See Clay and Sand	ditto						
Silt	Soft	Readily excavated. Easily moulded in the fingers. Silt particles are not normally visible to the naked eye. Slightly gritty. Moist lumps can be moulded with the fingers but not rolled into threads. Shaking a small moist pat brings water to surface which draws back on pressure between fingers. Dries rapidly. Fairly easily powdered	0.002 to 0.06 mm	<75	450	650				
Clay			See above	<75						
Sandy clay			See Sand and Clay	May need to be assessed by test						
Silty clay			See Silt and Clay	ditto						
Silt	Very soft	A natural sample of clay exudes between the fingers when squeezed in fist	See above	ditto	600	850				
Clay			See above	ditto						
Sandy clay			See Sand and Clay	May need to be assessed by test						
Silty clay			See Silt and Clay	ditto						
Chalk	Plastic	Shattered, damp and slightly compressible or crumbly	—	—	Assess as clay above					
Chalk	Solid	Requires a pick for removal	—	600	Equal to width of wall					

Table 4.2 *Soil characteristics and bearing capacities*
Based on information in the Building Regulations 1985, BS 8004: 1986, BS 5930: 1981 and Building Research Digests 64 and 67

In built-up areas local knowledge of the types of soil and their characteristics and properties can frequently be drawn on through local authorities and others which may be sufficient to make an investigation of the soil unnecessary, although it is usually advisable to link this with the findings from at least one trial pit.

Foundations are invariably formed in concrete, the durability of which, in this context, can be affected by sulphate salts in the soil, particularly in the ground water, and by frost action.

The nature of any ground water should, therefore, be checked and where sulphate salts are known to be present precautions must be taken to prevent destruction of the concrete by the use of a suitable cement.[1] Since concrete is attacked by sulphate salts only when they are in solution, adequate density and impermeability of the concrete is essential. During the early hardening period all concretes are more prone to sulphate attack than when matured and in severe cases asphalt or polythene sheet, as a tanking, is used to isolate ground water from the concrete during this early period.

Damage by frost action caused by the expansion on freezing of water in the pores of the concrete is unlikely in Great Britain since, as indicated earlier, for reasons other than protection of the concrete foundations should be placed below the level at which frost is likely to penetrate into the soil (see page 53).

4.1. FOUNDATION TYPES

Foundations, either of mass or reinforced concrete, range from a simple strip to a deep, piled foundation.

The many forms of foundations used in building work may be divided broadly into *shallow* foundations and *deep* foundations. Shallow foundations are those which transfer the load to the soil at a level close to the lowest floor of the building and include the spread foundations: strips, pads and rafts. These, of course, may be formed at great depth below ground where there is a basement. Deep foundations include piles and various types of piers which transfer their loads to the soil at a considerable distance below the underside of the building (see figure 4.1).

4.1.1 Choice of foundation

Unless conditions make the use of deep foundations essential, shallow foundations are always used as these

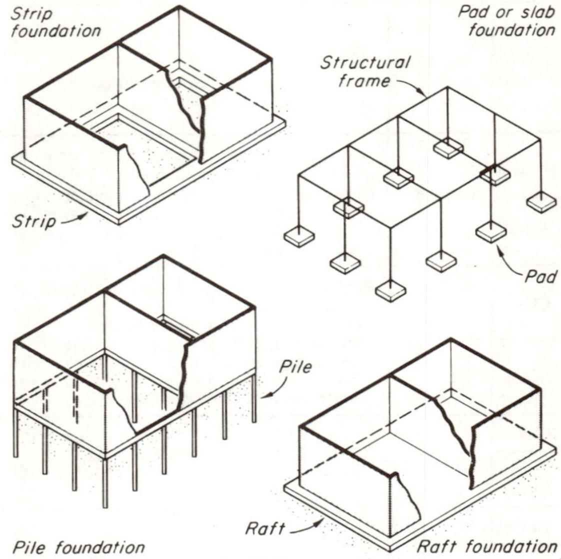

4.1 *Foundation types*

are nearly always the cheapest. An exception is possibly the use of short bored piles instead of strip foundations in shrinkable clays.

Strip foundations under continuous walls and pad or slab foundations under isolated piers or columns are used on sites where a sufficient depth of reasonably strong subsoil exists near the surface of the ground or, in the case of a building with a basement at the level of the proposed basement floor.

Raft foundations, by which the whole of the building area is covered, are used where no firm bearing strata of soil exists at a resonable depth below the surface and a maximum area of foundation is required to bring the imposed pressures within the low bearing capacity of the weaker soils and of some made-up ground. Rafts are also used on firm soils in circumstances described later in this chapter and also in other circumstances which are discussed in Part 2.

Piers and piles may be viewed as columns passing through weak soil to transmit the building load to lower strata where the pressure can be safely resisted. They may also be used to transfer loads to the soil below the level likely to be affected by moisture movement.

The choice of a foundation, as already indicated, must take account not only of the soil but also of the superstructure of the building.

[1] See *MBS: Materials.*

A stiff rigid building, one with plain monolithic concrete walls for example, will be adversely affected by differential movement to a greater extent than one with brick or block walls. Within certain limits distortion can be accommodated in the latter by fine cracks distributed throughout the joints whereas in the former the distortion will rapidly cause large cracks to develop.[1]

Small scale structures imposing small loads on the soil will cause only small settlements in most soils and the movement transferred to the superstructure will be very small and will have little effect upon it. The nature of the structure in this respect is, therefore, less important than with large-scale buildings under which consolidation may be considerable and settlement is likely to be significant. Foundation design then requires an appreciation of the ability of the structure to withstand movements without dangerous over-stress and damage to the structure.[2] Small-scale buildings affect the soil at shallow depths only as shown above so that provided the small loads on the soil are satisfactorily related to the strength of the soil near the foundation bearing by the use of an appropriate form of foundation, account need be taken only of soil movement due to causes other than loading. Where the soil is stable, that is where no general soil movement is likely, the possible causes of movement will be changes in moisture content of the soil and frost action, movement due to both of which can be avoided by placing the foundation at an adequate depth as already described.

On soils which, for some reason, are unstable special measures must be taken whatever the scale of the building.

Overall or total settlement must be limited so that services and drains connected to the building are not damaged; alternatively, provision must be made for flexible connections.

Table 4.3 indicates the suitability of the foundation types described above to the various types of soil. In this volume the types of foundations suitable for small-scale work will be considered in detail. Those required to solve the problems involved in large-scale works and in unstable soils are described in Part 2.

4.1.2 Spread foundations

Spread foundations, that is strips, pads and rafts, must be designed so that the soil is not over-stressed and so that the pressure on the soil under them is equal at all points in order to avoid unequal settlement under the actual foundation. The former is ensured by providing sufficient area of foundation (see below) and the latter by arranging the centre of gravity of the applied loads to coincide with the centroid of area of the foundation. In the case of strip foundations and isolated column slabs this requires the foundation to be placed symmetrically with the wall or column it supports (figure 4.2a).

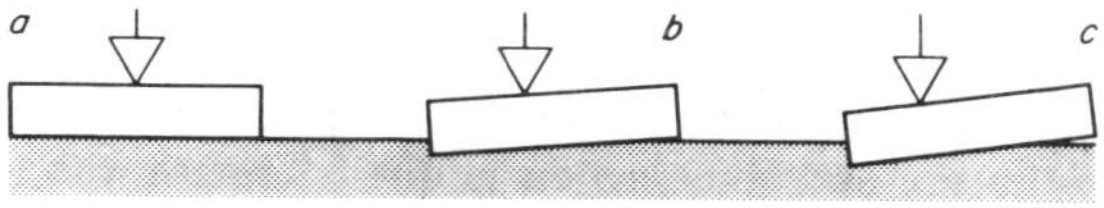

4.2 *Effect of eccentric loading on foundations*

If the load from a wall or column is applied eccentrically to a spread foundation the pressure on one side will be greater than the average pressure, causing greater consolidation of the soil on that side of the foundation (figure 4.2b). When the eccentricity is great the increased stress could, in fact, exceed the safe bearing capacity of the soil, even though the average stress might be well below it. When the eccentricity is greater than one-sixth of the foundation width tensile stress occurs and causes the foundation to rise off the soil, since there is no tensile resistance between the two, thus concentrating the pressure on a reduced area of soil (figure 4.2c) and resulting in very high stresses. Stress distribution in the soil will be similar to that in an eccentrically loaded wall and will similarly vary according to the degree of eccentricity as shown in figure 3.14. The determination of the stresses in the soil is carried out as for walls (page 40).

Strip foundations

These consist of a strip of concrete under a continuous wall carrying a uniformly distributed load (figure 4.3). The required area, as in the case of all spread foundations, is related to the imposed load and the bearing capacity of the soil. As the imposed load is considered as a load per metre along the wall the width of the strip is made such as to give sufficient area per metre run of foundation (figure 4.4). Thus, if the loading is

[1] Tests have shown that brickwork with openings for windows can withstand a distortion of about 25 mm in 25 m without serious cracking occurring. See also chapter 3 Part 2 for further reference to permissible distortion.

[2] The relationship of soil, foundations and superstructure is covered more fully in Part 2.

Soil type and site condition	Foundation	Remarks
Rock, solid chalk, sands and gravels or sands and gravels with only small proportions of clay, dense silty sands	Shallow strip foundations, pad foundations, (as appropriate to the load-bearing members of the building) Surface raft See Table 4.2	Keep above water wherever possible. slopes on sand liable to erosion. Foundations to be 450 mm below ground level on ground susceptible to frost heave (see text)
Uniform, firm and stiff clays: 1 Where vegetation is insignificant	Strip or pad foundations at least 1 m below ground level Bored piles See Tables 4.2 & 4.4	With these soils downhill creep may occur on slopes greater than 1 in 10. Unreinforced piles have been broken by slowly moving slopes
2 Where trees and shrubs are growing or to be planted close to the site	Bored piles See Table 1.4	
3 Where trees are felled to clear the site and construction is due to start soon afterward	Reinforced bored piles of sufficient length with top 3 m sleeved from the surrounding ground and with suspended floor Thin reinforced rafts supporting flexible superstructure Basement rafts See Part 2	
Soft clays, soft silty clays	Strip foundations up to 850 mm wide if bearing capacity is sufficient See Table 4.2 Rafts See Part 2	Settlement of strips or rafts must be expected. Services entering building must be sufficiently flexible. In soft soils of variable thickness it is preferable to pile to firmer stratum
Fill (made up ground) Peat	Pier foundations Piles driven to firm stratum below Special raft foundations with or without flexible superstructure See Part 2	If fill is sound, carefully placed and compacted in thin layers, strip foundations are adequate
Mining and other subsidence areas	Special raft foundations with or without flexible superstructure See Part 2	

Table 4.3 *Suitability of foundation types to various soils*
Based on information in Building Research Digest 67

30·00 kN/m and the soil is to be stressed not more than 50·00 kN/m² the minimum width should be 0·60 m (load divided by stress, see page 35). This means that in every metre run of foundation the load of 30·00 kN will be distributed over 0·60 m² of soil, resulting in a pressure of 50·00 kN/m².

In cases of light loading on reasonably strong soils a strip no wider than the wall it carries may suffice. In practice, however, with masonry walls some spread is usually provided on each side to give a minimum width of 450–500 mm to allow working room for bricklayers building the lower courses of walls (figure 4.3). This also provides some stability to the wall before it is tied in by floors and roof. For hand excavation of the soil a minimum width of 610 mm to 760 mm, depending on the depth, is needed to give sufficient working space for this operation.

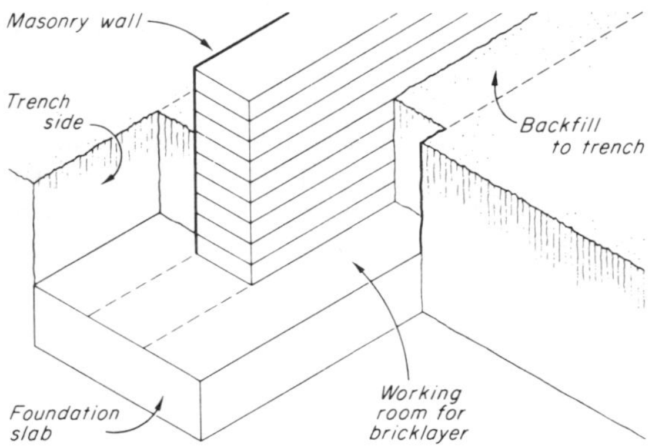

4.3 *Strip foundation*

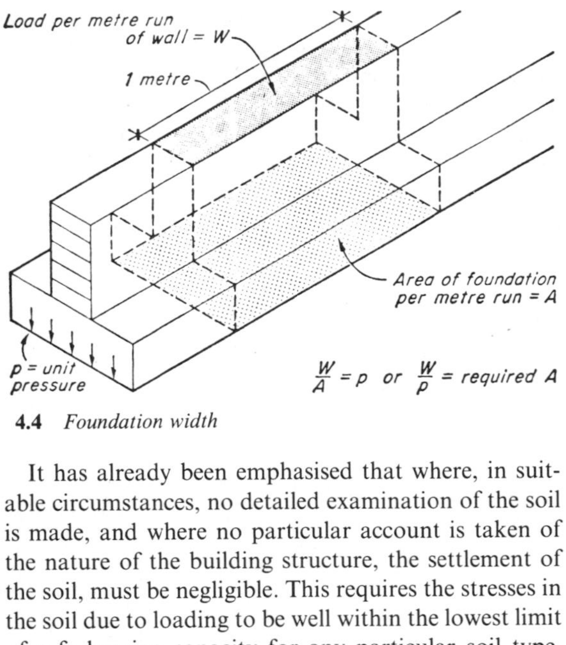

4.4 *Foundation width*

It has already been emphasised that where, in suitable circumstances, no detailed examination of the soil is made, and where no particular account is taken of the nature of the building structure, the settlement of the soil, must be negligible. This requires the stresses in the soil due to loading to be well within the lowest limit of safe bearing capacity for any particular soil type, resulting in wider foundations than those which would fully stress the soil. This is the basis of the minimum widths laid down for strip foundations in the Building Regulations and given in table 4.2.

Where the edges of a foundation project beyond the faces of the wall it supports, bending due to cantilever action will occur as a result of the resistance of the soil, causing bending and shear stresses in the foundation (figure 4.5 *A, B*). The tensile strength of unreinforced concrete is low, and in order to keep these stresses within the capacity of the concrete the strip must be of adequate thickness. Concrete fails under a compressive load usually by tensile shear failure along planes lying at an angle of about 45 degrees to the horizontal (*B* and see chapter 3). BS 8004: 1986 recommends an angle of spread of load from the wall base to the outer edge of the foundation of not more than 45 degrees which results in the thickness being not less than the projection of the base beyond the face of the wall it carries and provides sufficient area at the shearing planes to keep tensile stresses within the capacity of the concrete (*C*). Very wide strips are reinforced to keep their thickness within economic limits.

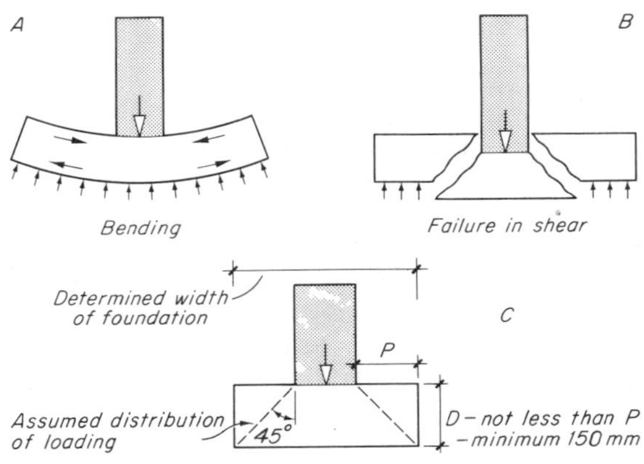

4.5 *Foundation thickness*

Heavy loads concentrated at points in the run of a wall carrying an otherwise uniformly distributed load will result in greater loads on the foundation at these points than on the remainder. In order to ensure equal stress at all points in the soil these extra loads must be distributed to the soil through larger foundation areas. When, for example, a beam bears on a wall its load may necessitate a thickening of the wall by a projecting pier as in figure 4.6. The beam load will normally be distributed over the combined area of the projection and the wall immediately behind and the foundation, therefore, must be extended symmetrically with this area of pier and wall by projecting the foundation in front of the pier the same extent as that to the wall in order to maintain the centroid of area of the foundation under that of the combined pier and wall. The foundation projection is returned the same extent on each side of the pier to ensure equal distribution of pressure on the soil. A similar return is made at a stopped end of a wall (see page 79).

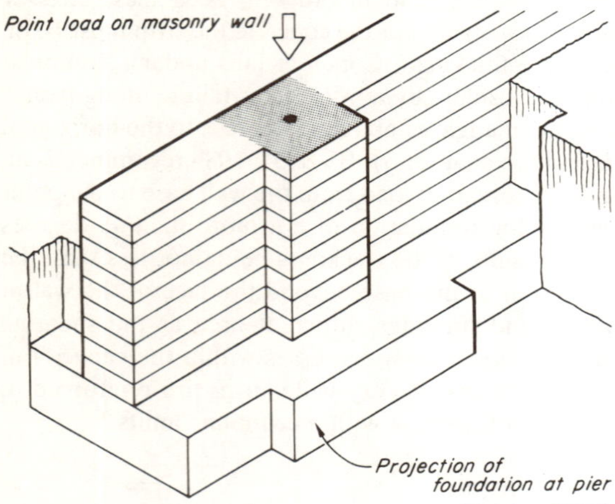

4.6 *Foundation to wall pier*

For mass concrete foundations, that is unreinforced, a 1:3:6 mix is commonly used, with a fairly large aggregate — say 38 mm to 50 mm. Concrete should be poured as soon as possible after excavation of the trenches. This is particularly important in clays and chalk which deteriorate on exposure to water and frost, losing strength when they become wet. In the case of clay drying-out causes shrinkage which is followed by expansion subsequent to further wetting after the foundations have been completed.

If the concrete cannot be placed on completion of excavation the bottom should be protected by 50 mm of weak concrete *blinding* or, alternatively, 75 mm to 100 mm of soil should be left for excavation immediately prior to concreting.

In granular soils, particularly if loose, it is good practice to put building paper in the trench to prevent leaching of the water and cement into the soil which results in poor quality concrete.

Trench fill foundations

Firm or stiff shrinkable clays are strong and when carrying light loads necessitate quite small foundations, possibly no wider than the wall carried. These soils, however, move considerably with changes in moisture content and, as already emphasised, the bottom of the foundation should be 1 m below ground level. In these circumstances a deep, narrow excavation is required, perhaps only 300 mm wide but 1 m or more in depth (figure 4.7).

Such a deep, narrow trench cannot be dug by hand, nor can brickwork be built up from the bottom. It can, however, be dug quickly by mechanical means and if the trench is filled with concrete to within a short distance of the ground surface the difficulties of bricklaying are overcome. Much less soil has to be excavated and moved than with a wider strip foundation and backfilling is eliminated.

This form is called a *trench fill foundation* and where conditions are suitable it is cheaper to construct and quicker to complete than the wider (often called 'traditional') strip foundation taken to the same depth.

The conditions necessary to make it economic are:

(1) a self-supporting soil to avoid timbering — firm, shrinkable clays possess this characteristic, and
(2) adequate runs of straight trenching with a minimum amount of corner trimming, to justify the cost of a mechanical excavator of one type or another.

Condition (2) generally necessitates a reasonably large contract of suitable types of building, that is without a large number of small breaks in the runs of wall. In the case of a single small building the limited amount of work to be done, the short straight lengths to be excavated and the relatively large proportion of trimming required usually make mechanical excavation uneconomic and the use of other types of foundation necessary.

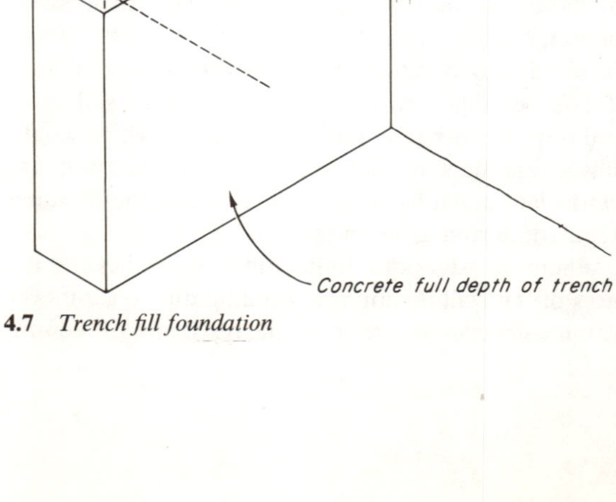

4.7 *Trench fill foundation*

Stepped foundations

Except in certain types of structure transferring inclined thrusts to the ground[1] all foundations must bear horizontally on the soil. If strip foundations to a building on a sloping site are at the same level throughout, those on the higher side will be a greater distance below ground level than the remainder, necessitating deeper trenches and a greater amount of walling in the soil (figure 4.8 *A*). On slight slopes this is of little consequence but when the slope is steep excavation and the amount of walling below ground become excessive.

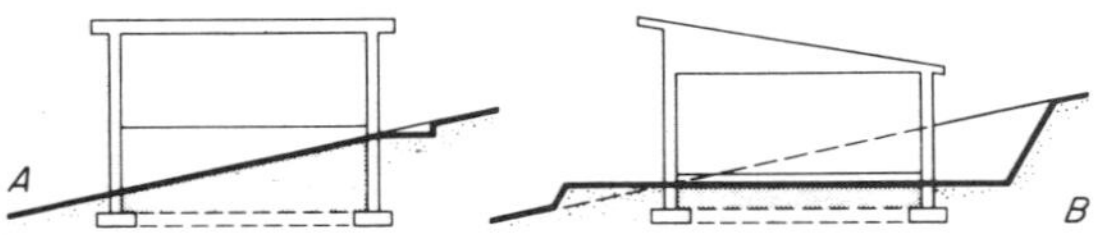

4.8 *Foundations on sloping sites*

There are two ways in which this excessive building into the soil may be reduced:

(i) by cut or cut-and-fill to provide a horizontal plane off which to build (*B*);[2] the choice of this method will depend on factors other than foundations such as the nature of the soil, the plan form of the building, the proximity of a tip for the spoil,
(ii) by stepping down the slope the foundations to those walls parallel to the slope (figure 4.9). These are known as *stepped foundations*.

The steps should be relatively short in length and they should be sufficient in number along the length of the foundation to keep their heights small and uniform. If the steps are too great in height the considerable difference in load on each side of a change in level, due to the varying heights of wall supported, may cause differential settlement. There is also the added shrinkage of the mortar joints in the wall below each step which, together with differential settlement of the soil, may cause cracking in the structure above the steps. Unless special precautions are taken to deal with this possibility the height of step should not exceed the thickness of the foundation. The lengths of the steps need not be uniform but should be varied where necessary to keep the heights as uniform as possible. At each step the higher foundation must lap over the lower for a distance at least equal to the thickness of the foundation or twice the height of the step or 300 mm whichever is the greater except in the case of a trench fill foundation with a thickness of 500 mm or more where the overlap should be twice the height of the step or 1 m whichever is greater.

The foundations should be so arranged that a step occurs at any intersection with a cross wall, the step being on the side where the ground level is highest. As the depth of foundations to internal walls is often less than to external walls a step may be required at the junction of internal and external foundations as shown in broken lines in figure 4.9, whether or not the external foundation is stepped.

On sloping sites it is advisable to lay sub-soil drainage, in the form of land drains, across the slope on the up-hill side of the building, in order to divert the flow of surface water away from the foundations.

Isolated column foundations

Isolated piers or columns are normally carried on an independent slab of concrete, commonly called a *pad foundation*, the pier or column bearing on the centre point of the slab. The area of foundation is determined by dividing the column load by the safe bearing capacity of the soil and its shape is usually a square. Its thickness is governed by the same considerations as for strip foundations and is made not less than the projection of the slab beyond the face of the pier or column or the edge of the baseplate of a steel column. It should in no case be less than 150 mm thick. As in the case of strip foundations when a column base is very wide a reduction in thickness may be effected by reinforcing the slab (see Part 2).

In a framed structure where loads on different columns vary, the sizes of the bases must vary in order to maintain equal soil pressure under each and thus eliminate differential or unequal settlement.

Light surface raft foundations

A raft foundation is a large slab foundation covering the whole building area, through which all the loads from the building are transmitted to the soil. These foundations have been referred to on page 56 and when used in low-rise buildings for the purposes described here they are laid on, or just below the surface of the ground and are termed *surface* rafts.

[1] See figure 3.11 and Part 2 — *Rigid Frames*.

[2] See also Ground Floors page 190.

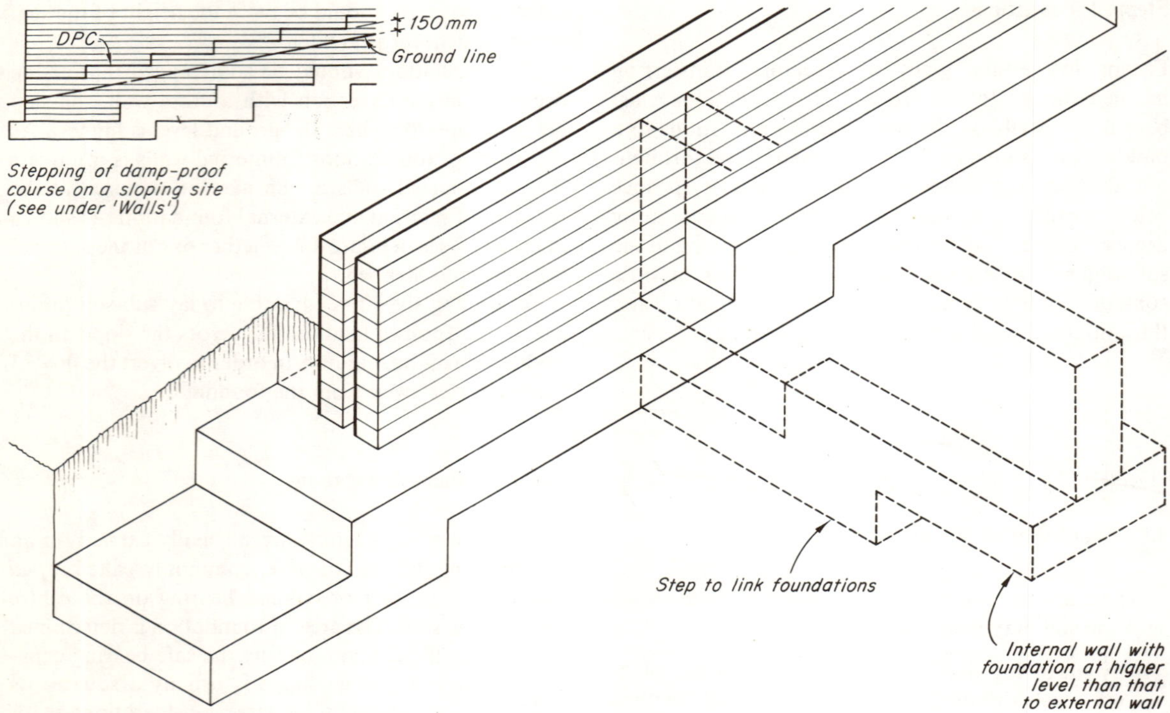

4.9 *Stepped foundation*

Solid concrete ground floor slab construction is normal today (see page 188). This slab, if about 150 mm thick and lightly reinforced, may be used as a light raft on all types of firm soils. Reinforcement is required at the top for crack control with some steel at the bottom under walls or columns to resist tensile stress in these zones (figure 4.10 *A*). On soils of moderate to low compressibility reinforcement at both top and bottom is required throughout the slab. The raft should be extended about 300 mm beyond the perimeter walls to spread the load and to protect the soil under the walls from possible frost action. On sands it is preferable to form a 'downstand' edge all round to prevent erosion of the soil under the perimeter of the slab (*B*). If used on shrinkable clays the soil under the external walls should be protected from moisture changes and consequent movement by an extension of the slab 1·25 m to 1·5 m beyond the walls (*C*). In this case reinforcement is generally as for rafts on other soils but top and bottom reinforcement must be provided under the external walls and in the extension to resist the tensile stresses at the top due to loads on the extension when the soil has shrunk under the slab edge and at the bottom due to the pressure of the clay when it swells.

On more compressible soils a stiffer raft is required and the type shown in (*D*) is suitable. The wide stiff edge beam is reinforced and is bonded to the slab by linking with the top slab reinforcement by bonding bars from the beam.

When rafts are used on fill material containing chemicals harmful to concrete or on sulphate bearing soils a damp-proof membrane should always be placed beneath the raft to prevent chemical bearing moisture from the soil reaching the concrete of the raft. (See page 56).

Light surface rafts can also be used to carry lightly loaded structures of certain types on soils subject to general earth movement and these are discussed along with other raft types in Part 2.

As in all spread foundations the centre of gravity of the loads should coincide with the centroid of area of the raft. This is facilitated when the building has a simple regular plan form with load-bearing elements such as walls, columns, stacks, disposed symmetrically about the axis of the building. Heavy elements such as

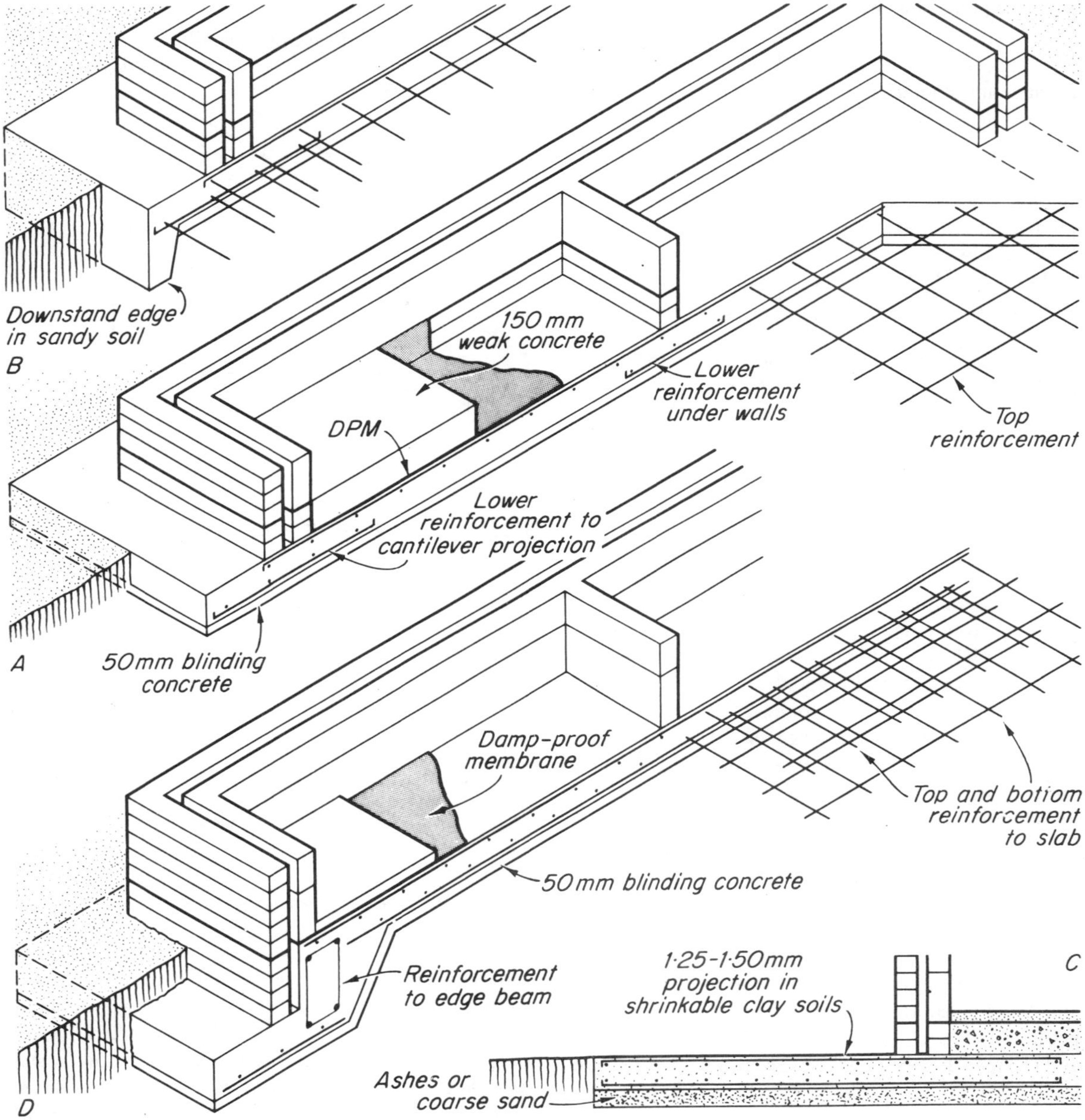

4.10 *Surface raft foundation*

stacks are best situated near the centre of the plan. Excessive variation of loading results in problems which need careful consideration in the design of the foundation. These are discussed in Part 2.

4.1.3 Pile foundations

Piles are often used to transmit loads through soft soils or made-up ground. In such circumstances, unless large in diameter, the piles will normally need to be reinforced. Piles of relatively short length can, however, be used economically in firm shrinkable clay as means of founding below the zone of moisture movement. Such piles require no reinforcement because the diameter is large relative to length so that the piles are stiff, in addition to which they receive considerable support from the firm soil through which they pass. In this type of soil the piles can be easily and quickly formed by boring. This particular form of pile is, therefore, called a *short bored pile*.

Short bored pile foundations

In shrinkable clays this foundation has a number of practical advantages over strip foundations: a reduction in the amount of excavated spoil, a cleaner site, faster construction and the fact that work can continue in weather which would make trench digging impracticable. When mechanically bored in sufficient numbers this type of foundation is competitive in cost with a trench fill foundation and cheaper than a traditional strip foundation of appropriate depth. For a single building it may be slightly dearer than a trench fill foundation, although against this must be placed the advantages of the piles. Generally speaking, the stiffer the clay the cheaper will this type of foundation be relative to strip foundations.

In order to obtain the advantage of greater speed and economy relative to strip foundations the clay must be suitable for easy boring. If many tree roots are present and the soil contains a great number of stones, especially if large, trench digging is likely to be quicker and cheaper than boring for piles, although if mechanical boring can be used, augers larger than hand boring will permit can be adopted which cope more easily with stones.

Mechanical boring is much quicker than hand boring, especially when the holes must be large, but to be economic requires a sufficiently large contract of work on one site[1] and, as for any mechanical plant, requires adequate preparation of the site and the programme of work to be carefully planned in advance to avoid idle time.

This type of foundation consists of a series of short concrete piles which, in the case of load-bearing wall structures, are spanned by a shallow reinforced concrete beam on which the wall is built (figure 4.11).

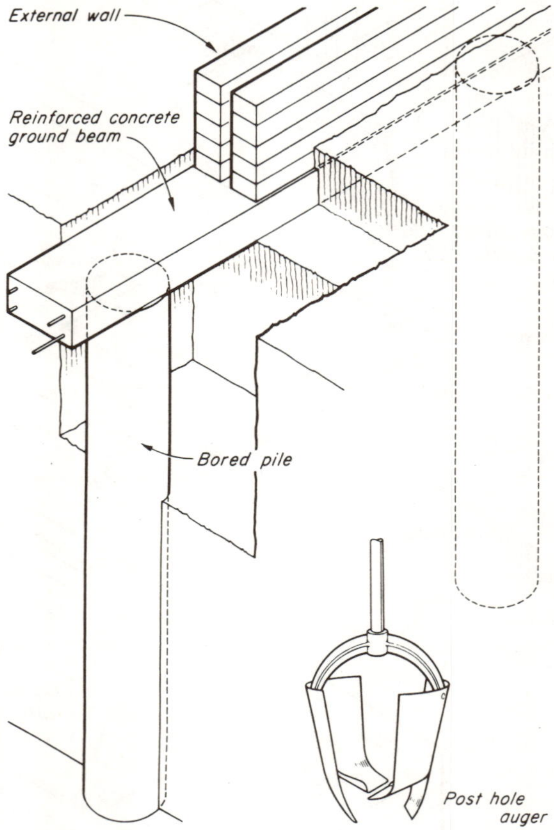

4.11 *Short bored pile foundation*

Holes for the piles are bored manually or mechanically on the centre line of the beams to the required depth and diameter (see table 4.4). When hand bored a bucket type post-hole auger is used rotated by two men, extension rods being added as the depth increases (see figure 4.11). Small stones and layers of gravel present no problem but large stones must be broken up by a heavy chisel on extension rods. Larger augers cope with stones more easily than smaller ones but above 350 mm diameter the weight of the spoil is too great for easy hand boring. A 250 mm diameter hole can be sunk 2·4 m in about 60 minutes, including rest periods, in soil free from stones.

[1] Not less than 100 piles.

Soil strength classification	Diameter of pile (mm)	Length of pile (m)			
		2·4	3·05	3·66	4·27
		kN	kN	kN	kN
Stiff — cannot be moulded with fingers	254	40	50	60	70
(unconfined shear strength	305	50	60	75	90
more than 72 kN/m² — see Part 2)	356	65	80	95	110
Hard — brittle or tough	254	55	65	80	90
(unconfined shear strength	305	70	85	100	115
more than 143 kN/m² — see Part 2)	356	95	110	125	140

The figures are for clay which increases in strength with depth to the 'stiff' and 'hard' classifications near the bottom of the piles. The figures should not be applied to piles in other situations.

Table 4.4 *Permissible loads on short bored piles*
Based on information in BR Digest 67 (Second series)

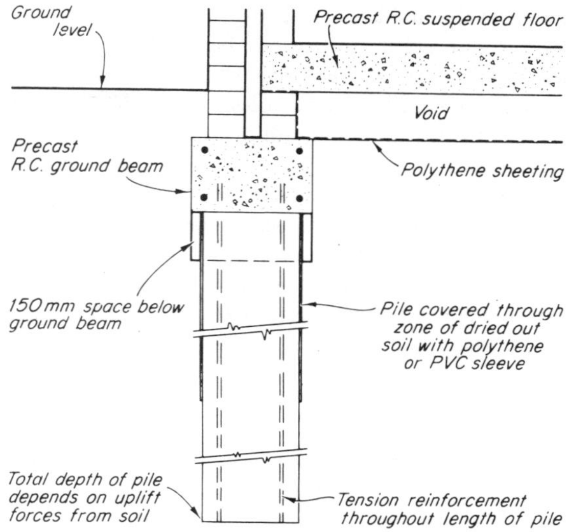

4.12 *Sleeved bored pile foundation*

In framed structures a pile or group of piles is placed under each column. In load-bearing wall structures piles are placed at the corners, at wall junctions and under stacks with further piles distributed between, sufficient to carry the imposed load, spaced as far as possible to produce uniform loading and to bring ground floor door and window openings centrally between piles.

The shallow reinforced concrete ground beams should have a depth/span ratio of 1/15 to 1/20. Reduced 'equivalent bending moments' are used in their design taking account of the fact that the brickwork on the beam tends to act with the beam and as an arch tending to concentrate the load towards the supports. Top reinforcement is placed over the pile positions to take up the negative tensile stresses at these points (see chapter 3 page 46).[1]

A 1:2:4 mix concrete is used for the work with a minimum water content to prevent excessive wetting and thus weakening of the clay. This is placed immediately each hole is bored, using a hopper to prevent soil entering the hole, each 305 mm to 610 mm lift being thoroughly tamped. The beams are normally cast in a trench to avoid shuttering. If this is excavated before the holes are bored the concreting of piles and beams can be done simultaneously. If the beams are to be poured after the piles have set 9·5 mm diameter steel bars should be cast in the tops of the corner piles, set 610 mm in the pile and projecting 610 mm and bent over for casting in with the beams.

Where trees exist on shrinkable clay soil closer to a building than their mature height or, in the case of groups or rows of trees, one and a half times their mature height, this type of foundation should always be used in order economically to take the foundations below the zone of drying-out of the soil (see page 53). In view of the fact that trees may well be planted close to a building during its life, even though not existing and not planned at erection, it can be held that all buildings founded on this type of soil should be built on this type of foundation.

When trees growing in this soil have been felled to permit building special foundations must be used for the reasons given on page 53. A bored pile foundation, generally similar to that shown in figure 4.11 but with certain modifications, should be used. The pile must be sleeved over the upper part through the dried-out zone of soil (see table 4.3, page 58 and figure 4.12) in order to

[1] See BRE Digest 67 together with Digests 240–242 for fuller details of this type of foundation and its design.

to reduce the uplift forces acting on the pile from the swelling soil and it must extend below this zone sufficient to produce the required frictional resistance to any residual uplift force. Tension reinforcement must be provided for the full depth of the pile to resist this force.[1]

In order to avoid pressure from the swelling soil on the underside of the ground beams they must be cast on a thick layer of compressible material, such as 150 mm of expanded polystyrene, or precast concrete beams may be used with a space left underneath. For the same reason the ground floor must be suspended out of contact with the ground.[2]

4.1.4 Pier foundations

These are frequently used on made-up ground where ordinary strip or pad foundations will often be inadequate to prevent excessive and unequal settlement, especially when the fill is poorly compacted. They can be economic up to depths of about 3·5 m to 4·5 m and consist of piers of brick, stone or mass concrete in excavated pits taken to the firm natural ground below. They are usually square and the size is dependent on the material used and the strength of the bearing soil below, but the smallest hole in which hand excavating can be carried out is about 1 m square. The foundation size is calculated as for a column base.

When this type of foundation is used the structure is carried on reinforced concrete ground beams spanning between the piers as shown in figure 4.13.

Piles may be used in similar conditions but will need to be reinforced. Some form of driven pile may be used in loose fills; on firmer fills bored piles may be used, but as boring is not suitable through many types of fill on made-up ground piers provide a useful alternative within the economic limits of depth given above.[3]

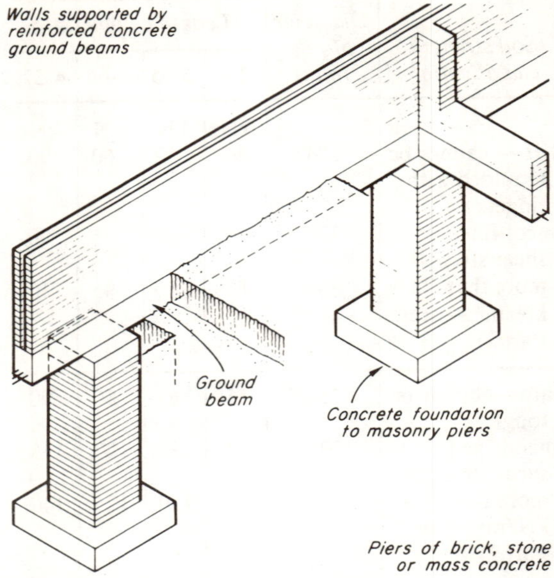

4.13 *Pier foundations*

[1] For reinforced bored piles see Part 2 chapter 3.

[2] See BRE Current Paper 61/78 and BRE Digests 241 and 298 for a consideration of trees on shrinkable clay soils.

[3] See Part 2 for piling and for ground treatments which may be applied to fills on made-up ground.

5 Walls and Piers

Walls are the vertical elements of a building which enclose the space within it and which may also divide that space. Together with the roof they form the 'environmental envelope' referred to in the first chapter and when the form of construction is based on a solid or surface structure (see chapter 1) the walls become also the basic supporting elements.

Types of wall

Walls may be divided into two types, *loadbearing* which support loads from floors and roof in addition to their own weight and resist side pressure from wind and, sometimes, from stored material or objects within the building, and *non-loadbearing* which carry no floor or roof loads. Each type may be further divided into external, or enclosing walls, and internal dividing walls.

The virtue of the loadbearing wall is that it is capable of fulfilling at one and the same time the dual functions of loadbearing and of space enclosure and division. In many circumstances, therefore, it is for this reason a most economical form of construction. Nevertheless, it suffers certain inherent disadvantages. As a loadbearing element it can become thick and heavy at the base of a very tall building although, unless the building is exceptionally tall, with modern materials and methods of design the wall will not be unduly thick. This is probably in many cases less of a disadvantage than the fact that loadbearing wall construction is restrictive when an open plan is required (see Part 2 *Masonry walls*).

The external non-loadbearing wall, related to a framed structure, is termed a panel wall if of masonry construction, an infilling panel if of lighter construction or a cladding when applied to the face of the frame (page 68).

The term partition is applied to walls, either loadbearing or non-loadbearing, dividing the space within a building into rooms. Internal walls which separate different occupancies within the same building or divide the building into compartments for purposes of fire protection are called respectively separating walls and compartment walls (see Part 2).

In addition there are retaining walls, the primary function of which is to support and resist the thrust of soil and, perhaps, subsoil water on one side. The most important functional requirement of the retaining wall is strength and stability.

Functional requirements

The primary function of the wall is to enclose or divide space but in addition it may have to provide support. In order to fulfil these functions efficiently there are certain requirements which it must satisfy. They are the provision of adequate:

- strength and stability
- weather resistance
- fire resistance
- thermal insulation
- sound insulation.

It is not possible to place these functional requirements in order of importance since this will vary with the main function of the wall. For example, all, with the exception possibly of sound insulation, must be considered in the external loadbearing wall, whereas in the case of the loadbearing separating wall only strength and stability, fire resistance and sound insulation need usually be considered. In the case of panel walls the same considerations will apply as to external walls but compressive strength will be of less importance.

In studying the functional requirements of walls it is necessary to have regard to the forms of construction which may be employed. These are described by the following terms and are illustrated in figure 5.1.

Masonry wall, in which the wall is built of individual blocks of materials, such as bricks, concrete blocks or stone, usually in horizontal courses, cemented together with some form of mortar. Masonry construction may be in the form of either a single thickness of wall known as solid or solid wall construction, or two thicknesses with a space between known as cavity or cavity wall construction.

Monolithic wall, in which the wall is built of a material requiring some form of support or shuttering in the initial stages. The traditional earth wall and the modern concrete wall are examples of this. Monolithic concrete walls may be either of plain concrete or reinforced concrete.

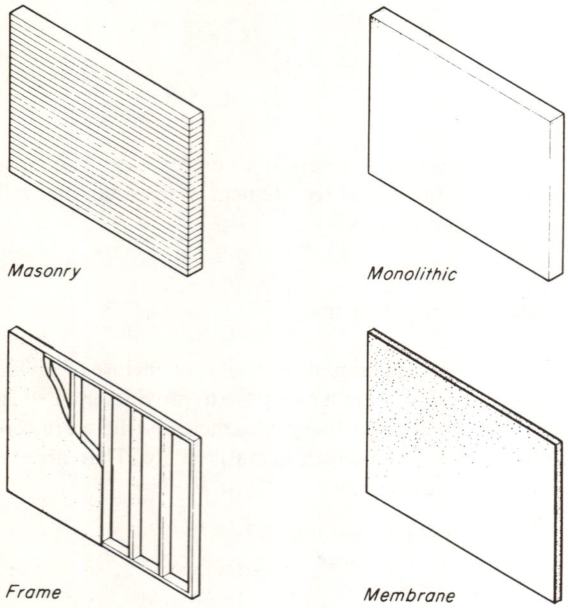

5.1 *Wall construction*

Frame wall, in which the wall is constructed as a frame of relatively small members, usually of timber, at close intervals which together with facing or sheathing on each side form a complete system. It should be noted that this is a *wall* construction and should not be confused with a structural frame of a building.

Membrane wall, in which the wall is constructed as a sandwich of two thin skins or sheets of reinforced plastic, metal, plywood or other suitable material bonded to a core of foamed plastic or to internal ribs to produce a relatively thin wall element of high strength and low weight.

These forms of construction can all be used for loadbearing or non-loadbearing walls or for panels to a structural frame. Another form of construction adopted for framed buildings consists of relatively light sheeting, or precast concrete slabs, secured to the face of the frame to form the enclosing element. These are generally termed *claddings* and are discussed in Part 2. A particular form of cladding consisting of a light framework and infilling panels is called curtain walling.

Strength and stability

The strength of a wall is measured in terms of its resistance to the stresses set up in it by its own weight, by superimposed loads and by lateral pressure such as wind; its stability in terms of its resistance to overturning by lateral forces and buckling caused by excessive slenderness.

The mode of failure of a wall by overloading, overturning or by buckling is described in chapter 3 and as shown there the provision of adequate thickness and, possibly, lateral support are necessary in order to attain sufficient strength and stability.

In small-scale buildings of solid masonry construction the external wall thickness is rarely determined by strength requirements alone. The load on the wall of a two-storey domestic building pierced with average size window and door openings is quite small and well within the bearing capacity of a normal half-brick wall. This results in functional requirements other than that of strength being the determining factors as far as thickness is concerned. For small-scale buildings of this type the wall thickness is, therefore, not normally calculated in terms of strength, but is established by an alternative, empirical, method described on page 72.

For domestic loading and storey heights external frame walls can be very much thinner than masonry or some monolithic walls. This is due to the nature of the materials used and to the fact that the functional requirements of weather resistance and thermal insulation and the strength requirements are not satisfied by the same component parts of the wall. The thickness of the structural component does not, therefore, depend upon the requirements of weather resistance and thermal insulation.

Weather resistance

The external walls of a building, whatever their form, are required to provide adequate resistance to rain and wind penetration. The actual degree of resistance required in any particular wall will depend largely upon its height and upon the locality and degree of exposure to the elements.

Wind force and rainfall vary considerably throughout the British Isles so that a form of construction adequate for one locality may not be satisfactory in another. Within any locality there can also be variations of exposure: for example, a site near the coast is likely to present greater problems of rain exclusion than one a mile or two inland. Such factors must be borne in mind.

Generally speaking the problems of wind penetration rarely presents difficulties in masonry wall con-

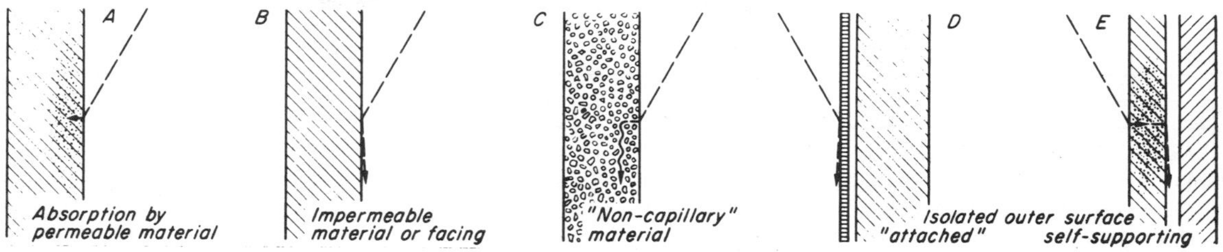

5.2 *Water resistance of walls*

struction. Tests by the Building Research Station on solid and cavity walls have shown that, provided these are plastered internally, there is a negligible penetration of wind. The possibility of wind penetration does arise with some types of modern walling of dry construction consisting of external cladding or sheathing and dry internal linings on some form of frame. Here, some barrier may be required similar to the layer of building paper or bituminous felt normally placed under weatherboarding on a timber frame wall. Reference to variations in wind pressure for variations of locality and height is made in the sections on *Loading* in chapter 4 of Part 2.

Wind, of course, has considerable influence on rain penetration, forcing the water through pores and cracks which otherwise it might not penetrate. This is especially so on high buildings. Careful design of external joints is, therefore, essential and often necessitates shaping of adjacent edges and sometimes the inclusion of wind baffles or the use of sealants as discussed in chapter 2. BS 5618 and BS 5628: Part 3 both give methods for assessing the degree of exposure of a particular wall to wind driven rain. That given in BS 5628 is a later method and being based on worst expected, rather than average annual, conditions is to be preferred. This Code of Practice gives information on the suitability of various forms of solid wall construction for different exposure conditions, on the lines of table 5.1, and gives some constructional factors affecting rain penetration of cavity walls, which should be taken into account when choosing a construction suitable for a particular exposure category.

Rain penetration through walls can be resisted in three ways (i) by ensuring a limited penetration only into the wall thickness (ii) by preventing any penetration whatsoever through the outer surface (iii) by interrupting the capillary paths through the wall. In the first the water will be absorbed by a permeable walling material and held, as in a sponge, near the outer surface until dry weather conditions permit it to evaporate (figure 5.2 *A*). In the second the use of an impermeable walling material, or an impermeable facing, will force the water to run down the wall face without entering the wall thickness (*B*). Both methods present difficulties which are discussed later. The alternative to either is the third method, the breaking of the capillary paths being accomplished by the use of a solid wall structure in which no capillary paths exist (*C*), such as no-fines concrete (see page 70), or by the provision of an outer surface which is isolated from the inner surface by a continuous gap or cavity. The outer surface or skin may be non-loadbearing in the form of traditional tile or slate hanging or of suspended cladding panels (*D*), or it may be loadbearing or self-supporting as in cavity wall construction (*E*).

The practical aspects of these methods of resisting rain penetration are discussed below and later in this chapter.

In addition to protection against lateral penetration of rain a wall must be protected at its base against ground moisture which can enter and rise by capillary attraction. The function and the provision of horizontal and vertical damp-proof barriers are discussed later. Protection may also be necessary against the entry of subsoil water under pressure through basement walls. Methods of dealing with this problem are discussed in Part 2.

Monolithic concrete walls are discussed in Part 2. At this point some practical aspects relating to their resistance to rain penetration are considered.

A well-graded and carefully mixed and placed cement concrete wall can be impervious to water. Small areas of such walls can be quite waterproof, but with larger areas problems of cracking arise due to shrinkage and thermal movements and to possible settlement. The dense monolithic nature of the wall tends to produce a few large cracks, possibly penetrating its full thickness, rather than many fine ones. These, together with the considerable volume of water streaming down

Construction of wall			*Exposure*		
Brickwork	*Concrete blockwork*	*Stone rubble*	*Sheltered**	*Moderate**	*Severe**
Cavity wall, unfilled Solid wall covered externally with slate, tile or other hanging	Cavity wall, unfilled Solid wall covered externally with slate, tile or other hanging Rendered wall of hollow blocks of dense or lightweight aggregate concrete not less than 190 mm thick with shell bedding Rendered wall of solid aerated concrete blocks not less than 250 mm thick	Cavity wall, unfilled Solid wall covered externally with slate, tile or other hanging Solid wall battened and lined internally	Suitable	Suitable	Suitable
Rendered solid wall not less than 215 mm thick	Rendered wall of lightweight aggregate concrete or solid aerated concrete blocks not less than 190 mm thick Unrendered wall of hollow blocks of dense or lightweight aggregate concrete not less than 190 mm thick with shell bedding	Rendered 406 mm solid wall	Suitable	Suitable	Not suitable
Unrendered 328 mm solid wall	Unrendered 328 mm solid wall	Unrendered 406 mm solid wall	Suitable	Not suitable	Not suitable
Unrendered 102·5 mm wall	Unrendered solid wall 190 mm thick or less on full horizontal mortar bed. (Generally, but some open-textured blocks would be suitable in sheltered conditions)	Unrendered wall less than 406 mm thick	Not suitable	Not suitable	Not suitable

*See BS 5628: Part 3: 1985 for definitions of these exposure categories, for their sub-division into six categories arising from improved meteorological data and for a similar Table based on these extended categories.

Table 5.1 *Suitability of masonry walls for various exposures*

the impermeable face of the wall, can result in serious water penetration. Precautions against such cracking are taken by controlling shrinkage and moisture movement by steel reinforcement, by allowing for thermal movement by means of expansion joints and by the careful detailing and execution of construction joints, all of which are considered in Part 2.

Walls constructed with no-fines concrete[1] do not, by

[1] This is composed of cement and coarse aggregate alone, the fine aggregate being omitted.

their nature, resist water penetration by shedding the water off the surface in the manner of dense concrete walls. The omission of the fine stuff from the aggregate results in the formation of relatively large inter-connected spaces round the pieces of aggregate. Thus although water enters the surface of the wall is it unable to pass through it by capillary attraction. It tends, as in traditional 'dry' walling in various parts of the country, to fall within the wall near the outer face and run out at a lower level (figure 5.2 *C*). Damp-proof barriers, or 'courses' must, therefore, be placed over the heads of all openings, except immediately under an eaves, and be laid to conduct moisture to the outer face as in the case of cavity walls. Such walls must be finished externally with a rendering in order to prevent water being forced through by wind pressure. The rendering must be a suitable porous type with, preferably, a rough surface. No-fines concrete walls 203 mm to 229 mm thick, rendered and satisfactorily detailed at openings, are quite resistant to moisture penetration.[1]

Fire resistance

A degree of fire resistance adequate for the particular circumstances is an essential requirement in respect of walls which, like upper floors, are often required to act as highly resistant fire barriers. They are used to compartmentalise a building so that a fire is confined to a given area, to separate specific fire risks within a building, to form safe escape routes for the occupants and to prevent the spread of fire between buildings.

The term fire resistance is a relative term applied to elements of structure and not to a material. It is not to be confused with non-combustibility. An element may incorporate a combustible material and still exhibit a degree of fire resistance which will vary with the way in which the material is incorporated in the element. The degree of resistance necessary in any particular case depends on a number of factors which are discussed in Part 2 chapter 10.

Thermal insulation

The external walls of a building, together with the roof, must provide a barrier to the passage of heat to the external air in order to maintain satisfactory internal conditions without a wasteful use of the heating system. They should also serve to prevent the interior heating up excessively during hot weather.

Adequate thermal insulation is attained in a variety of ways. Reliance upon the thickness of normal solid structural masonry and concrete necessitates impractical thicknesses of wall and it is necessary to incorporate in such construction cavities and materials with high insulating values in order to keep the thickness within reasonable limits. Frame walls of timber, which is a good insulating material, by their nature incorporate cavities and with appropriate internal linings they provide good insulation with a relatively small thickness of wall. Practical ways of incorporating insulation in wall construction are discussed in the following pages. Heat transmission values for various forms of construction are given in *MBS: Environment and Services* where the principles of thermal insulation are fully discussed.

Sound insulation

Only in exceptional circumstances are the sound insulation qualities of an external wall a significant factor in its design since the other functional requirements which must be fulfilled usually necessitate a wall which excludes noise sufficiently well in most circumstances. Windows, of course, provide weak points in this respect and in some circumstances these may have to be treated as double windows as described in *MBS: Environment and Services*, in order to attain a satisfactory degree of insulation. Sound insulation is, however, often a significant factor in the design of internal walls. Weather exclusion and, generally, thermal insulation are not functional requirements of these walls but the prevention of the passage of sound from one enclosed space to another is often an important function they must fulfil. Since the strength requirements of an internal wall, especially if it is non-loadbearing, may result in a relatively thin wall, the requirements of sound insulation can be the critical ones because the efficiency of a solid wall in preventing the transmission of air-borne sound depends upon its mass. As with thermal insulation, however, an adequate degree of sound insulation can sometimes be attained only with an excessive thickness and weight of solid wall. In such cases discontinuous construction, that is construction with two leaves with a cavity between, must be adopted. Reference should be made to *MBS: Environment and Services* where this subject is fully discussed.

[1] See Part 2 for further details of no-fines wall construction.

5.1 MASONRY WALLS

The term masonry is used today to mean bricks or blocks of any material laid one on another usually with mortar as a binding material to form building elements, rather than simply stonework to which it was once limited. Apart from certain forms of stone walling referred to later, all masonry consists of rectangular units built up in horizontal layers called *courses*. These units are laid in certain specific ways relative to each other for reasons which apply irrespective of the material used and which are discussed below.

Masonry units at the present time consist of bricks of various types and materials, blocks, which are units larger in size than bricks, and stone. The mortar in which they are laid is a mixture of sand or other fine aggregate with cement or lime as a binding material, its function being (i) to bind together the walling units, (ii) to distribute pressures evenly throughout the wall from unit to unit and (iii) to fill the joints between the units in order to prevent wind and rain penetration and to maintain the overall thermal and sound insulating characteristics of the wall.

Loadbearing masonry work has for many years provided the cheapest structure for building types such as blocks of flats up to five storeys in height and, broadly speaking, loadbearing brick or block wall construction will still produce the cheapest structure for small-scale buildings of all types where planning requirements are not limited by its use. For buildings with cellular plan forms and using modern methods of structural design very tall structures can be built economically with walls of this type.

For many small traditional types of building and for residential buildings up to three storeys in height the Building Regulations provide means for determining wall thicknesses other than by a process of calculation. Although the method, which may be called empirical, has limitations it is, nevertheless, simple to use. Strength and stability are ensured by limiting the width of openings in order to provide sufficient bearing area of wall, by requiring lateral support up the height of the wall by adequately braced floors and roof (see Part 2, 4.1.3) and by relating the thickness to the height and to the length of the wall between adequate lateral supports in the form of cross or buttressing walls.[1] The design of loadbearing masonry walls is covered by BS 5628: Part 1: 1978 and the application of its provisions in the calculation of wall thicknesses is given in Part 2, chapter 4, where further reference is made to the empirical method and to the relative economics of masonry walling as a structural and enclosing medium.

5.1.1 Bonding of masonry walls

A masonry wall may be constructed as shown in figure 5.3 *A* with the units or blocks laid lengthwise along the wall, those in each course lying directly on a block below. A load applied to a block at the top of the wall will be transferred to those immediately below it and thus to the foundation, the pressure being concentrated on a narrow band one block wide. This concentration of pressure could lead to unequal settlement in the wall due to greater consolidation of the mortar joints on this narrow band (*B*). Should the wall undergo lateral pressure at one point as indicated in (*C*) the narrow band of wall sustaining the pressure would tend to overturn.

If, however, the blocks are laid to overlap those in the courses below as shown in (*D*) the effect of loading will be different. A vertical load on one block will then be distributed to the two blocks below on which it bears and from those to an ever-increasing number as indicated. This results in a rapid distribution of the load over a greater area of wall with a consequent reduction in the stress in the masonry and less likelihood of unequal settlement. By this means also the pressures

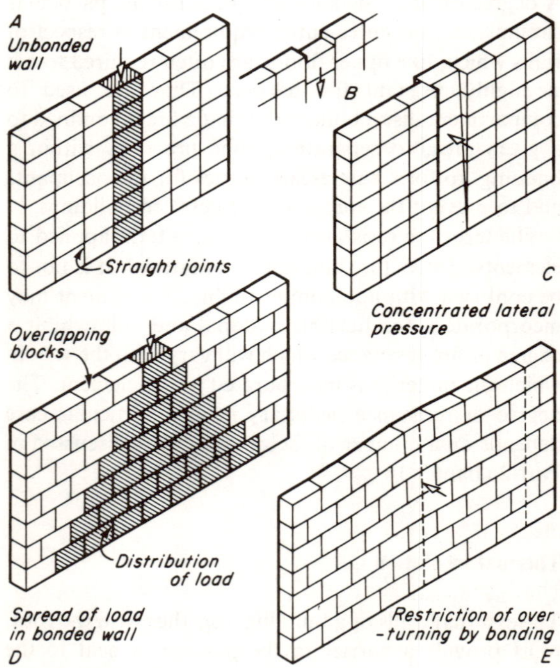

5.3 *Bonding of masonry walls*

[1] The conditioins which must be met when this method is adopted are laid down in Approved Document A Part C.

from a number of point loads on the wall will overlap and produce a reasonably even stress over the base of the wall. Under the application of lateral pressure at one point the tendency of the wall to overturn at that point will be restricted by the masonry on each side to which it is connected by overlapping blocks (*E*). To this extent the wall will be more stable. This effect would be apparent even without mortar joints, the weight of the blocks alone providing considerable resistance to movement.

This overlapping of the units is termed *bonding* and is always adopted, except occasionally for non-loadbearing panel walls or applied facings.

When thicker walls are required these may be built with wider units, as in the case of blockwork or ashlar stonework, laid in the manner described, or they are built in multiples of narrow units, as in the case of bricks. If constructed in two thicknesses both laid with straight joints and placed side by side to produce the required wall thickness as in figure 5.4 *A* the distribution of applied loads will similarly be concentrated as before in narrow widths of wall. Furthermore, the individual sections are liable to buckle separately under load as shown in (*B*). In order to avoid this the units could be laid across the wall as in (*C*) but as this would still result in concentrations of load on narrow portions of the wall a spread of load is obtained by overlapping the units as in (*D*).

This latter method is very little used and the generally adopted solution is to introduce into a wall built up basically as in 5.4 *A* units laid across the wall either at intervals along each course or as alternate complete courses as shown in (*E*). These units act as transverse ties to prevent buckling of the separate half-sections of the wall, ensure a spread of load across the thickness of the wall and serve to produce the overlapping or bonding of the units along its length. In cavity walls, the leaves of which are laid as in figure 5.3 *D*, metal wall ties across the cavity fulfil the same functions as these transverse units in resisting buckling of the relatively thin leaves.

In comparing figure 5.3 *D* with 5.4 *E* it will be seen that in the former the units overlap each other by half their length and in the latter by only a quarter of their length due to the use of transverse bonding units. In both cases the overlap is adopted in order to avoid continuous vertical joints up the wall. These overlaps are obtained by the introduction of a half-unit at the end of each alternate course in the first case and a quarter unit in each course of transverse units in the second. The amount which one unit overlaps another is called the *lap*.

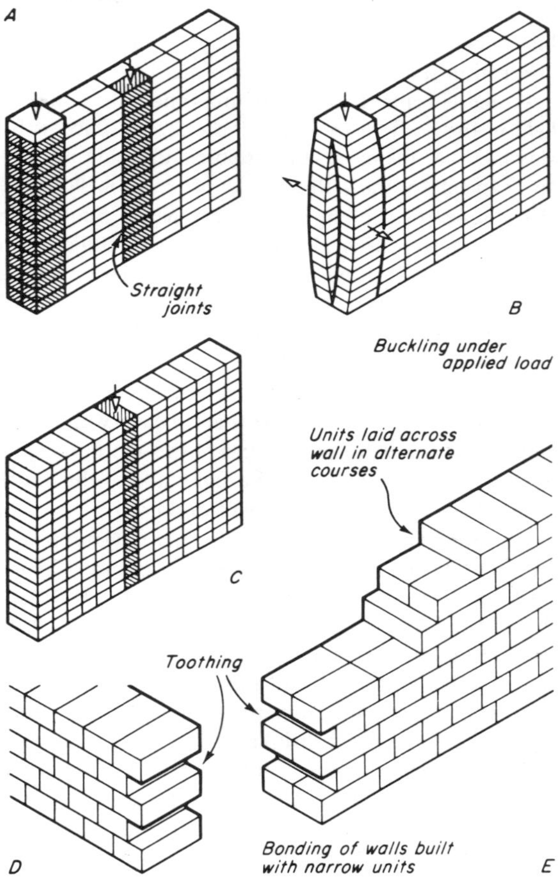

5.4 *Bonding of masonry walls*

5.1.2 Weather resistance of masonry walls

Solid walls

If water is to be prevented from getting to the inside of a wall by means of absorption it is essential that the mortar and the walling units should have similar absorptive qualities. Strong dense mortars should be avoided in order to ensure sufficient porosity in the joint and to reduce shrinkage so that cracking between mortar and units is kept to a minimum. Penetration occurs more often by capillary attraction through cracks between the mortar and units than through the units themselves.[1] The use of insufficiently cured concrete blocks increases the risk of cracks at the joints since drying shrinkage will continue after the blocks

[1] See *Mortars for Jointing* in *MBS: Materials*.

have been built into the wall. Complete curing of such blocks before use is essential. The units should have an absorption similar to that of a normal facing brick.

Water will enter the pores of units and mortar and be held in the body of the wall (figure 5.2 *A*). Success by this method, therefore, presupposes adequate absorptive capacity of the wall to make penetration slow and the absence of prolonged and very heavy rainfall. In such circumstances a 215 mm wall may be sufficient in very sheltered positions, but for all normal exposures a thickness of at least 328 mm is required. On exposed sites in districts of heavy prolonged rainfall excessively thick walls are required. Table 5.1 gives some indication of the suitability of masonry walls under different conditions of exposure.[1] The thick, heavy wall essential in most cases for the success of this method is one of the reasons which has brought about the more general use of cavity wall construction.

The difficulties in producing a barrier to water penetration by means of an impermeable wall (figure 5.2 *B*) of small bonded units are considerable and are centred round the joints. Impervious units, such as engineering bricks, are usually smooth-faced and do not assist adhesion between block and mortar. The dense mortars required to provide impermeability in the joints undergo a large initial shrinkage. There is, therefore, a marked tendency for cracks to develop at the joints and for rain to penetrate these joints by capillary attraction. Since rain will not be absorbed by the impervious walling units it will stream down the face of the wall and rapidly enter any such cracks. Apart from the danger of complete penetration of water through cracks to the inner face of the wall there is the danger of water remaining in the centre of the wall, because in fine weather it is unable easily to evaporate through the dense face. Ultimately it will penetrate to the inner face. In frosty weather this trapped water may freeze and in expanding cause disintegration of the wall. In order to minimise these dangers a high standard of workmanship is essential to ensure that all joints are flushed up solid. To reduce shrinkage the use of a high proportion of cement should be avoided, workability in the mortar being obtained by the use of a carefully graded sand or by the addition of a small proportion of lime or other plasticiser. The mix should be no wetter than that required to permit the joints to be thoroughly filled and consolidated.[2]

When the ends of cross walls[3] are exposed on elevation between thin infilling panels some positive weatherproofing is required. In such cases the distance between the exposed and internal faces of the wall is small and insufficient to permit the normal process of absorption of rain and subsequent evaporation to occur before moisture penetrates to the interior. Painting of the reveals or the use of impervious facing bricks alone at the wall ends will not necessarily prevent water penetration. This is because the joints will still be exposed and water may pass through cracks by capillary attraction. A number of alternative methods of dealing with the problem are shown in Part 2 figure 4.8.

Renderings, used as a means of reducing water penetration through any type of solid wall, can act either as an impervious or as an absorbent skin. For the former, necessary in severe exposures, a dense mix must be used with which, because it is strong, considerable shrinkage cracking in the render can develop, into which can flow the water streaming down the non-absorbent surface. In order to minimise such cracking the adhesion to the background must be good. The practical difficulties in attaining a surface free of cracks have brought about the general use, wherever possible, of weaker renderings based on cement: lime mixes which crack less due to shrinkage and produce an absorbent finish which inhibits the flow of water over the surface. Ways of providing the necessary edge protection to prevent penetration of water behind the rendering and other precautions are described in Part 2, chapter 4.

The alternative to these methods of resisting rain penetration, and one which overcomes the problems inherent in both absorbent and impermeable solid wall construction, is that of breaking the path of moisture through the wall by a continuous gap or cavity as mentioned earlier. One way of doing this is by the application of an outer surface in the form of traditional tile or slate hanging or weather boarding, or of suspended cladding panels of various materials. These are fixed to the wall in such a way that separation by a cavity is ensured (figure 5.2 *D*) so that, with the incorporation of some form of wind or moisture barrier

[1] Based on tables in Code of Practice 121: Part I: 1973 *Brick and block masonry*, BS 5390: 1976, *Code of practice for stone masonry* and *Concrete Block Walls* published by the Cement and Concrete Association.

[2] See *MBS: Materials* for suitable mixes for normal brick and blockwork.

[3] These are internal walls carrying all floor and roof loads and running at right-angles to the length of the building. The external walls often consist of infilling panels between these walls thus exposing their ends (see Part 2 chapter 4).

where considered necessary, any water which may pass through the joints in the outer surface does not penetrate to the wall itself. In tile or slate hanging rain penetration is prevented by hanging the tiles and slates to break joint as in roofing and in weatherboarding by overlapping the boards and placing a moisture barrier behind as described in Part 2 under *External facings and claddings*. With suspended cladding panels water penetration is prevented by sealants or gaskets in the joints or is controlled by open-drained and ventilated joints (see page 28) or by open joints acting in conjunction with a cavity behind the cladding which forms a ventilated air space by virtue of the open joints. This method is commonly termed rain screen or open rain screen construction (figure 5.5).

Rain screen construction In this method water penetration is controlled by ensuring that the inner side of the air space is airtight so that under the action of wind blowing through the open joints there is no air flow across the space, thus permitting a build-up of air pressure in the space equal to that on the external face of the cladding, as described on page 28 in relation to the open-drained joint. The open-jointed external cladding is called the *rain screen*.

Vertical joints in the rain screen not exceeding 2·5 mm and horizontal joints not exceeding 5 mm in width will reduce rain driving across the air space to a very small quantity. Where joints are wider than this water penetration can be limited by restricting the flow of water across the surface of the screen to the joints by a rough external surface or vertical profiling or by reducing or preventing the flow of water into the joints by profiling the returns into the joint as in figure 5.5 *A* and figure 2.4*A*, by forming lipped edges (*B*) or a lapped joint (*C*) or by incorporating a channel barrier as in (*D*) and in figure 4.35 (*G*) in Part 2. Since it must be assumed that some water will enter the air space drainage should be provided by means of Z-flashings in some horizontal joints to limit the amount of water retained in the space and to prevent it coming into contact with the air barrier (figure 5.5). To this end the rain screen is preferably fixed to vertical supports to permit free drainage down to these Z-flashings.

In order to ensure the essential pressure equalisation in the air space the latter should be not less than 25 mm in width from screen to air barrier and should be divided by airtight baffles into areas the largest dimension of which should be about 5 m, except within 25 per cent of the top or corners of the construction where it should be limited to about 1·5 m. This sub-division is necessary because of the considerable variation in air pressure over the face of a building which, as a consequence in an undivided cavity, could result in air flowing through the cavity from a point of high pressure to an area of low pressure, thus preventing the necessary equalisation of pressure at that point and resulting in increased penetration of rain.

Cavity walls

In cavity wall construction the wall is built in two leaves or skins with a space between (figure 5.2 *E*) and is the other means by which the outer surface of the wall may be isolated from the inner surface by a continuous gap, as in the methods described above. In this case, however, the outer leaf is self supporting and may be loadbearing. Provided all details are well designed, particularly around openings, and the work carefully executed, it has proved to be a most reliable method of avoiding moisture penetration through walls.

The successful functioning of a cavity wall depends upon the cavity being continuous without bridging of any kind capable of transferring moisture to the inner leaf. There should be no projections on the inside of the outer leaf extending into the cavity as these can collect mortar droppings, and water trickling down the inside

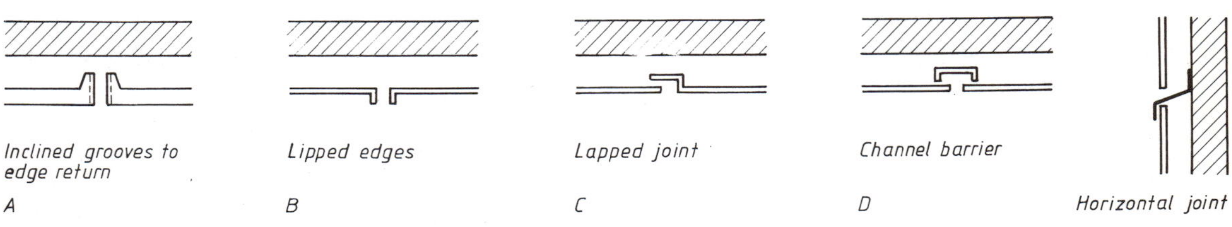

5.5 *Edge treatment to rain screen joints*

face may drop from one projection to another and splash across to the inner leaf. In walls incorporating thermal insulation filling the cavity any such projections will reduce the thickness and thus the effectiveness of the insulation at those points.

Horizontal damp-proof barriers, known as damp-proof courses, must be provided over all openings and vertical damp-proof courses at all points of contact between inner and outer leaves. All such points of contact should be minimised. Where possible it is preferable so to detail around openings that the cavity is not closed by returning one leaf on to the other (see figure 5.20 *A, C*).

Methods of damp-proofing round openings and details of the construction of cavity walls are given later in this chapter.

5.2 BRICKWORK

Bricks and brickwork generally

Bricks are walling units made of burnt clay or shale, sand or flint and lime (calcium silicate) or of concrete moulded in various ways to form blocks of suitable and defined dimensions.[1] The method of production varies with the material and the characteristics required.

There are three varieties of clay brick known as common, facing and engineering bricks, the difference between the first two being mainly that of appearance. The third is used primarily for its high strength although some of the common and facing bricks have a relatively high strength as well. Calcium silicate and concrete bricks are classed according to their strength and drying shrinkage.

The size of the most common form of clay brick, the standard brick, is 215 mm × 102·5 mm actual size, with a height of 65 mm. Calcium silicate bricks and concrete bricks are also manufactured to these dimensions.

It will be seen that the length of the standard brick is rather more than twice its width. This is to ensure that two bricks laid side by side with a 10 mm joint between are equal to the length of the brick, this being necessary because of the manner in which these bricks are built up to form a wall. Bricks and blocks are normally referred to by their designated size which is the actual size plus a 10 mm joint width to each dimension.

Moulded and pressed bricks are formed with a depression, termed a 'frog', on one or both bedding faces, the function of which is to reduce the weight of the brick and to form a key for the bedding mortar. Some bricks are perforated right through for the same reasons. Bricks may be specially moulded to shapes required for particular purposes (see *MBS: Materials*).

5.2.1 Solid wall construction

Bonding, as already shown, avoids as far as possible the coincidence of the vertical joint between any adjacent wall units with vertical joints in the courses above and below. Such coinciding, continuous vertical joints are termed *straight joints*. The manner in which these are avoided by the use of units laid across the wall thickness is explained in principle on page 72. In brickwork those bricks laid lengthwise in the wall are called *stretchers* and the course in which they occur, or which commences with a stretcher, a *stretching course*. Bricks laid across the wall thickness are called *headers* and the course in which they occur or which commences with a header, a *heading course*. The vertical sides, or faces, of a brick are named accordingly. The bottom face, which is bedded in mortar on the course below is called the *bed* face (figure 5.6).

Bonding of brickwork

Bricks may be arranged in a wide variety of ways to produce a satisfactory bond and each arrangement is identified by the pattern of headers and stretchers on the face of the wall. These patterns vary in appearance, resulting in characteristic 'textures' in the wall surfaces, and a particular bond may be used primarily for its surface pattern rather than for its strength properties. Many so-called decorative bonds result in a number of straight joints within the wall but this is usually of no great significance unless high strength brickwork is required. In order to maintain bond it is necessary at some points to use bricks cut in various ways, each of which has a technical name according to the way it is cut. These are illustrated in figure 5.6.

The two simplest arrangements, or 'bonds' as they are called, are *stretching bond* and *heading bond*. In the former each course consists entirely of stretchers laid as in figure 5.11 and is only suitable for half-brick walls

[1] For the definition of a *brick* in terms of size and for a consideration of the characteristics and properties of bricks of all types see *MBS: Materials* chapter 6.

such as partitions and the leaves of cavity walls, since thicker walls built entirely with stretchers, even though laid in stretching bond, suffer the defects already described.

In the latter each course consists entirely of headers laid as in figure 5.4 *D* with a 56 mm lap and is only used for curved walls.

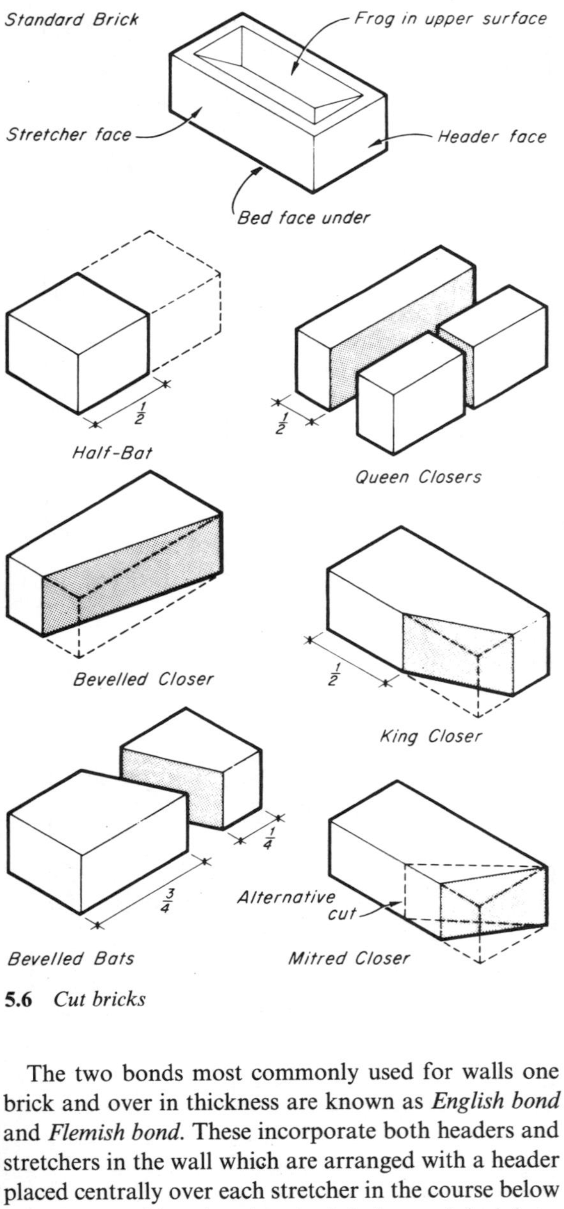

5.6 *Cut bricks*

The two bonds most commonly used for walls one brick and over in thickness are known as *English bond* and *Flemish bond.* These incorporate both headers and stretchers in the wall which are arranged with a header placed centrally over each stretcher in the course below in order to achieve bond and minimise straight joints.

English bond

This consists of *alternate courses* of headers and stretchers as in figure 5.7 *A, B, C* where it will be clear that each stretcher has a header immediately over it, the intervening spaces in the heading courses being filled with headers. An increase in the thickness of a wall necessitates a variation in the arrangement of the bricks in each course in order to meet the requirements of a good bond. It will be seen from the illustration of the one-and-a-half brick wall that on plan each course is, in fact, made up a series of units one brick wide and the wall thickness in depth, so laid over each other that a header lies centrally over the stretcher below. This bond is entirely free from straight joints.

Flemish bond

This consists of *alternate bricks* laid as headers and stretchers in each course as in (*D, E, F*) each stretcher having a header immediately over it.

Double Flemish bond means that the typical face pattern of alternating headers and stretchers shows on both sides of the wall as illustrated. As in English bond each course on plan is made up of units of bricks so laid over each other that a header lies centrally over a stretcher below; the units in this case, however, are one-and-a-half bricks wide on the face, as will be seen from the illustration.

Because of the large numbers of straight joints which occur this bond is considered to be less strong than English bond but it is sufficiently strong for all general purposes.

Single Flemish bond, in which Flemish bond shows on the face only with English bond on the opposite side, is used to economise in expensive facing bricks but is not applicable to walls less than one-and-a-half bricks thick.

Garden wall bonds

In building walls one-brick thick which are to be exposed on both faces, such as garden or boundary walls, the fact that the headers pass through the thickness of the wall creates difficulty in obtaining a fair-face on both sides due to the variations in the lengths of bricks. Garden wall bonds are designed to reduce the

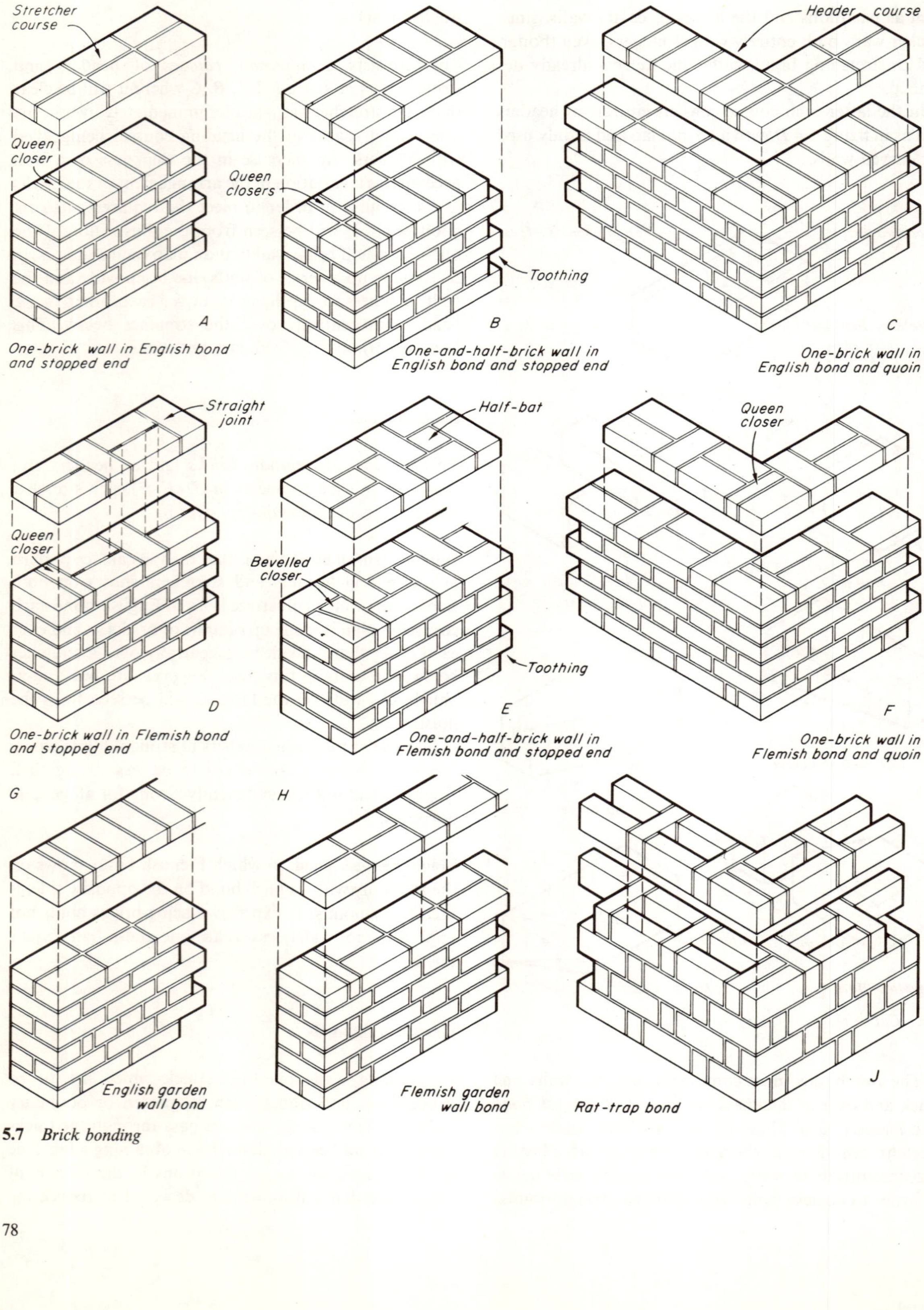

5.7 *Brick bonding*

number of headers in the wall and thus simplify the task of selecting headers of uniform length. There are two forms of this bond, *English garden wall bond* with one course of headers to three or five courses of stretchers and *Flemish garden wall bond* with one header to three or five stretchers in each course (*G*, *H*). They are deficient in strength because of the large amount of longitudinal straight joints which occur, but are sufficient for non-loadbearing walls such as these.

These bonds are sometimes used instead of stretcher bond for the outer leaf of cavity walls (see page 81) in an attempt to improve the face appearance by the introduction of some headers all of which are snap headers. The half-bats for these must be cut carefully so that they do not protrude into the cavity and provide lodgement for mortar droppings.

Brick-on-edge bonds

In these bonds bricks are bedded on edge so that the courses are a half-brick high. Brick-on-edge, in the form of *Rat-trap bond*, may usefully be adopted for walls which are to be clad with tiles or similar covering. In this the bricks are laid in Flemish arrangement but with an 85 mm cavity between each pair of stretchers (*J*). This saves about 25 per cent of brickwork relative to a 215 mm solid wall and gives a spacing of horizontal joints at 112·5 mm which is a suitable gauge for tile hanging and permits the tiles to be nailed directly to the brick face (see Part 2 *Tile hanging*).

Bonding at stopped ends and quoins

The termination of a run of wall is called a *stopped end*. A *quoin* is the external angle formed at the return of a wall. Where the angle formed is greater or less than 90 degrees the term *squint quoin* is used, qualified as obtuse or acute respectively.

Stopped ends At the end of a run of wall the toothing formed by the bonding of the bricks (figure 5.7) must be closed to form a smooth square end. This is accomplished by filling in the 56 mm space in each heading course with part of a brick which is termed a *closer* since it 'closes' the bonding of the wall.

In positioning the closer in the heading course the last header is transferred from its normal position over the centre of the stretcher below to the end of the wall, the closer being set in the space left as shown in figure 5.7. This is done to avoid the possibility of the narrow closer being dislodged from a position at the extreme end of the wall. Closers of different shapes (figure 5.6) are used according to the requirements of bonding to avoid straight joints (figure 5.7*E*). Queen closers are cut from whole bricks but it is difficult to cut a full-length closer 51 mm wide so they are usually formed from two halves.

Quoins As at stopped ends the toothing at the ends of the return walls must be closed and closers are used in a similar manner. In fact, in the bonding of quoins the heading course in each wall is formed as a stopped end and is carried through to the angle and forms the beginning of the stretching course on the return face, the latter course butting square against the former as will be seen in (*C*, *F*). Thus the heading courses overlap at the quoin and become the stretching courses on the return faces. In walls thicker than one brick slight variations are sometimes necessary to preserve face bonding.

Bonding of junctions

Where two walls meet to form a *tee-junction* the normal method is to butt against the main wall the stretching courses of the cross-wall and to bond in the heading courses 56 mm making use of closers and bats as necessary (figure 5.8 *A*). At a *cross junction* or *intersection* every alternate course in each wall passes across the other wall, the bonding being arranged to avoid straight joints.

Piers

The term *pier* is used broadly to mean a column of masonry either freestanding or attached to a wall, the function of which is to support concentrated loads from beams, roof trusses or arches. Piers also may be used solely to increase the lateral stability of a wall, in which case they are called *buttresses* (see chapter 3). Freestanding piers have been referred to as isolated or detached piers but BS 5628: Part 1 defines these as *columns*. Piers bonded to a wall are referred to as such.

Piers are arranged so that the headers of the pier bond into the stretching courses of the wall, the stretchers lying against the wall (figure 5.8 *B*, *C*). The need for queen and king closers and for bats will vary with the projection and width of the pier and with the

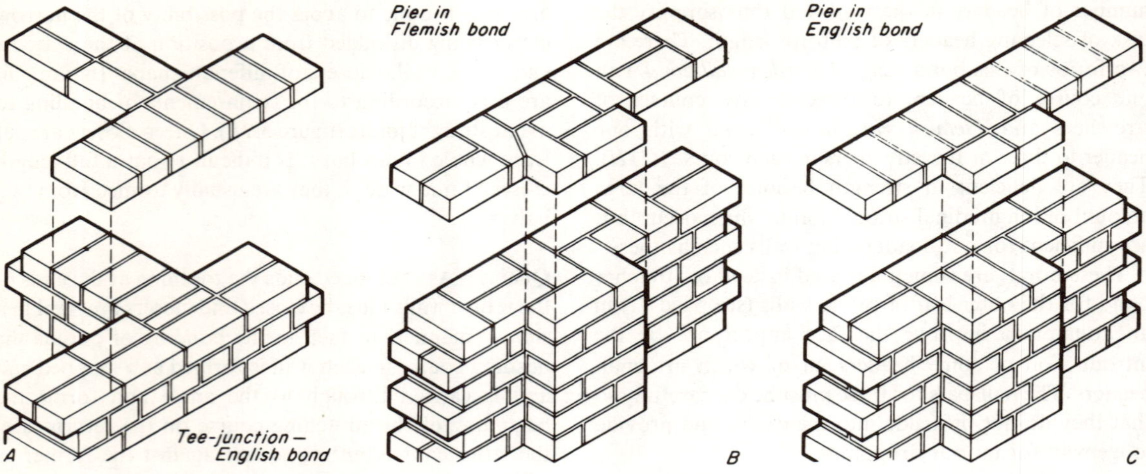

5.8 *Brickwork: bonding of junctions and piers*

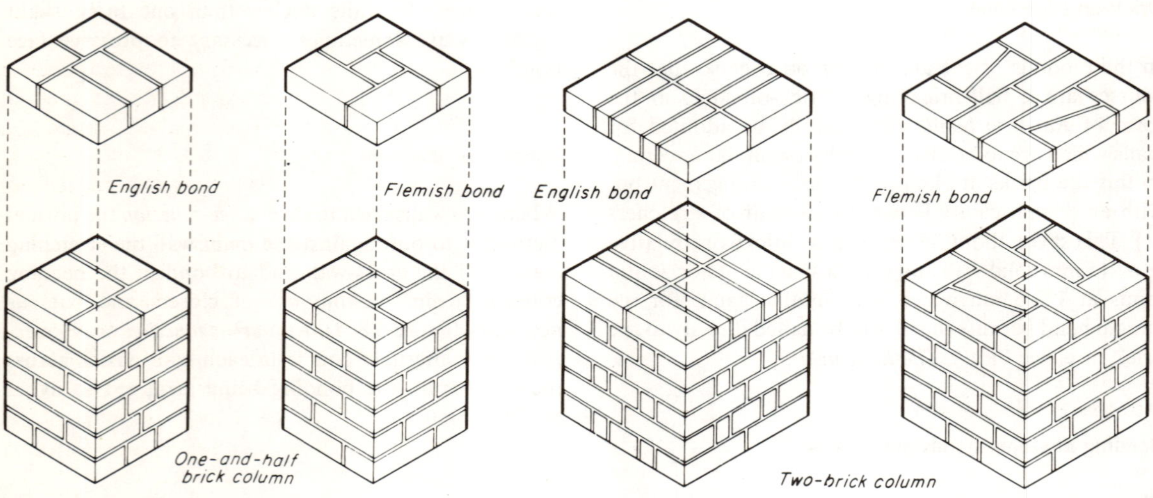

5.9 *Brickwork: bonding columns*

position of the pier relative to the normal bonding of the wall.

The bonding of brick columns is simple. It will be seen in figure 5.9 that the arrangment in each course is basically the same but alternate courses are either reversed or turned through 90 degrees to obtain a satisfactory bond.

Thermal insulation of solid walls

Thermal insulation must be applied to a solid brick external wall in order to achieve the standards of insulation required by the Building Regulations and it may be applied on either the internal or external face of the wall.

Internal insulation is applied direct to the wall face by adhesive or mechanical fixing (figure 5.10 *A*) or is separated from it by an air space not less than 25 mm wide, the latter providing enhanced insulation and permitting a reduction in thickness of the insulating material (*B*). Suitable materials are to be found in the many forms of insulating slabs or boards available and in dry linings incorporating insulation.[1]

[1] For the various forms of insulating materials see *MBS: Materials*.

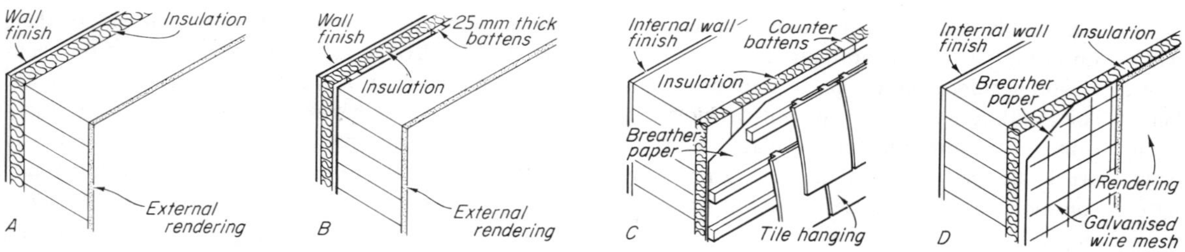

5.10 *Thermal insulation of solid walls*

With some of the materials a vapour barrier (see page 116) will be required on the warm side of the insulation to prevent interstitial condensation[1] on the inner face of the brickwork which would have the effect of reducing the insulating value of some materials such as fibreboard. Others, such as some expanded or foamed plastics, may provide in themselves sufficient resistance to the passage of water vapour to prevent such condensation.

Internal insulation provides a surface which warms up quickly and results in heating economies in buildings which are only heated intermittently.

External insulation may be in the form of stiff boards or slabs or of flexible mats or batts, protected by a waterproof covering of rendering or cladding. The one-brick wall 215 mm thick, at present widely displaced by the cavity wall, requires some form of external covering in order to provide adequate resistance to the penetration of damp (see Table 5.1, page 70) and it can, therefore, usefully be insulated in this way.

As in the case of internal linings the stiff forms may be attached to the wall by adhesive or mechanical means and are covered by rendering or a cladding such as tile or slate hanging (figure 5.10 *C*). Flexible insulation is attached for ease of installation to a breather paper (see page 116) and galvanised open wire mesh by means of which it is mechanically fixed to the wall and which forms a key for a protective rendering (*D*). In fixing to the wall the insulation must not be compressed thereby reducing its insulation value.

External insulation should extend below the level of the ground floor in order to avoid a potential cold-bridge by which heat loss could take place through the wall. Below DPC level this should be in the form of non-absorbent insulation board taken down at least 300 mm below ground level and protected from damage on its outer face by 50–75 mm concrete slabbing taken up to the termination of the main covering to the insulation at DPC level. At roof level where the insulation terminates at eaves or verge the roof insulation must be so arranged as to avoid such gaps between it and the wall insulation as would form cold-bridges.

External insulation permits the thermal capacity of the wall, that is its capacity to absorb and store heat, to be utilised. This is economically advantageous where heating is continuous.[2]

5.2.2 Cavity wall construction

The functional advantages of an external wall built in two leaves or skins with a space between have been discussed earlier in this chapter. The better resistance to rain penetration and the greater degree of thermal insulation it provides compared with a solid wall of the same thickness and material, results in this form of construction (figure 5.11) being almost universally adopted for external walls.

The tendency for a wall to buckle under load is normally avoided by providing adequate thickness or by buttressing (see chapter 3). By connecting the two leaves of a cavity wall they may be made to stiffen each other by acting together under load and the thickness of each may, therefore, be thinner than if they acted separately. Nevertheless, the lateral stiffness of the whole wall is less than that of a solid wall equal in thickness to the sum of the thicknesses of the two leaves and this must be taken into account in the determination of wall thicknesses (see Part 2 chapter 4).

The outer leaf is usually a half-brick thick in stretching bond[3] and the inner leaf the same or, more commonly for reasons given below, of lightweight concrete blocks 100 mm thick.[4] The width of the space, or cavity, between the leaves may vary from 50 mm to 150 mm.

[1] That is condensation within the fabric of the building rather than on the surface. See Part 2 chapter 4.

[2] Thermal insulation and various methods of achieving the required standards of insulation in the building fabric are discussed more fully in *MBS: Environment and Services*.

[3] See page 79 regarding the use of snap headers.

[4] See part 2, section 4.3.1 where reference is made to the manner in which the required thickness of the leaves of a cavity wall is related to the height and length of the wall.

50 mm is normally considered the minimum desirable width to prevent inadvertent bridging, for example by mortar droppings lodging on projecting bed joints below. The Building Regulations permit cavity widths from 50 mm to 100 mm and for calculated walls BS 5628: Part 1 suggests a maximum of 150 mm.[1] The latter is considered a maximum desirable width in order to restrict the free length of the ties connecting the leaves and thus reduce their tendency to buckle under compressive forces.

The two leaves are connected by metal *wall ties*. These must be strong enough to develop mutual stiffness in the leaves and they must be so designed that water cannot pass from outer to inner leaf and, further, that mortar droppings cannot easily lodge on them during the building of the wall and bridge the cavity. BS 1243: 1978 specifies three types of tie which satisfy these requirements (figure 5.12). Each of these ties is formed with a 'drip' at the centre which prevents water passing across; the wire ties should be laid with the twisted ends or the crimp hanging down. BS 5628: Part 1 recommends and the Regulations require that where the cavity width exceeds 75 mm, only vertical-twist strip or similar ties should be used. Adequate stiffness is provided by the ties if they are placed at intervals not exceeding those shown in table 5.2, the ties in each row being staggered relative to each other as shown in figure 5.11. At the jambs of openings the greater loads due to the reactions from lintel or arch over results in a greater tendency of the leaves to buckle. The number of ties at the jambs should, therefore, be increased by placing them 300 mm apart vertically within 150 mm of the opening. To ensure a satisfactory bond with the leaves a cement-lime mortar at least should be used with the ties embedded at least 50 mm in each leaf.

[1] BS 5628: Part 1 also suggests that it should not exceed 75 mm where either of the leaves is less than 90 mm thick.

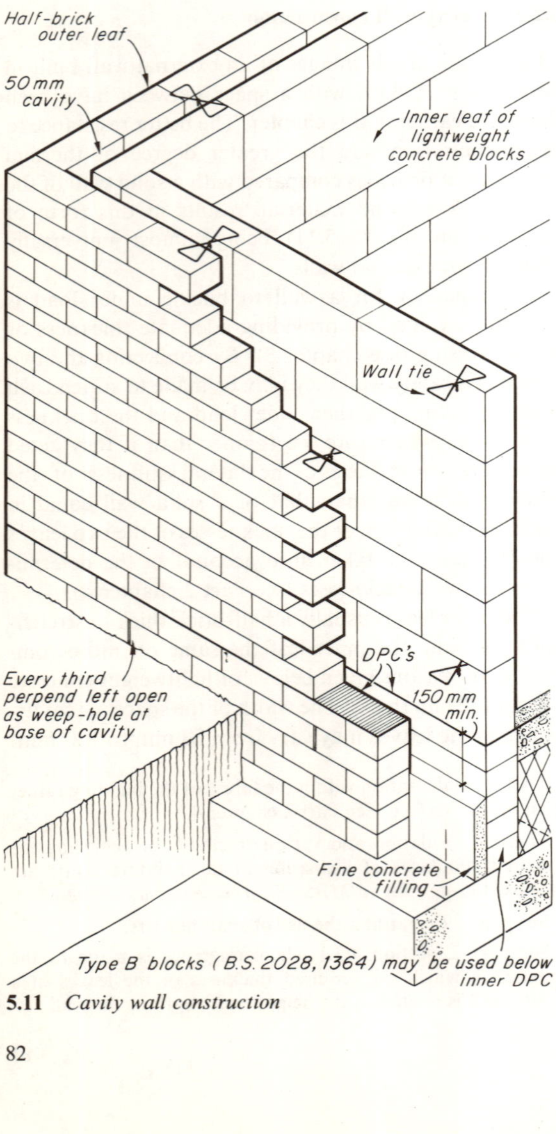

5.11 *Cavity wall construction*

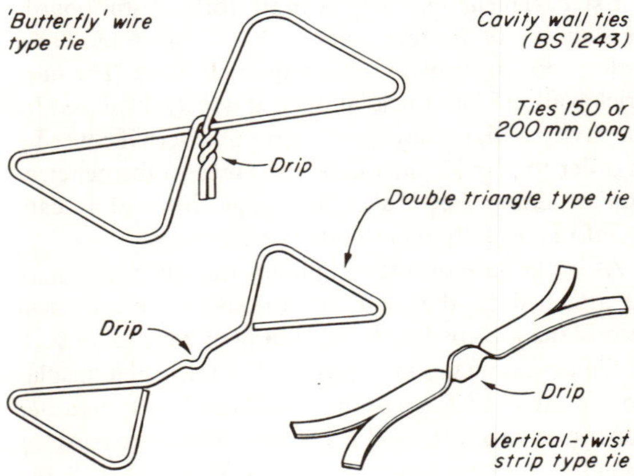

5.12 *Cavity wall ties*

Least leaf thickness (one or both) mm	*Cavity width mm*	*Spacing of ties*	
		Horizontally mm	*Vertically mm*
75–90	50–75	450	450
90 or more	50–75	900	450*
90 or more	75–100	750	450*
90 or more	100–150	450	450

Table 5.2 *Spacing of wall ties*
The figures in this Table are those recommended in BS 5628: Part 1.
* The figures in these lines are those permitted for leaf thickness and cavity width and required for tie spacings by the Building Regulations AD A1/2 Part C.

When used in conditions of severe exposure the ties should be made of stainless steel or non-ferrous metal.

Unless brick is required for loadbearing reasons (which is normally not the case in domestic work up to two storeys high) or in order to obtain a fair-face brick finish internally, the inner leaf is usually built in lightweight concrete blocks. The reasons for this are (i) the thermal insulation of the wall is increased and (ii) building in blockwork is quicker and cheaper than in brickwork (see page 100).

Type B loadbearing blocks with a strength of not less than 2·8 N/mm² are normally used and for a two-storey house the inner loadbearing leaf need not be greater than 100 mm thick. Although BS 5628: Part 1 allows thicknesses down to 75 mm and the Regulations down to 90 mm such thin leaves are now unlikely to be used in view of the thermal insulating values required by the Building Regulations, since a reduction in thickness results in a reduction in the thermal insulating value of the wall. 100 mm blocks are, therefore, commonly used. This type of lightweight block may be used below the damp-proof course.

Wherever possible the roof load should be distributed to both leaves of a cavity wall by suitable detailing at the bearing as in figures 7.11 and 7.28. The reduction in eccentricity of loading in this way reduces the stress in the wall and makes the wall more stable. For the same reason floor joists, which are carried entirely by the inner leaf, should preferably not be supported by metal hangers but should bear directly on the leaf on a MS bearing bar (figure 8.8). The use of a timber wall plate in this position is not desirable for reasons given on page 197.

The detailing at the head of the wall referred to in the previous paragraph also provides the cavity barrier against flame and smoke movement which is required at this point by the Building Regulations.[1]

The base of the cavity is filled with fine concrete the top of which must be kept at least 150 mm below the level of the lower of the damp-proof courses (see figure 5.11). This provides a space as a precaution against moisture rising above the damp-proof course. Every third vertical joint in the outer leaf at the base of the cavity is left open as a means of discharge for any water which might collect at this point. The risk of mortar droppings forming a bridge at the base of the cavity is also minimised by this extension of the cavity below the horizontal damp-proof course. By bedding a number of bricks in sand at the quoins raking of the cavity on completion of the wall can be carried out, after which the bricks are finally bedded in mortar.

Thermal insulation of cavity walls

In order to achieve the standards of thermal insulation required by the current Building Regulations it is necessary to use added insulation in brick or brick-block cavity walls or to use greater thicknesses of lightweight concrete blocks. A variety of methods has been developed and some are illustrated in figure 5.13. In other illustrations in this chapter the application of these methods of insulation has been omitted in order to maintain the clarity of the structural detailing.

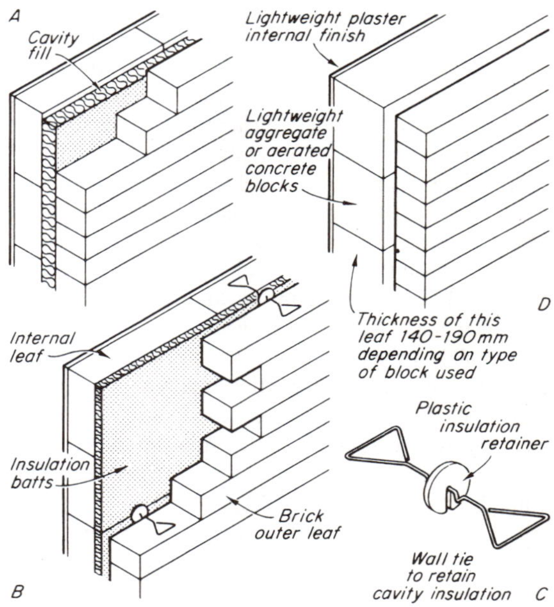

5.13 *Thermal insulation of cavity walls*

In addition to applying insulation to the interior face of the wall, as for solid walls, it will be seen from figure 5.13 that apart from increasing the thickness of the block inner leaf (*D*) the methods consist of either filling the cavity with suitable material or applying a suitable material to the cavity face of the inner leaf.

Cavity fills may be blown or injected into the cavity of the wall after construction (*A*). These may be in the form of injected urea-formaldehyde[2] foam or of blown-in mineral or glass fibre or beads or granules of

[1] See Approved Document B Appendix G.

[2] Approved Document D lays down specific requirements covering the use of materials which give off formaldehyde fumes.

polystyrene or perlite. The cavity may also be filled during construction with mineral or glass fibre slabs or quilt or expanded polystyrene slabs.

With bead and granule fillings detailing must be such that the fill, because of its free-flowing property, is not able to flow out of the cavity into voids such as the spaces between floor joists. Although good workmanship is always desirable in cavity wall construction this is especially so when fill is to be blown-in or injected since it is difficult to check that the wall is, in fact, in a satisfactory state to take the fill and function efficiently afterwards. With the use of slab or quilt fill placed in the cavity during construction work can be inspected as the insulation is placed in position.[1] The 50 mm cavity in general use has hitherto been wide enough in most circumstances to provide sufficient cavity fill to meet the required standard of insulation for the wall, but ever-increasing standards are likely to necessitate a wider cavity to provide an adequate thickness of fill.

All fills bridge the cavity and thus increase the risk of rain penetration through the wall. Although rain rarely penetrates by capillary action through the pores of the material it may pass through wide gaps in the joints between slabs or shrinkage cracks in dried foam. Tests by the Building Research Station involving blown and injected fills indicate that in exposed situations under severe weather conditions water may be expected to pass across cavities filled with any of these materials although in widely differing degrees.

Blown-in beads and blown-in man-made fibres such as mineral or glass fibre and batts of man-made fibres may be used in any exposure conditions provided they are not subjected to conditions more severe than those recommended for the equivalent unfilled wall. Injected insulants such as urea-formaldehyde should be subject to further restrictions related to the local exposure conditions and type of wall construction and BS 5618 gives criteria as to the suitability of walls for filling with this foam together with maximum exposure conditions for different materials and finishes for the outer leaf.[2]

A method of assessing the degree of exposure of a wall, given in BS 5628: Part 3, is referred to on page 69.

Filling the cavity results in the outer leaf being not only slightly colder, but remaining wetter after rain than with an open cavity. Research by the Building Research Station suggests that these differences in temperature and moisture content do not have a significant effect on the durability of the outer leaf as was previously considered possible.

When the preservation of a cavity is desired partial fill insulation may be employed using insulation batts attached to the cavity side of the inner leaf leaving a residual cavity in front of the batts (figure 5.13 *B*). For reasons already given 50 mm is considered the minimum desirable width of cavity in an unfilled cavity wall. Where batts are used in this manner on the inner leaf the Building Regulations (AD L Section A) imply that a minimum residual cavity of 25 mm can be satisfactory for a dwelling, but BS 5628: Part 3 recommends a general minimum of 50 mm. In situations of severe exposure in areas prone to heavy, driving rain it is wise to retain a clear residual cavity of this greater width and it is essential in walls over 12 m high. The batts must be securely fixed to the inner leaf to prevent them falling forward to bridge the cavity. This is accomplished by the use of retaining clips or by wall ties produced to ensure this, an example of which is shown in (*C*). Where the latter are used the layout of the ties may need to be changed in order to provide adequate support to the batts. For example, 1200 mm long batts would require ties at 600 mm horizontal centres in vertical rows instead of in a staggered layout.

Where cavity insulation of any type stops short of the top of the wall, for example at ceiling level in a gable wall, it should be protected at the top by a cavity tray with stop ends unless it is partial fill with a residual cavity not less than 50 mm wide. In buildings taller than the 12 m provided for in British Standards (see footnote[2] below) up to a height of 25 m, cavity trays should be used within 12 m of the base of the cavity and at 7 m intervals above except, again, with partial fill and a residual cavity not less than 50 mm.[3]

As alternatives to the preceding methods the necessary insulation may be applied to the interior face of the wall, as described for solid walls on page 80, or thicker insulating blocks may be used to achieve the desired insulation value for the wall. When using thicker blocks for the inner leaf (*D*) the overall thickness of the wall may be minimised by the use of blocks

[1] On this aspect reference should be made to BS 8208: Part 1: 1985 *Assessment of suitability of external cavity walls for filling with thermal insulants.*

[2] It should be noted that the recommendations in the Standards concerned with the installation of cavity fill insulation, BSs 5618, 6232, 6676 and 8208, are limited to buildings with a height of wall from lowest DPC to top of wall not exceeding 12 m that is, normal three-story construction. See above for reference to heights greater than this.

[3] For a useful overall consideration of cavity insulation see *Cavity insulated walls: Specifiers guide* published by Brick Development Association, Cement and Concrete Association and others.

specially developed for this purpose which have better insulating characteristics than the normal lightweight aggregate or aerated concrete block.

When brick rather than insulating block is required for the inner leaf to achieve a fair face finish internally the necessary insulation must be provided by cavity fill or partial fill, both of which will necessitate a constructional cavity greater in width than would be necessary with a block inner leaf in order to accommodate the thicker insulation required and, in the case of partial fill, still to preserve a residual air space of adequate width.[1]

Whatever the method of wall insulation adopted it is essential that at the head and foot of the wall the insulation be taken at least to the level of the roof insulation and to the level of the ground floor in order to avoid such gaps between lateral and vertical insulation as would form potential cold bridges.

In order to maintain the thermal insulation value of an unfilled cavity it should not be ventilated other than by open vertical drainage joints which provide sufficient ventilation for any moist, humid air in the cavity. For this reason when it is necessary to provide airbricks in the wall for, say, the ventilation of a hollow timber ground floor the air should pass through slate or pipe ducts as shown in figure 8.6. Similar detailing is necessary when cavity fill is used for insulation to avoid obstruction of the airways by the fill.

5.2.3 Openings in walls

Openings are required in walls for reasons of lighting, access or ventilation.

Dimensions and position of openings

A number of considerations determine the dimensions of openings and their position in a brick wall — functional, structural and economic.

In the case of windows the area and height of the openings must relate to the size and, particularly, the depth of the room to be lit (see *MBS: Environment and Services*). Similarly requirements of ventilation have a bearing in this respect.[2] Door openings must be sufficiently wide to permit the passage of furniture and equipment as well as persons, both able and disabled,[3] and in the case of fire escape doors, to comply with the minimum widths laid down in regulations.

Structural considerations are concerned with the relative widths of openings and adjacent walling. The greater the width of opening the greater the weight of wall over transferred to the walling on each side which must be strong enough to carry it. This is particularly important where a series of closely spaced openings leaves relatively narrow piers of brickwork between. The narrower these are the greater is the significance of the height of the openings because of the effect upon the stability and bearing capacity of the piers (see chapter 3). In cavity wall construction most of the load over an opening, including floor loads, is usually carried by the inner leaf and when openings are wide the ability of the jambs to support these loads must be carefully investigated. At these points it may be necessary to use a spreader or thicker blocks or to form a pier.

Economic considerations are concerned with the width and position of openings relative to the normal bonding of the wall in which they occur. In order to avoid irregular bond above and below the opening its width should be a multiple of brick sizes: for English bond a multiple of one brick, for Flemish bond a multiple of one and a half bricks, with a minimum width of 450 mm in the latter. By this means the correct face appearance is maintained throughout and the perpends are kept true[4] above and below the opening without the labour and wastage in cutting bricks. The dimensions of the brickwork between the openings should be based on brick sizes wherever possible in order to maintain bond and avoid the cutting of bricks. For economic building brick sizes should thus be applied to both openings and intermediate walling. In respect of openings in particular this is not always convenient and the setting of windows and doors in storey-height infilling panels between areas of brickwork offers the advantage of eliminating brick cutting and broken face bond above and below the openings if these are not of brick dimensions as well as that of disassociating the work of different trades (see chapter 2).

[1] See footnote index[2] to page 81.

[2] See, for example, the Building Regulations 1985: AD F1. It should be noted here that for reasons of thermal insulation AD L2/3 sets a limit to the total window area in any wall.

[3] See BS 5810: 1979 *Access for the disabled to buildings.*

[4] *Perpends* are the vertical joint lines on the face of the wall. Perpends are 'kept true' when the vertical joints in alternate courses lie on vertical lines.

Each part of the wall round an opening is defined by a particular name as shown in figure 5.14 *A*. The wall immediately adjacent to the side of the opening is called the *jamb* and that to the top, the *head*. The return face to the former is the *reveal* to the opening and to the latter, the *soffit*. The bottom of a window opening is the *cill* and of a doorway, the *threshold*, both being horizontal planes requiring protection from the weather. At the head provision must be made to support the wall over.

5.2.4 Openings in solid walls

Jambs of openings

These may be square or rebated. In a *square jamb* the reveal is flat as in figure 5.14 *A* and is in fact, identical to a stopped end and is bonded in the same way. In a *rebated jamb* the reveal is recessed, or rebated, as in (*B*). The bonding of this form of jamb varies with the width of the stop, which is usually 102·5 mm but may be more in thick walls, the depth of the recess, which may be 56 mm or 102·5 mm and with the bond and thickness of the wall. In all cases the use of various types of closers and bats is necessary in order to avoid straight joints and to maintain the face appearance.

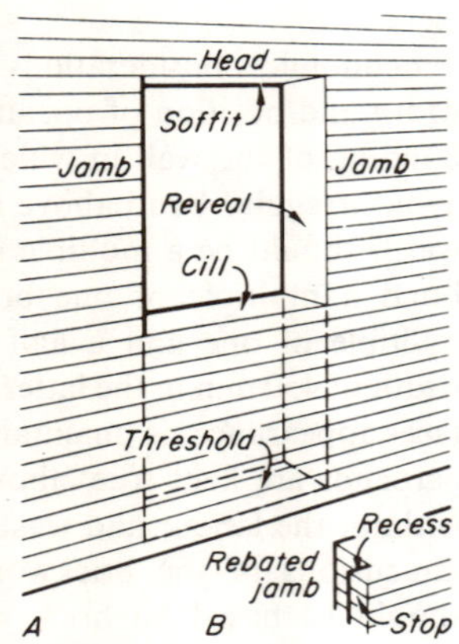

5.14 *Openings in walls*

Heads of openings

Support to the wall above an opening is provided by a horizontal beam called a *lintel* or by some form of arch, producing a *square-headed opening* and an *arched opening* respectively.

Lintels

The most common materials used for lintels are reinforced concrete and steel. Alternatively, it is possible to dispense with a separate lintel and to reinforce the brickwork to span the opening itself.

Reinforced concrete lintels are made of 1:2:4 mix concrete normally reinforced with one steel bar to each 102·5 mm in the width (figure 5.15 *A*). For reasonably short spans over door and window openings the 'arching' action of normal well-bonded brickwork due to the overlapping of the bricks may be taken into account and it may be assumed that the lintel will carry only that brickwork enclosed by a 45 degree isosceles triangle with the lintel as its base, the walling above the triangle being supported by the jamb brickwork on each side through the 'arching' action of the bond. This method of assessing the amount of brickwork carried by the lintel should be used only where there are no wall openings within this triangle and for spans up to about 3·5 m in two to three storey domestic buildings. BS 5977: Part 1: 1981 *Lintels* uses this method but with a base to the triangle equal to 1·1 times the clear span of the lintel. It also gives methods of dealing with loads, other than that of the brickwork, applied within or near the load triangle.

For spans up to 2·4 m the sizes of lintel and the amounts of reinforcement shown in figure 5.16 may be used. The steel bars should have 25 mm cover of concrete and the bearings on the wall should be at least equal to the depth of the lintel. Lintels above 2·4 m span should be calculated.

Long span concrete lintels may be cast in situ in formwork erected at the head of the opening, but where suitable lifting tackle or crane is available for hoisting into position or where the lintel is short enough to be man-handled conveniently by two men, precasting is usually adopted. This has the advantage that it may be matured long enough to permit the initial shrinkage to take place before building into the wall and that the walling above may be proceeded with immediately after bedding in position. If precast lintels are to be manhandled the weight must be kept within the limit for two men which is usually considered about 68 kg. This restricts the size to about 1219 mm × 215 mm × 102·5 mm.

Detailing at the head of the opening will depend on whether or not the brickwork is to carry over the lintel and on how the face brickwork is supported. Methods of providing support are shown in figure 5.15. The depth of the lintel at the face may be reduced to a 75 mm toe which will carry the outer 102·5 mm of brickwork as at (*B*), forming what is known as a *boot lintel* and minimising the amount of concrete exposed on the wall face. Alternatively, the outer 102·5 mm of brickwork may be carried by a 75 mm × 75 mm MS angle rag-bolted back to the lintel behind as in (*C*). The angle, which should be galvanised or painted with bituminous paint to prevent corrosion, is less obvious than the concrete toe. The face brickwork itself may be

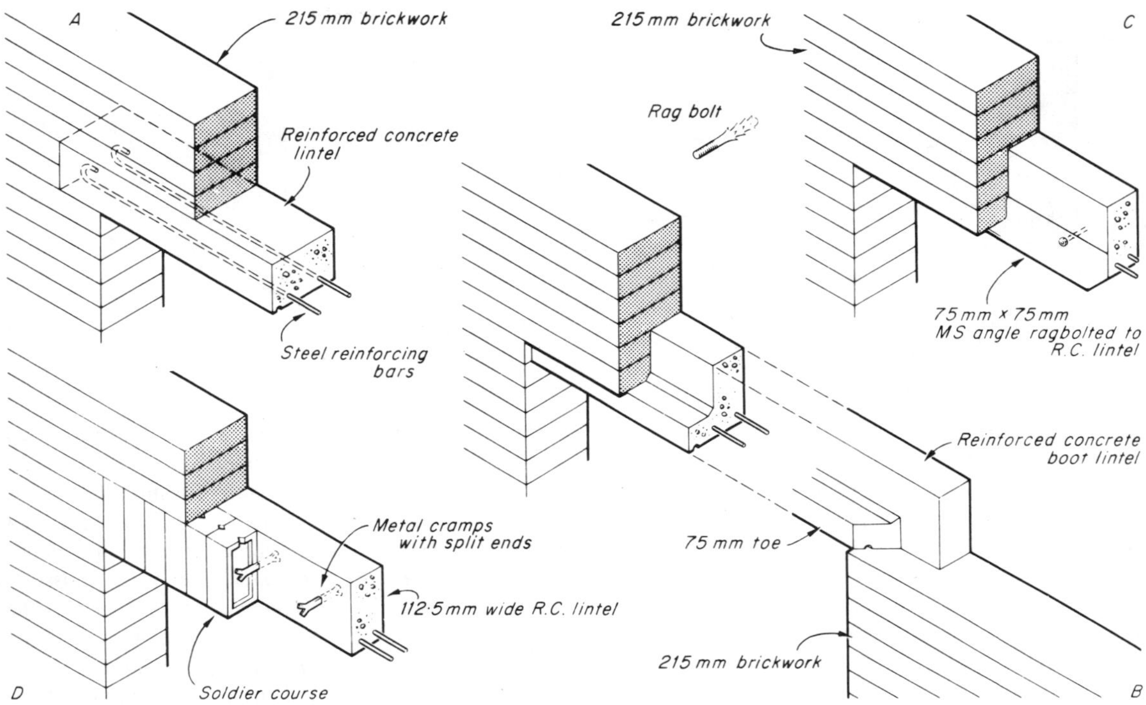

5.15 *Lintels to openings in solid walls*

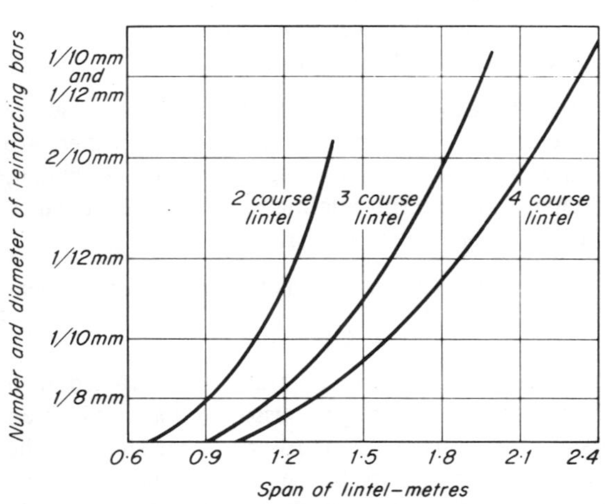

This graph relates span of opening and lintel depth in brick courses to the required amount of reinforcement per half-brick thickness of wall carried by lintel

5.16 *Reinforcement to concrete lintels*

tied back to the lintel by metal cramps but this necessitates the use of bricks bedded on end[1] as shown in (*D*).

Over large openings steel lintels are in the form of rolled steel joists which must usually be protected from the effects of fire by concrete casing or some other form of protection (see Part 2). For short spans the outer face of the brickwork may be supported solely by a steel angle acting separately from the concrete lintel and thus relieving it of the weight of the brickwork. In this case the vertical leg of the angle must be deep enough to give sufficient stiffness, similar to that in the cavity wall example in figure 5.21 *D*.

Brickwork may be made to act as a beam or lintel by the inclusion of steel reinforcement, the manner of incorporating this depending upon whether the bricks are bedded on end as a soldier course or run through in normal courses.

A method involving a soldier course is shown in figure 5.17 *A* where it will be seen that longitudinal steel bars are set in the central vertical joint with bent wire stirrups bedded in every third vertical joint along the length of the course. The longitudinal bars act primarily in tension due to bending and the stirrups resist shear stresses. The longitudinal bars must extend 150 mm to 229 mm into the jambs and all the vertical

[1] Known as a *soldier course*. This type of lintel has been called a *soldier arch* but the term should be avoided as it is not a true arch.

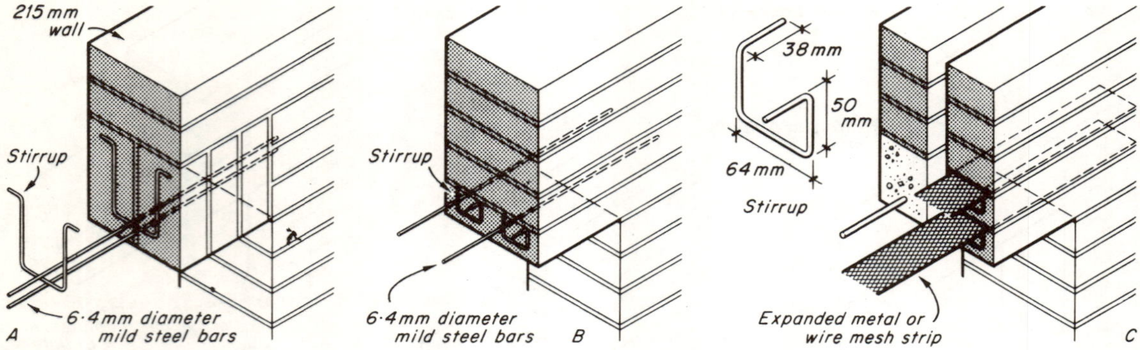

5.17 *Reinforced brick lintels*

joints must be solidly grouted up. For this work the mortar used must be cement mortar or cement gauged with a small proportion only of lime.

With normal coursing the reinforcement, either rods or expanded metal, is placed in the first and, perhaps, second bed joint with 3·2 mm to 6·4 mm diameter wire stirrups in every vertical joint of the reinforced courses, the ends of these being hooked over the tops of the bricks as shown in (*B, C*). To give sufficient effective depth there should be at least four courses of bricks in the lintel with a greater depth for wide spans. As in the previous example reinforcement must extend 150 mm to 229 mm into the brickwork at each bearing.

Reinforcing bars should be galvanised and, being set in the joints of the brickwork, should not exceed 6 mm in diameter and should have a cover of not less than 15 mm from the exterior face of the brickwork. The design of reinforced masonry such as this should be based on recommendations given in BS 5628: Part 2: 1985.

Arches

The arch is a technical device used to span an opening with components smaller in size than the width of the opening. It consists of wedge-shaped blocks which, by virtue of their shape, mutually support each other over the opening between the supports, or *abutments*, on each side. It exerts a downward and outward thrust on the abutments which must, therefore, be strong enough to resist this thrust in order to ensure the stability of the arch (see chapter 3).

The terminology relating to the parts of the arch is indicated in figure 5.18.

The wedge-shaped blocks are called *voussoirs*, the centre one being the *key* (key-brick, keystone, according to the material) as it is the last voussoir laid and locks the remainder in position. Until it has been bedded the arch is not self-supporting.

The circular course of voussoirs forming the arch is called a *ring* and brick arches are often built up with a number of rings to obtain sufficient structural depth and for other reasons given later. The lowest voussoirs in the arch are the *springers* and these rest on the sloping surfaces of the abutments which are termed the *skewbacks*. *Extrados* and *intrados* are the terms applied to the external and internal curves of the arch respectively and the *soffit* is the inner curved surface.

The bed joints of an arch are those between the voussoirs, normal to the curve of the arch.

Arches may be classified in two ways:

1 According to the amount of labour involved and the degree of finish of the arch. In this respect there are three classes of arch; rough, axed or rough-cut, gauged, these terms relating to the manner in which the voussoir shapes are formed.

2 According to shape, such as flat or camber, segmental, semi-circular.

Rough arch This form of arch is constructed with rough voussoirs. A rough voussoir is one in which an uncut brick is used, the necessary wedge shape of the voussoir being achieved by the wedge-shaped bed joints between adjacent bricks in the arch (figure 5.18 *A, B*). The arch should be constructed in half-brick rings in order to avoid the very thick joints which occur with a 215 mm voussoir (*B*), although it is usual to use a full brick at the skewback in order to distribute the thrust from the rings on to the abutment as shown. The use of perforated bricks or those with a frog on each bed ensures a good key between the mortar and the bricks. The rough arch is relatively cheap and is used where the work will be hidden by plaster or other finish.

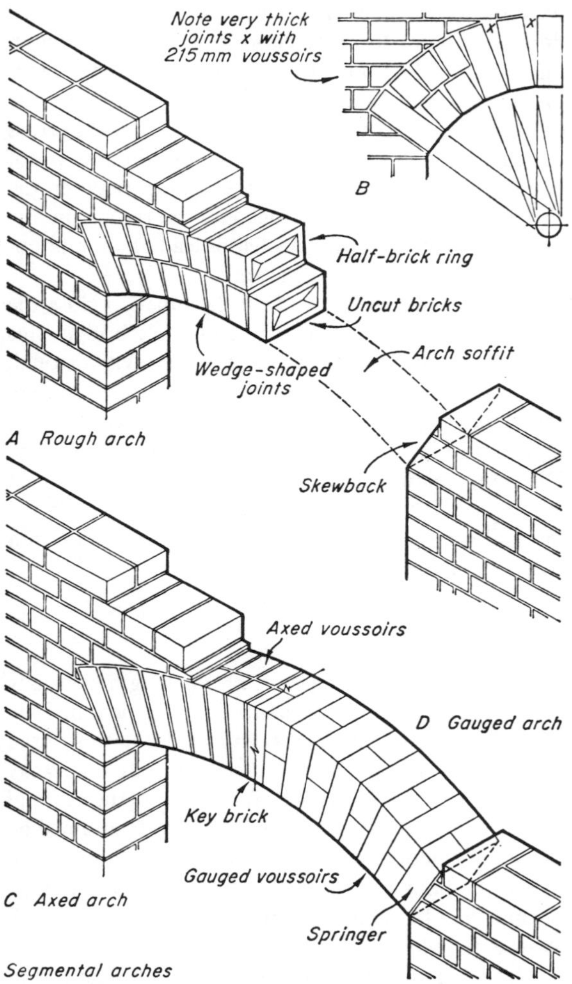

5.18 *Arch construction*

Axed or rough-cut arch This arch uses axed voussoirs which are ordinary facing bricks cut to a wedge shape on site. The joints in this form of arch are rather thick and irregular as in ordinary brickwork. Full brick voussoirs may be used since they are cut to shape and the joints are of uniform thickness as shown in (*C*).

Gauged arch This form of arch uses gauged voussoirs which are bricks so accurately cut to shape (hence the term *gauge* which means measure) that an arch may be formed with bed joints as thin as 0·8 mm to 1·6 mm (*D*). Such accuracy is possible only with special bricks called *rubbers* which are sufficiently soft to be sawn and rubbed to shape. In order to obtain the very thin joints pure slaked lime putty is used for bedding the voussoirs. This form of arch is now rarely used.

Segmental and semi-circular arches Segmental arches (figure 5.18) may be constructed in any of the forms described above, but semi-circular arches should not be constructed with rough voussoirs unless of large radius, because of the wide joints which result when these are laid to a sharp curvature (see *B*).

Flat or camber arch This arch is constructed in gauged or axed form but it is not very strong and should not be used for spans exceeding 1·5 m. The extrados is horizontal and the intrados is given a slight upward curve, or camber, to correct the illusion of sagging produced by a horizontal intrados (figure 5.19).

If the brick length from which the voussoirs are formed is limited to 215 mm as in axed work, and if no cross joints are desired, the depth of the arch must be less than 215 mm as in figure 5.19 *A*. For greater depths of arch either purpose-made bricks must be used or cross-joints must be incorporated (*B*).

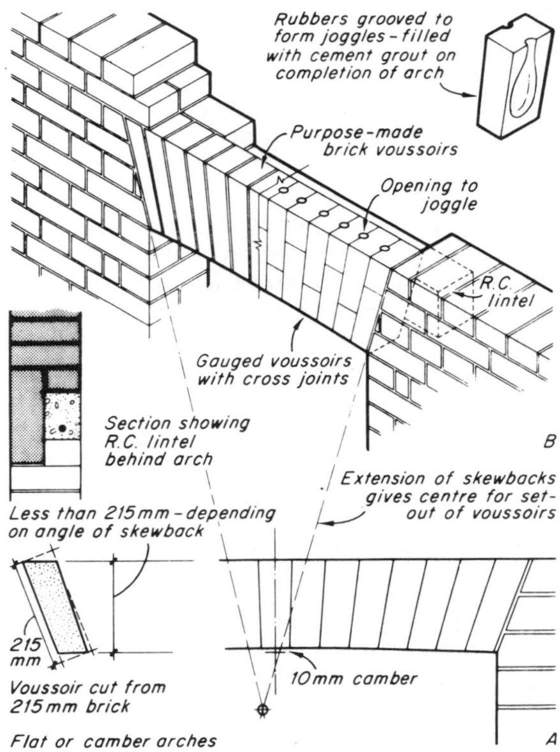

5.19 *Arch construction*

Arches of all types require some temporary support until they have been completed and the mortar has set and hardened. This normally takes the form of a framework of timber known as *centering* on which the voussoirs are laid to the required curve (see page 239).

Cills and thresholds
The detailing at cills and thresholds for openings in solid walls is basically the same as in cavity wall construction and is discussed in the following section.

5.2.5 Openings in cavity walls

Careful detailing round openings in cavity walls is essential. It is mainly at these positions that, apart from the ties, the cavity is bridged and damp may penetrate to the inner leaf.

When cavity thermal insulation is adopted detailing round openings must be such that the insulation does not stop short of the door or window frames thus reducing the insulation value of the wall at those points and giving rise to potential cold bridges. Although the Building Regulations permit lintels, jambs and cills to be regarded as part of the openings, with a lower standard of insulation at these points than for the wall generally, taking advantage of this relaxation could result in such cold bridges, leading to internal surface condensation round the opening.

Jambs of openings
The opening may have square or rebated jambs as in solid walls. In either case the cavity must be closed to preserve its integrity as a thermal insulator and to form a finish at the jamb. The closure must be carried out in such a way that moisture cannot penetrate to the inside wall and the simplest and cheapest way is to use the door or window frame to close the cavity as shown in figure 5.20 *A*. When the frame is set nearer the internal or external wall face than will permit it to close the cavity one or other of the leaves may be returned across the cavity for this purpose and a vertical damp-proof course incorporated to break the contact between the two leaves (*B, D*). Sheet metal or more commonly, because cheaper, bituminous felt or lead or aluminium-cored felt is used for this purpose.

The returning of one leaf on the other to close the cavity in this way produces a potential cold bridge. This could be avoided in (*B*) by carrying the cavity through and closing it with a proprietary cavity closer to take the internal plaster finish (*C*). In the case of (*D*) insulation may be interposed as shown between the DPC and the inner leaf to break the cold bridge. In either case, if thermal insulation is positioned on the internal wall face the cold bridge could be dealt with without the adoption of these details.

Heads of openings
The leaves above the head of an opening are usually supported by lintel rather than arch construction. They may be carried in various ways all similar to those used for solid walls. A single reinforced concrete lintel may carry both leaves, showing its full depth on the external face as in figure 5.21 *A*, or a reduced depth, attained by the use of a boot lintel, as in (*B*). Instead of the concrete nib of the boot lintel a steel angle bolted to the lintel behind may be used to carry the outer leaf, and will show externally to a lesser extent (*C*).

All these lintel forms constitute a cold bridge and to avoid heat loss and internal surface condensation at these points it is essential to provide insulation on the

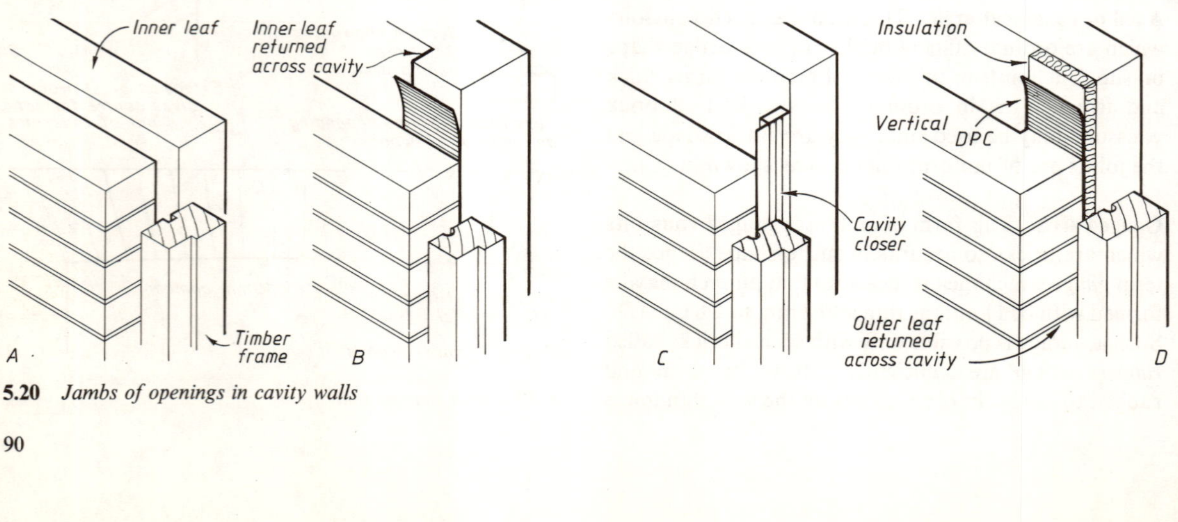

5.20 *Jambs of openings in cavity walls*

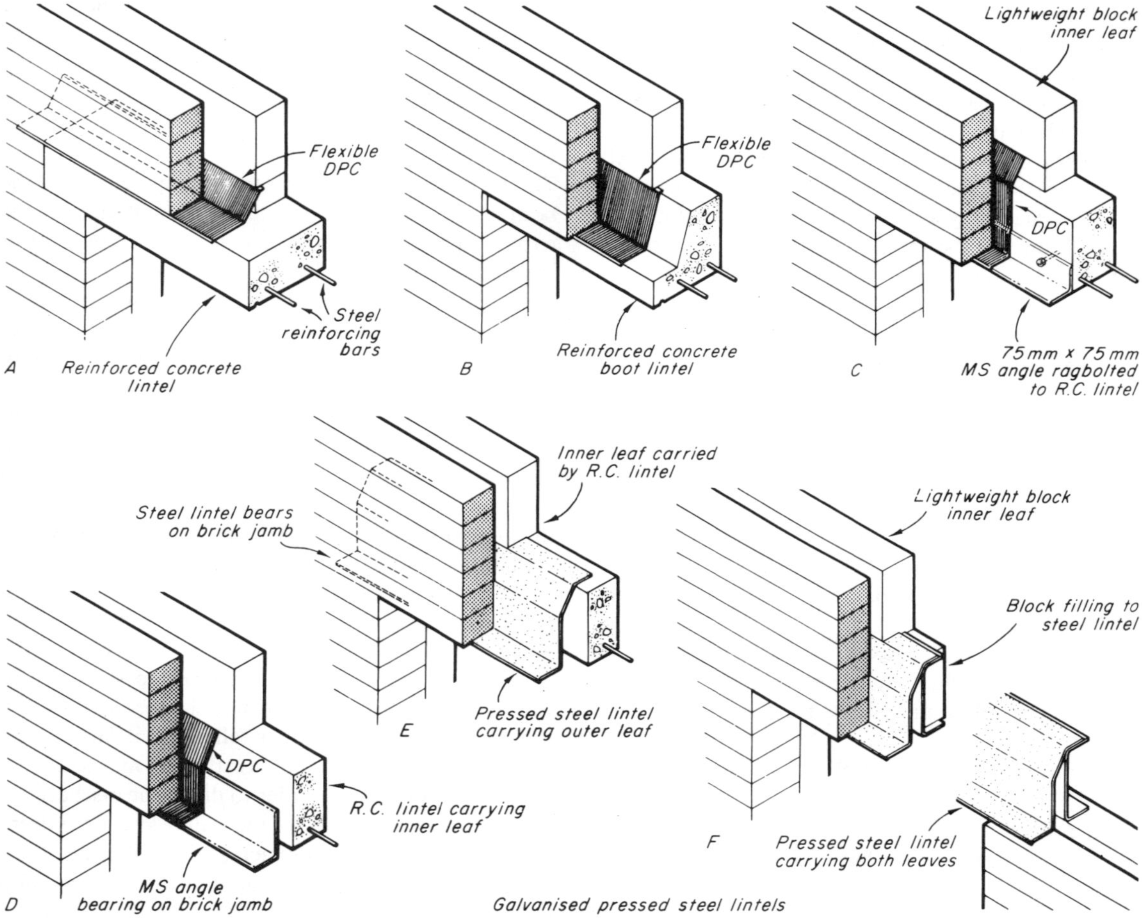

5.21 *Lintels to openings in cavity walls*

internal face of the lintel or to use lightweight aggregate or aerated concrete for the lintel itself.

Alternatively, each leaf may be supported separately. This has the advantage of breaking the cold bridge. For short spans a small steel angle can be used for the outer leaf (figure 5.21 *D*). For larger spans there are available galvanised pressed steel lintels one form of which is shown in (*E*). In both cases lightweight aggregate or aerated concrete may be used for the inner lintel in order to maintain the insulation value of the inner leaf at this point in the wall. A variation of the steel lintel in (*E*) incorporates a pressed steel channel which supports the inner leaf as shown in (*F*), and which permits blockwork immediately over the opening thus overcoming the difficulty sometimes encountered in drilling dense concrete for fixings for curtain rails and pelmets.

The outer leaf may be made self-supporting by bedding rods or expanded metal in the courses of brickwork immediately over the opening to form a reinforced brick lintel as already described (figure 5.17 *C*).

It is essential that the bridging of the cavity by the lintel or the frame head be protected by a suitably formed damp-proof course, called a *cavity tray* in this position, so that moisture falling down the cavity will not lodge and percolate through the inner leaf. The tray should be shaped to drop at least 75 or preferably 150 mm towards the outer leaf as shown in the illustrations, so that any moisture is conducted away from the inner leaf, and it should be continued down so that the outer edge is as near the opening as possible to prevent the penetration of rain to the interior at this point. The ends of the cavity tray should project 150 mm beyond the sides of the opening in order to discharge any water well clear of the jambs except where there is cavity fill insulation or batts with a residual cavity of less than 50 mm, when the tray should be provided with stop ends. A few open vertical drainage joints should be left in the bottom course of bricks or, where the brickwork is to act structurally as in figure 5.17, short lengths of copper drainage tube should be bedded at the bottom of the normal joints.

When an external cavity wall becomes an internal wall at a lower level due to the extension of a lower floor beyond the wall (figure 5.22) cavity trays must also be used to prevent moisture falling down the cavity to the interior. If the roof to the extension is pitched a stepped DPC will be necessary in the form of overlapping trays with stop ends, following the profile of the roof as in figure 5.22. Each tray at its inner end will link with the stepped apron flashing to the roof. Where the roof to such an extension is flat a horizontal tray will be required as for the cavity parapet wall shown in figure 5.25 *C*.

The material used for cavity trays should be capable of being formed to shape and dressed over sharp bends: sheet metals and commercially pre-formed trays are suitable.

The galvanised pressed steel lintels shown in figure 5.21 *E, F*, are usually further protected by chlorinated rubber primer and are shaped to perform the function of a cavity tray as well as that of a lintel.

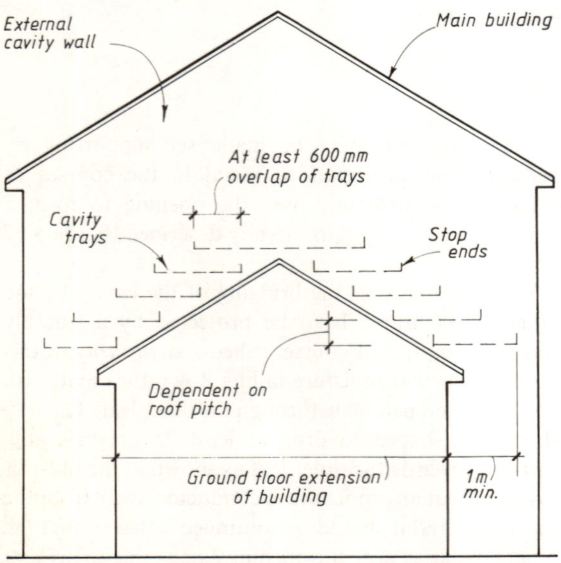

5.22 *Cavity walls—change from external to internal*

Where cavity insulation is used, either in the form of fill or batts on the inner leaf, the insulation must be continued in any space below the cavity tray or below or within a pressed steel lintel.

Cills

The term *cill* has already been broadly defined as the bottom of a window opening. More particularly, the term is applied to the external protective covering of some form applied to the wall at this point.[1]

The wall at the bottom of a window opening is particularly vulnerable to the penetration of water since it is immediately below the impervious glass surface of the window down which all the rain falling on it flows. The function of the cill, therefore, is to protect this part of the wall from the penetration of considerable quantities of water. It should be so designed that in fulfilling this function it also prevents driving rain penetrating the joint at the seating of the window frame on the cill.

Suitable materials for the construction of cills are stone, concrete, brick or quarry tiles laid in cement mortar. roofing tiles laid to break joint in cement mortar, and metal, all of which are widely used.

The top surface of the cill is made to slope downwards and outwards and is then said to be *weathered*, in order to discharge rainwater falling on it and the cill itself is made to project not less than 25 mm to 38 mm beyond the wall face in order to direct the discharge of water away from the face of the wall below. To prevent the backward flow of water across the underside of this projection through wind or capillary attractions a *drip* is formed at the bottom front edge of the cill projection beyond which the water will not pass. In stone, concrete and clayware cills, this is formed by the provision of a groove or *throating* on the underside of the cill. A half-round groove 12 mm in diameter is satisfactory. Bricks and roofing tiles are bedded at an angle to form a weathered top surface and the inclined underside produces a drip at the bottom edge. The front edge of a metal cill is turned down to form a drip well away from the wall face. These cill characteristics are shown in the details in figure 5.23.

The joint between cill and window frame is normally sealed with mastic. A further barrier to water penetration may be incorporated in the form of a strip of galvanised steel called a *water bar*, 19 mm × 3·2 mm to 32 mm × 6·4 mm, bedded half its depth in cement mortar in a groove formed in stone, concrete or moulded

[1] This term is also applied to the bottom member of a timber window frame, see *MBS: Components*.

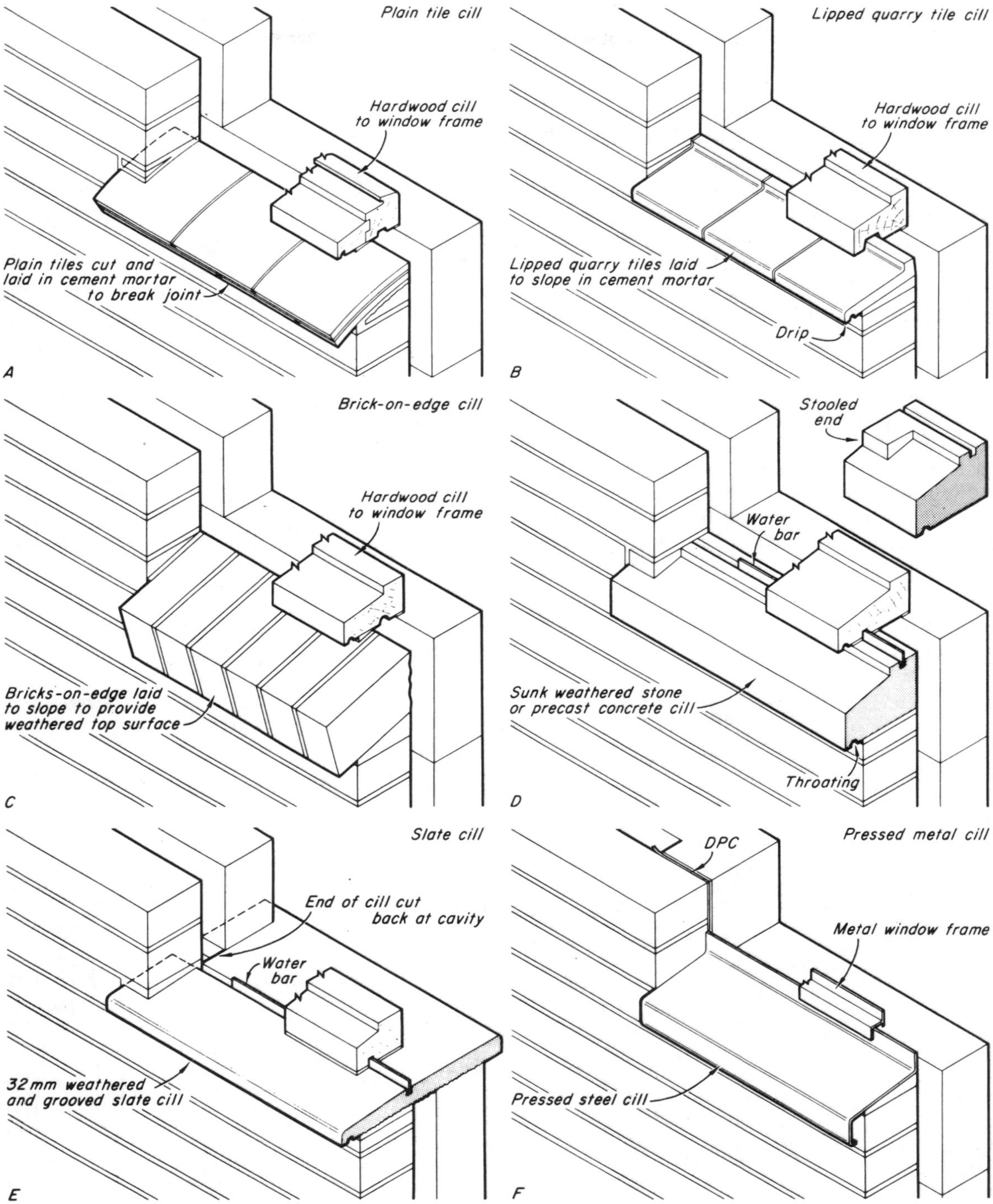

5.23 *Cills to window openings*

clay cills, the upper projecting half engaging in a similar groove in the underside of the window frame which is filled with white lead and oil or a mastic before the frame is bedded on to the cill (*D, E*). In the case of cavity walls a water bar is not essential if the cavity is maintained at this point and provided an anti-capillary groove is formed on the underside of the window frame to prevent the passage of water to the inner leaf as shown in *A* and *C*. Where, however, the cill runs across the cavity as in (*E*) or the cavity is bridged solidly in some other way, and also in the case of solid walls, such a water bar is essential. In the latter if roofing tile or brick cills are used the water bar is set in the joint behind the cill as shown in the threshold detail in figure 5.24 *A*.

As an additional means of protecting this joint the weathered top surface of the cill may be sunk slightly at the top (figure 5.23 *D, F*). This has the effect of raising the joint above the water-retaining surface and serves to break the force of water blown back to the joint. Lipped quarry tiles may be used to achieve this and to fulfil the function of a water bar at the same time (*B*).

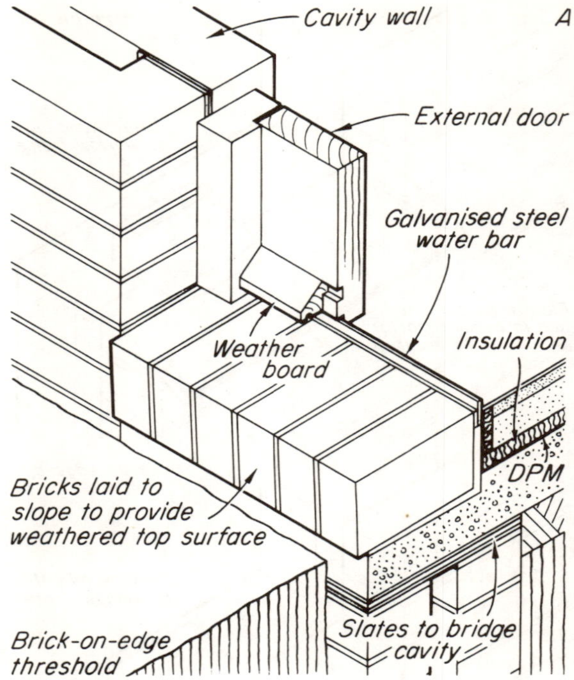

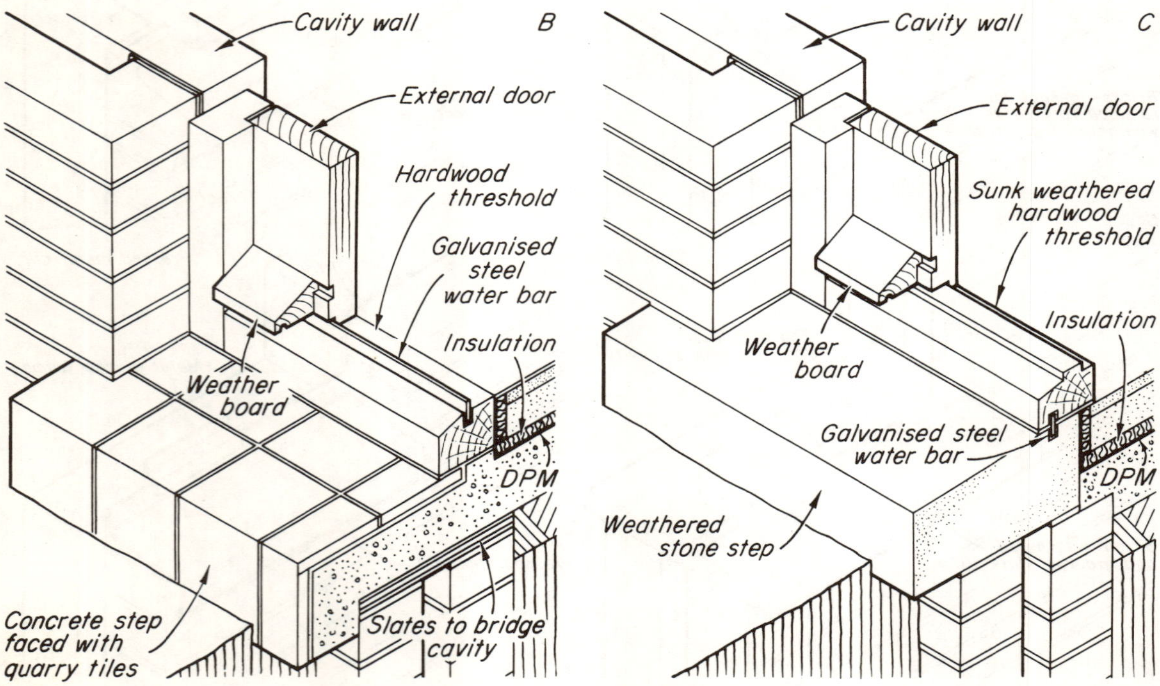

5.24 *Thresholds to door openings*

The weathering to a stone or precast concrete cill may be stopped short of the ends to provide a flat seating for the brick jambs. This is called a *stool* or *stooling* (see (*D*)). If this is omitted, as it may be if the slope of the weathering is shallow, the jamb bricks bedding on the ends of the cill must be cut wedge-shaped. Machine worked stone cills are cheaper to form without stooled ends.

Natural stone and precast concrete cills are similar in section and are normally not less than 75 mm thick with the depth varying according to the depth of the window reveal. The section shown in (*D*) is typical. Slate by its nature can be used in thin sections and slate cills are commonly from 25 to 50 mm thick. Because of its highly impervious character a slate cill may be safely carried across a cavity to form the internal cill as shown in (*E*), but because of the potential cold bridge thus formed this detail should be used only in centrally heated buildings and then only in those not liable to high internal humidity. Where, however, the cavity at this point is maintained as in the other examples and is bridged only by the timber cill of the window frame there will be no cold bridge since timber is a good thermal insulator.

Metal cills may be of cast metal or may be hand-formed out of sheet copper or zinc (figure 4.22 in Part 2) or pressed out of sheet steel (*F*) and secured to the wall by MS brackets.

Where possible the cill should not project into the cavity of a cavity wall, but where the window position makes this essential the cill should be isolated from the inner leaf by a damp-proof course and the ends projecting beyond the sides of the frame should be cut back to the face of the frame to avoid a lodgement for mortar droppings, as in the case of the slate cill (*E*).

When the window frame is set close to the outer face of the wall the main cill may be eliminated and its functions be fulfilled by a projecting timber cill to the window frame (see *MBS: Components*).

Thresholds

The term *threshold* has been defined as the bottom of a door opening. In the case of external doors it is applied particularly to those members the function of which is to form a firm and durable base to the doorway and to exclude water. Suitable materials are stone, concrete, brick, quarry tile and timber.

As the floor level is normally above the ground level outside the door an external threshold usually incorporates a step. This may be formed in various ways, either as an extension of the concrete floor slab or as a separate member of some other material as shown in the details in figure 5.24. The width of the threshold should be wide enough to accommodate the human foot and, preferably, be weathered on the top surface.

Apart from sheds and outbuildings external doors are hung to open inwards and the incorporation of a water bar in the threshold and a weather-board (or weather mould) on the door is essential in order to prevent the entry of water under the door. The water bar forms a barrier to water blown across the threshold and the weatherboard throws water away from the bottom edge of the door and prevents it running down behind the water bar (figure 5.24 *A, B*). A drip must be formed on the underside of the weather-board for the same reasons that one is formed on a cill.

A hardwood threshold is sometimes incorporated as part of the door frame, especially where the door is set within a prefabricated wall unit including door, window and infilling panel. This is useful when the distance between ground and floor levels is somewhat greater than a reasonable rise for a single step (*B*) or when clearance for a mat is required in order to avoid a mat well (*C*). In the latter detail a water bar is incorporated under the timber threshold to prevent the passage of water to the floor inside through the joint which lies above the damp-proof membrane in the floor. This detail shows a sunk weathered threshold which avoids the need for a top water bar.

The provision of a timber threshold results in a drop in level immediately below the weather-board and if the latter is made to project slightly beyond this, as in (*B, C*), water from the door does not fall on to the water retaining surface immediately in front of the water bar. People are less likely to trip over a water bar set in a relatively large and visible timber threshold than when it is set in a flush threshold as in (*A*).

The threshold of a door constitutes a potential cold bridge between the floor structure and the exterior of the building. This can be broken by the floor insulation if provided; if not, edge insulation in the floor as shown in figure 5.24 should be provided at the door opening, extending about 900 mm into the floor and about 300 mm on each side of the opening (see page 193).

5.2.6 Damp-proof courses

In addition to the lateral penetration of rain moisture may move vertically throughout a wall to the interior, either from the ground on which the wall bears or through the exposed head of a wall at a parapet or chimney. In certain circumstances it may also move laterally through a wall from adjacent ground against which it is built. Such movement is prevented by the provision of moisture barriers which, as indicated earlier, are called *damp-proof courses*, a term which is usually abbreviated to DPCs.

Horizontal damp-proof courses

These are used at the head and base of a wall to prevent vertical movement of moisture. Suitable materials for this purpose are sheet metal of which the most commonly used are lead and copper, bitumenised felt with or without a core of thin sheet lead or aluminium, polythene, asphalt, slates and engineering bricks. These are all covered by BS 743.

Lead or copper sheet is flexible and is laid in mortar with 75 mm lapped running joints and full-width laps at junctions and quoins. As fresh lime or Portland cement mortar may cause corrosion of lead this should be coated with bitumen as protection. Bitumenised felt is laid in mortar with running laps at least 100 mm wide and full-width laps at junctions and quoins: BS 743 includes seven types. Black low density polythene, like bitumenised felt, is a flexible DPC and is similarly laid in mortar with laps at least equal to the width of the DPC. These two materials are the cheapest of damp-proofing materials and are, therefore, widely used.

Asphalt is usually laid 12·5 mm thick. It is very durable but it does not withstand much distortion. Good quality roofing slates or engineering bricks laid in two courses form durable but rigid damp-proof courses which will accommodate only slight structural movements. They need to be very carefully laid in 1:3 cement mortar, the slate courses being laid to break joint and the vertical joints of the brick courses preferably being left unfilled. Slates for this purpose should be at least 229 mm long. The slate DPC because of its thickness and the brick DPC because of its colour and texture, will both be apparent on the face of the wall and for this reason may not always be acceptable.[1]

The damp-proof course at the base of a wall must be set at least 150 mm above the ground level in order to minimise the danger of an accumulation of soil and leaves building up against the wall to a greater height than the DPC and thus permitting moisture to by-pass it to the wall above. This also prevents the same thing occurring when heavy rain splashes up off adjacent paved areas.

The relationship of the horizontal DPC in the wall to that in the floor structure is important and is discussed on page 190 where it is noted that the DPC in the inner leaf of a cavity wall need not be at the same level as that in the outer leaf since the cavity acts as a vertical damp barrier. The need for horizontal DPCs over openings in no-fines concrete walls and in cavity walls has been referred to on pages 71 and 91 respectively.

The provision of DPCs at the head of a wall is considered in the section on *Parapets and copings* below.

Stepped and vertical damp-proof courses

When a building is set into a sloping site or 'steps down' the site as in figure 8.2 the horizontal DPC must be stepped in order to maintain it at least 150 mm above ground level at all points (figure 4.9). The vertical portions should preferably be not greater than three or four courses in height in order not to weaken the wall unduly. If deeper steps are essential the DPC should follow the toothing line of the brickwork down the step to ensure a degree of bond.

Where the floor level drops below that of the adjacent ground as in figure 8.2 (*B*) and the walls are of solid construction a vertical damp-proof course must be incorporated to join the damp-proof floor membrane with that in the walls. Apart from polythene film, any of the materials suggested for the floor membrane (page 192) are suitable and are applied to the inner face of the wall. As an alternative a waterproof cement rendering applied to the internal or external face of the wall may be used; if applied externally the floor membrane must pass through the wall to meet the cement rendering. Exposed areas of internal DPC require facing over in some way to take the wall finish. In the case of cavity walls, if the distance between the floor and ground levels is not great and there is no possibility of flooding of the floor, the cavity will fulfil the function of a vertical DPC (figure 8.1 *D*). If, however, the distance is considerable, in addition to the possibility of excessive pressure of the soil on the outer leaf, there is the danger of the base of the cavity filling with water and rising above the DPC in the inner leaf. In such circumstances the cavity should be filled and a vertical DPC provided.

The need for vertical DPCs at points of contact between the leaves of a cavity wall round openings has been referred to on page 90.

Whenever the floor level is placed below the known subsoil water level on the site, as is often the case with basements, the whole waterproofing system must be able to resist the entery of water under pressure and methods of providing for this are described in Part 2.

[1] See also BRE Digest 77 (Second series) *Damp-proof courses*.

5.2.7 Parapets and copings

A *parapet* is the upper part of an external wall carried above the level of a roof gutter or a roof plane.

This portion of wall tends to become saturated in wet weather by reason of its height and exposure on both sides and if it is more than a course or two high a horizontal damp-proof course should be incorporated at its base to protect the walling below. This should be placed level with the top of the upstand to the gutter or flat roof covering as the case may be, as shown in figure 5.25 *A*. In the case of a cavity wall the cavity may terminate at this point, and the remainder of the parapet be built in 215 mm brickwork as in (*B*), or, particularly in tall parapets, the cavity may be continued full height as in (*C*). The cavity here will prevent rain entering on one side from saturating the whole parapet, so that subsequent drying out will be quicker. This is important at times when frost follows soon after wet weather since the expansion on freezing of the water in a saturated parapet might lead to spalling of the brickwork. In the case of a rendered solid wall as in (*D*), a cavity parapet prevents water which enters at the back from reaching the brickwork immediately behind the rendering thus reducing the possibility of sulphate attack on the latter should the bricks contain soluble salts.

In the case of a cavity parapet wall the DPC must be carried across the cavity to prevent water running down the inside face of the back leaf and thus to the interior of the building. It may be stepped to discharge any water to the outer leaf; this may cause staining on the outer face of the wall. If stepped as shown in (*C*) moisture could travel down the underside of the slope and penetrate beneath the roof covering; this is possible in areas where the wall would be exposed to strong wind-driven rain and is a factor to be considered at design stage.

Minimum thicknesses of solid and cavity parapet walls to dwellings and small buildings are laid down in AD A1/2 C11 for different heights up to a maximum of 860 mm above the junction of structural roof and main wall.

The top of a parapet is protected by a capping of brick, stone, precast concrete or metal which is called a *coping* and which should, preferably, be designed to throw water clear of the wall below.

Brick copings The bricks for these should be hard and durable and should be bedded in cement mortar. The simplest form is the brick-on-edge coping with square or bull-nose bricks (figure 5.25 *A*). This may be improved by a projection on each side to throw water clear of the wall, formed either by two courses of clay tiles laid to break joint in cement mortar, called a *creasing* (*B*), or by oversailing courses of bricks projecting similarly on each side. A saddle-back coping (*D* for shape) formed with specially moulded brick or terracotta has the advantage of throwing off water more quickly than a normal brick-on-edge.

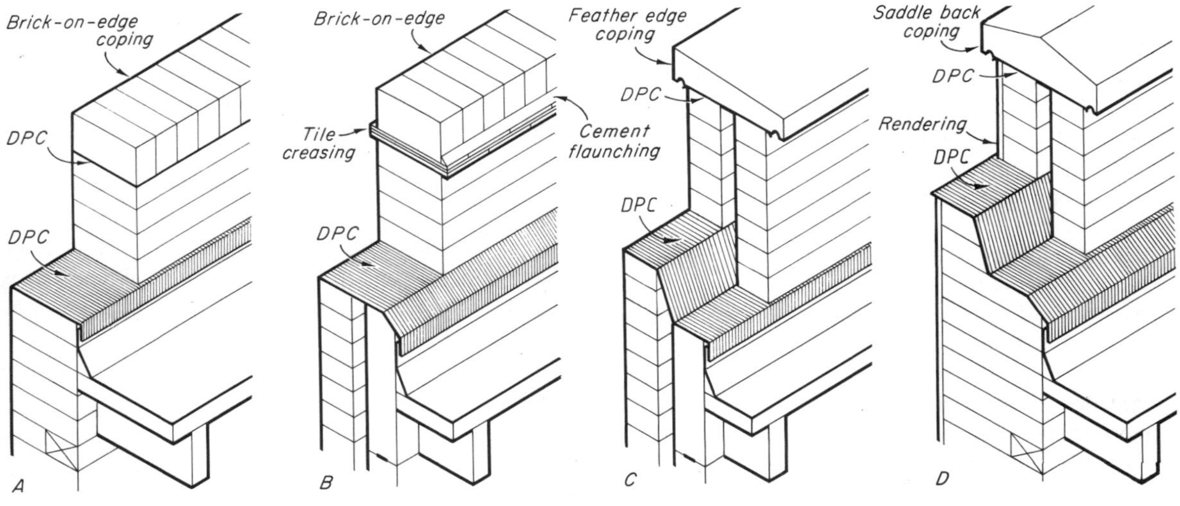

5.25 *Parapets*

The large number of vertical joints in a brick coping is a disadvantage and, even where tile creasing is used, a horizontal DPC should always be provided immediately beneath the coping.

Stone and precast concrete copings These are to be preferred because of the fewer vertical joints. The two most commonly used shapes are shown in (*C, D*) and these are designed to project 38 mm to 50 mm beyond the wall faces with *weather-grooves* on the underside to form drips on each side for the same purpose as on a cill. In spite of the reduced number of vertical joints penetration of water to the wall should be prevented by a horizontal DPC as for brick copings. When a flexible DPC is used over a cavity wall, as in these examples, some form of thin, rigid support should be provided to prevent the DPC sagging into the cavity. Where the wall below is rendered the DPC must be extended over the rendering since adequate edge protection to the latter is essential.

Parapet stones may be secured to each other as shown in figure 5.35 (*D, E*) in order to maintain alignment and to give mutual resistance against uplift due to heavy local gusts of wind.

Metal copings Copper, lead or zinc may be used to form an impervious covering to the wall head. This method is common on the Continent where zinc is widely used. Typical details are shown in figure 5.26. Where secured direct to the wall the metal should be laid on an underlay of building paper or thin bitumenised felt to permit free thermal movement and to prevent damage from any roughness of the wall head. Lead may be lead-burned at the cross-joints with provision for thermal movement in the form of a double welt at every 9 m at least; copper should be double welted at the cross-joints with a special expansion joint at the same intervals as for lead; zinc may be laid in continuous lengths up to 4·5 m to 6 m with welted joints at these intervals.[1]

Raking copings

Raking copings to gable[2] parapets need not be weathered as water quickly discharges down the slope. Some support is required at the base of stone copings to prevent them sliding off the wall head and this is provided by a *springer* or *footstone* which may be a block shaped to tail well into the wall and on which the mitre of the coping is worked as in figure 5.27 *A, B* or simply a mitred portion of the coping which is slate dowelled to the wall below. Long raking copings exceeding 3 m in length are provided with intermediate support by bonding stones called *kneelers* (figure 5.27 *A*).

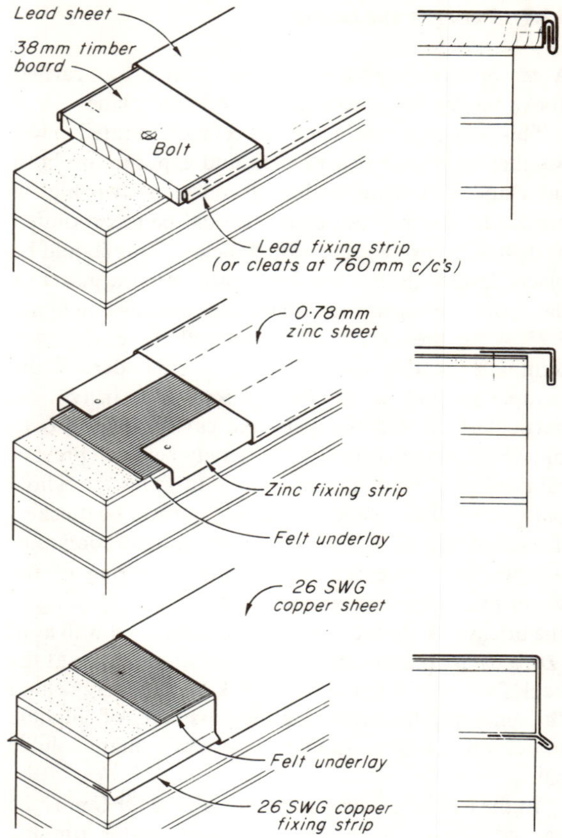

5.26 *Metal copings*

The stones are butt-jointed as for horizontal copings for slopes of 40 degrees and over. Below this rebated joints are often used to provide a barrier to water penetration.

Whether or not a gable terminates in a parapet some form of gable springer is often required to form a stop to a projecting closed eaves. In the case of a parapet the brickwork may be corbelled out as in (*B*). When there is no parapet a precast concrete or stone springer may be used as in (*C*).

[1] For details of joints in sheet metalwork see *MBS: Components* under *Roofings*.

[2] For definition of *gable* see page 177.

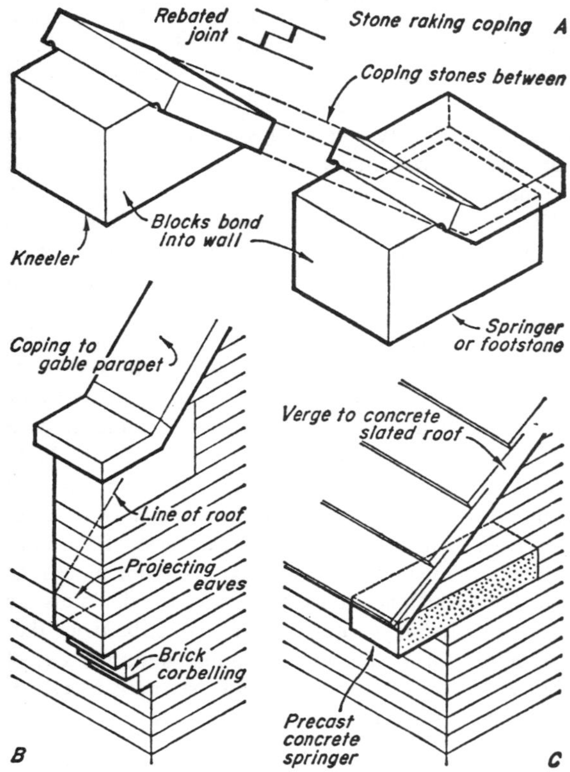

5.27 *Raking copings and springers*

5.2.8 Mortars, jointing and pointing

Mortars Cement mortars, although they set and harden quickly, are very strong and dense and offer no flexibility in a building which may be subject to stresses induced by settlement or by thermal and moisture movements. The effects of such movements tend to be concentrated in wide cracks which may pass through both bricks and mortar. Cement mortars should be used only with the strongest bricks, where imposed loads are heavy and a very strong structure is required.

Where loads are not unduly high and bricks of high compressive strength are unnecessary, lime or plasticiser should be introduced into the mix. This has a number of advantages. A mortar of adequate workability is obtained without an excessive proportion of cement, there is less tendency for the mortar to shrink away from the bricks and it retains water better than a cement mortar, thus preventing the mixing water being drawn out of the mortar prematurely by absorbent bricks. Cement-lime and cement-plasticiser mortars, being weaker than the bricks, also permit movement stresses to be relieved by the setting up of fine cracks throughout the joints rather than by the formation of a few wide cracks passing through bricks and mortar. This, together with the fact that these mortars are more absorbent than a cement mortar, results in brickwork of greater weather resistance (see also pages 73–74).

It has been shown that the strength of brickwork does not increase in direct proportion to the strength of the mortar used and that variations in the strength of mortar have only a relatively slight effect upon the strength of the brickwork.[1] Thus for most strength requirements it is possible to use a mortar other than a cement mortar and so gain the advantages described above.

Reference should be made to chapter 15 of *MBS: Materials*, which deals with mortars and where tables are given relating various mortars to types of masonry units and their strength and moisture movement characteristics and to conditions of exposure.

Jointing and pointing The face edges of the joints in brickwork may be finished in various ways in order to compress and smooth the exposed surface. If carried out as the work proceeds it is termed *jointing*, if after the brickwork is complete, *pointing*. Jointing is most commonly used at the present time and is to be preferred as it leaves the bedding mortar undisturbed. Different forms of joint are shown in figure 5.28.

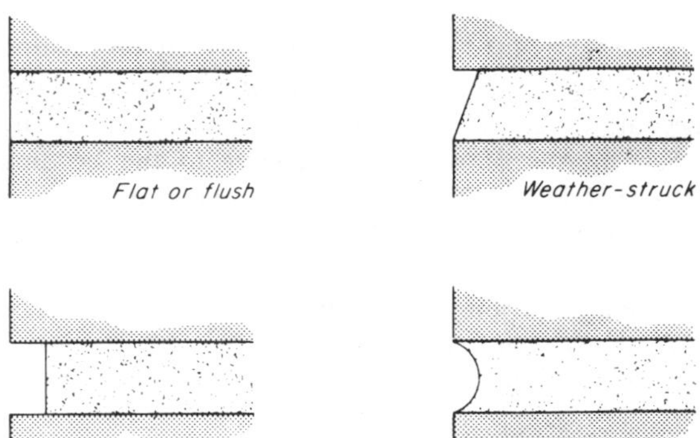

5.28 *Joints for brickwork*

The *flush joint* is usually formed by striking off the surplus mortar with the edge of the trowel. It may be rubbed in one direction to a smoother finish if required

[1] See Part 2, 4.1.1

when the mortar has sufficiently stiffened after the laying of a few bricks.

The following joints are tooled with a trowel or other appropriately shaped jointer which compacts the mortar and makes them more resistant to rain penetration than the flush joint.

The *weather struck joint* is formed by compressing the joint with the tip of the trowel as it is drawn in one direction at a slight angle along joint. The perpend joints are either bevelled in one direction or shaped to a 'vee' section.

The *keyed* or *bucket handle* joint has a curved face edge and the perpends are formed to the same shape.

The *recessed joint* has a recessed flat face edge. The perpends also may be recessed or they may be finished flush to emphasise the horizontal joints.

The type of joint finish should be carefully selected relative to the exposure of the wall. They may be placed in descending order of resistance to rain penetration as follows: 1 weather struck and keyed; 2 flush; 3 recessed. The weather struck and keyed joints are suitable for all exposure conditions and should always be used for cavity walls filled with insulation in severe exposure conditions. The recessed joint should not be used where there is danger of strong wind-driven rain and only with good, hard durable bricks, because of the water-retaining ledge which is formed. Where cavity fill insulation is incorporated this joint should be used only in sheltered conditions; where the fill is urea-formaldehyde foam BS 5618 limits both the degree of exposure and the height of wall.

Pointing is used at present more widely on existing rather than new brickwork when the joints have become defective. On new work it is normally used only when mortar of a different colour to the bedding mortar is desired. The mortar is raked out of the joints to a depth of 13 mm to 19 mm to provide a key for the pointing mortar, the brickwork is brushed clean and wetted and the joint recesses filled with the mortar. The pointing is then finished in any of the ways described above.

5.3 BLOCKWORK

Building blocks and blockwork generally

The term *building block* normally refers to a walling unit larger in size than that of a brick[1] but small enough to be handled by the blocklayer with one or both hands, although some of the larger size solid blocks require two men to handle.

Compared with brickwork the use of blocks results in great economy in construction time.[2] Tests have shown that blockwork walling can be built in about half the time required to build comparable walling in brickwork. The reason for this is the larger size of the unit which necessitates less activity in laying a given volume of wall since, for example, one 450 mm × 225 mm block represents six standard bricks each of which has to be laid individually. Linked with this economy in labour is an appreciable saving in bedding mortar because of the reduction in joints.

A further advantage arises from the different behaviour under load of walls built of units which are high relative to their thickness, which results in block walls having a greater strength relative to the strength of the blocks of which they are constructed than brick walls to the strength of the bricks of which they are built. This relationship, called the *strength ratio*, varies with the ratio of the height to thickness of the individual walling unit. Thus, for example, using blocks the height of which is not less than twice the thickness the design stress may, within certain limits, be twice that permitted for a wall built of similar units the size of bricks, so that for the same wall strength blocks of lower compressive strength may be used.[3]

5.3.1 Concrete blockwork

The techniques adopted for building concrete blockwork in respect of strength and stability, durability and weather resistance are similar to those used for brickwork but with some important differences, especially in relation to the prevention of cracking in the walls. As with brickwork general design considerations are laid down in BS 5628: Part 1: 1978.

Blocks are made from dense and lightweight aggregate concretes and from aerated concrete and may be solid, hollow or cellular in form. The range of sizes is considerable, the smallest standard block being 390 mm × 90 mm, which with a 10 mm joint gives a designated size of 400 mm × 100 mm. Thicknesses range from 60 mm to 215 mm.[4] In addition to the full

[1] For the definition of *brick* and *block* see *MBS: Materials*, chapter 6.

[2] National Building Study Technical Paper No. 1. 'A Work Study in Bricklaying' HMSO 1968.

[3] See BS 5628: Part 1: 1978.

[4] See *MBS: Materials* chapter 6, for information on concrete blocks generally.

standard units, some of which are illustrated in *MBS: Materials* chapter 6, half length blocks, quoin blocks and cavity closers are available.

Bonding of blockwork

The principles of bonding are the same as for brickwork but because of the range of thicknesses available the blocks are only bonded longitudinally, no cross bonding being required. Stretching bond is, therefore, normal although this may be varied for facing work as described later.

Half-lap bond is normal but where necessary to permit bonding at returns and intersecting walls this may be reduced to one-quarter of the block length. In short lengths of thin partition work the lap at such junctions may be reduced to not less than 65 mm.

When the block thickness is less than half the length of the block the use of quoin blocks avoids the presence of narrow edge faces on the extreme corner of the wall and the need to cut blocks (figure 5.32). The same advantages are obtained by the use of full and half-length cavity closers at openings in a cavity wall where the cavity is to be closed by returning one leaf upon the other (figure 5.32).

Except at quoins loadbearing concrete block walls should preferably not be bonded at junctions and pier positions as in brick and stone masonry. At tee-junctions and intersections one wall should butt against the face of the other to form a vertical joint which provides for movement in the walls at these points and thus controls cracking in the walls (see page 102). Where lateral support must be provided by an intersecting wall, and in the case of piers and buttresses, the walls and piers should be tied together by 6 mm × 32 mm wide metal ties with split ends, spaced vertically at intervals of about 1200 mm. When hollow blocks are being used the cores at these points may usefully be filled with mortar or concrete, especially in the case of loadbearing piers, and the ends of the ties be bent and embedded in the filling (figure 5.29 *A*). Non-leadbearing walls, for similar reasons, are tied together at intersections by strips of expanded metal or galvanised mesh bedded in alternate courses (*B*).

Apart from piers, the bearing area of a hollow block wall may be increased at points of concentrated load by filling the cores at such points with concrete for the full height of the wall and, if required for reasons of strength and stability, vertical reinforcing bars can be placed in the filling (*C*). Reinforced columns thus formed and reinforced piers can be linked with longitudinal reinforcement (*D*) or with horizontal bond beams (see below) to provide a frame system within the block walling.

The strength of concrete blocks either of dense or lightweight aggregate is sufficient for normal small-scale work but where loading is heavy only dense concrete blocks are suitable. Hollow blocks of dense concrete may be used for loadbearing walls but the courses directly supporting floor and roof structures should be built of solid construction in order to distribute the loading over the length of the wall and thus avoid the concentration of stresses. This may be accomplished by the use of solid blocks for such courses, by filling the cores of the hollow blocks in these courses with concrete, the wet concrete being supported by strips of expanded metal laid in the bed joint of the course, or by using lintel or bond beam blocks filled with concrete (figure 5.30). Apart from this particular use, bond beam blocks may be introduced at each storey height to form continuous reinforced concrete beams tying together the walls of a building when this is desirable for structural reasons, or at lintel and cill

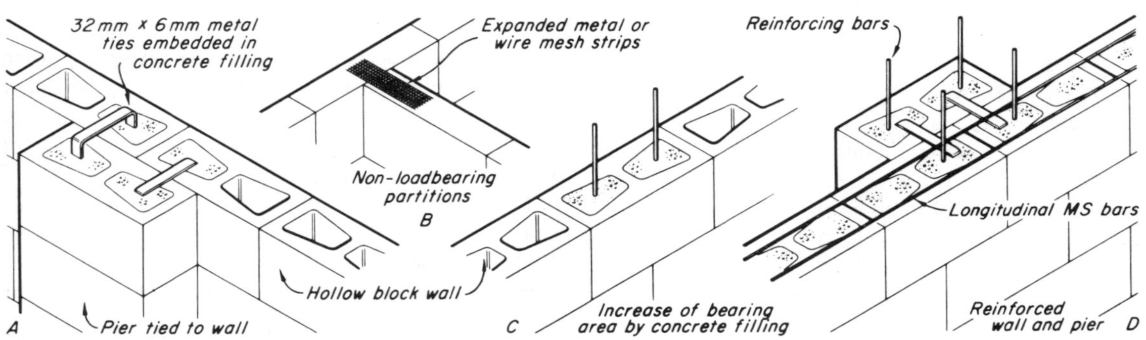

5.29 *Blockwork and reinforcement*

levels to provide crack control (see below). Bond beam blocks, without concrete filling, may be introduced to accommodate horizontal service runs linking with the vertical hollow cores in the block walling.

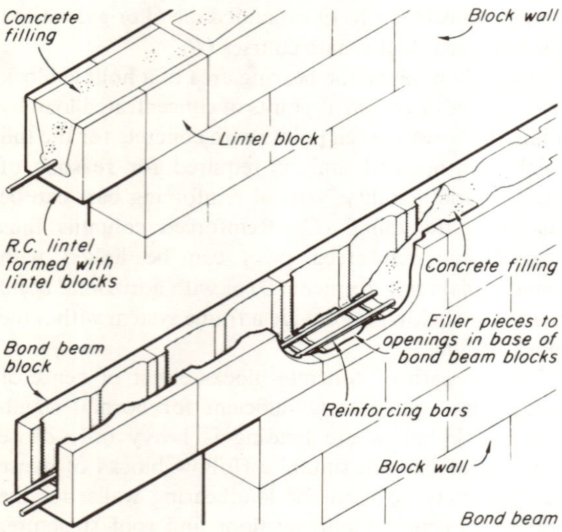

5.30 *Lintels and bond beams*

Crack control

Precautions against cracking in concrete blockwork must be taken because of the considerable moisture and shrinkage movement which occurs in concrete and in units made of concrete (see *MBS: Materials*). Except for very high loading the mortar for concrete blockwork should not be stronger than 1:1:6 cement, lime and sand mix, or its equivalent of cement and sand with a plasticiser. This ensures that shrinkage stresses are distributed through the joints and cracking in the blockwork is avoided. In addition to this other provisions should be made at the design stage. The methods adopted are based on restricting the length of wall in which movement will take place and on avoiding as far as possible narrow sections of wall over and under openings where stresses would concentrate and cause cracking.

In addition to control joints at piers and junctions referred to on page 101 long lengths of walling must be subdivided by control joints so that the length of each wall does not exceed one-and-a-half times to twice its height. Short lengths of wall are preferable although experience indicates that proportion is more critical than size.

Concentrations of stress above and below openings may be distributed through the wall, and cracking be thus controlled, by steel reinforcement either in the form of masonry reinforcement bedded in the courses immediately above and below the opening or by bond beams in the same positions. Narrow sections of wall above and below openings can be avoided by forming windows and doors in storey height infill panels thus limiting the wall to solid rectangular panels on each side.[1]

Weather resistance

Although a single leaf 190 mm to 215 mm block wall will often be sufficient for strength purposes, as with 215 mm brickwork this will not be watertight under all conditions of exposure and for severe conditions external rendering or other protection is essential (see Table 5.1). For walls with no more than a moderate degree of exposure unrendered 190 mm hollow blocks may be used bedded on 50 mm strips of mortar applied to the face edges only as shown in figure 5.31. This is known

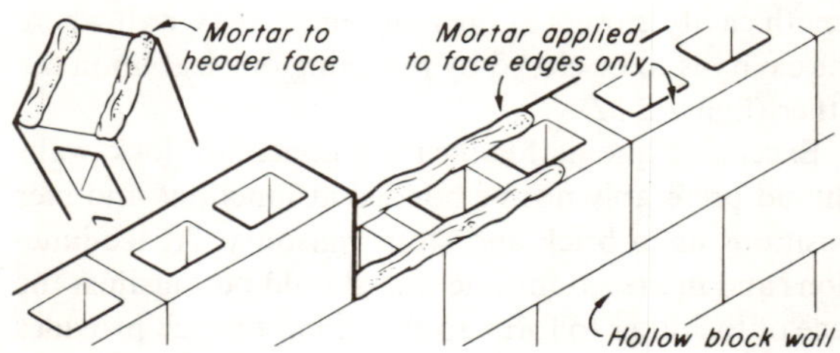

5.31 *Shell bedding of concrete blocks*

as 'shell' bedding which breaks the continuity of the mortar bed and thus reduces the danger of capillary passage of moisture across the wall thickness. This technique does, however, reduce the bearing strength of the wall to about three-quarters of that with normal bedding and, as stated above, is suitable only for situations where the risk of water penetration is not great. The only certain means of preventing water penetration under all conditions of exposure when external protection is not being applied is the cavity wall and for reasons of thermal insulation the inner leaf would be built in lightweight blocks, whatever type of block was used for the external leaf.

Methods of construction and detailing are generally similar to those for cavity walls in brickwork (figure 5.32). The use of cavity closer blocks at openings as mentioned earlier avoids the cutting of blocks and the

[1] See Part 2 chapter 4 under *Movement control* for further details of methods of crack control in block walls.

presence of narrow face edges at these points and simplifies work. In situations of severe exposure walls, the outer leaf of which is constructed of porous blocks, should be built with non-ferrous wall ties.

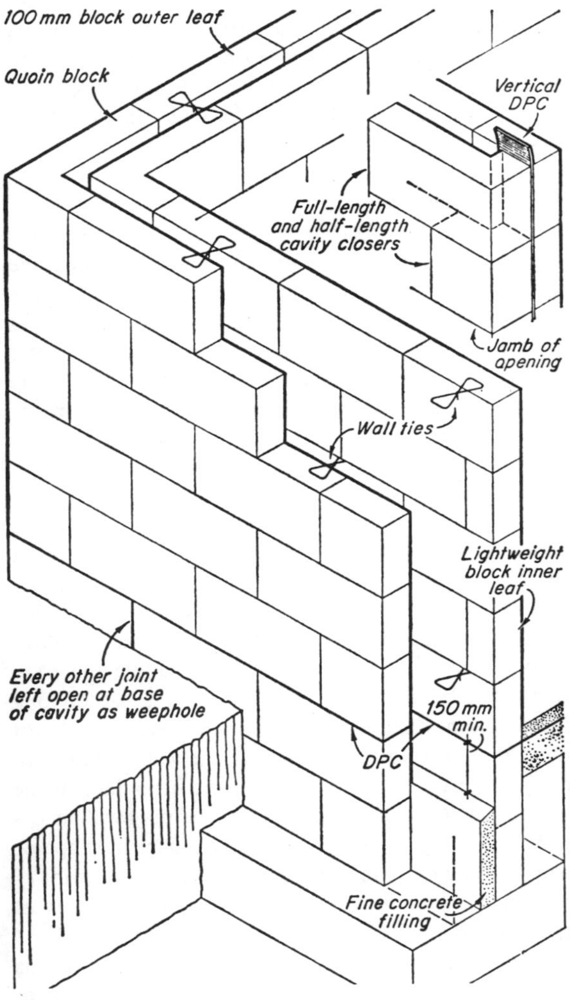

5.32 *Cavity wall construction in blockwork*

Thermal insulation

The methods of thermal insulation already described for brick solid and cavity walls may be applied to block walls. However, the possibility of using lightweight insulating blocks throughout the construction means that in achieving comparable standards of insulation the thickness of the applied insulating material may often be reduced or it may be omitted altogether.

Openings in walls

The remarks on page 85 relative to openings in brick walls apply equally to those in block walls. For economic building the widths of opening and the lengths of wall between openings and quoins should be such as to make use of full and half-length blocks in order to avoid the labour and wastage in cutting blocks.

Heads of openings in blockwork are normally formed with a lintel, usually of reinforced concrete, although steel lintels of various types may be used. Reference should be made to the section on *Brickwork*, where these are illustrated (figures 5.15 and 5.21). The use of lintel and bond beam blocks in blockwork (figure 5.30) permits in situ cast concrete lintels to be formed without the use of formwork, enables the bonding pattern to be maintained over the openings and avoids the necessity of handling large, heavy precast lintels. The openings in the base of a bond beam block are either filled with pieces of the inner cores which are knocked out, or expanded metal strips are placed below, in order to support the concrete while it is wet.

Cills and thresholds may be formed in any of the ways described under *Brickwork*.

Mortars, jointing and pointing

The requirements for mortar for blockwork are generally the same as for brickwork (see page 99). As with brickwork variations in the strength of the mortar have only a slight effect upon the strength of the blockwork. For reasons already given a mix not stronger than 1:1:6 or its equivalent is desirable except when high strength blocks sustaining heavy loads are used. In these circumstances a stronger mix is necessary in order to attain blockwork of a high bearing capacity. Reference should be made to chapter 15 of *MBS: Materials* which deals with mortars.

The methods of jointing and pointing suitable for brickwork are applicable to blockwork. Reference should be made to the section on brickwork in regard to the protection of walls against damp penetration by means of damp-proof courses and copings.

Facing work

Many types of concrete facing blocks are now available varying in texture, colour, shape and size. Colour is varied by variation of aggregates or by use of coloured cements or pigments; texture is varied by different mould patterns including bold profiled patterns producing geometrical shapes, by exposing the aggregate on the face or by splitting blocks so that the split faces have a texture resembling that of quarried stone.

Bonds other than the stretching bond may be used for facing work by using blocks of varying face sizes and include those used for squared rubble stone walling (see page 106)[1].

Mortar for facing work is the same as for common blockwork but colour may be incorporated. Jointing or pointing is carried out generally as for brickwork and where reinforcement is used in the joints, jointing is preferable to pointing.

For the provision of services and fixing of fittings to blockwork and the use of concrete blocks in partitions see section on *Partitions* page 118.

5.4 STONEWORK

Natural stone is durable but expensive and it is, therefore, used today mainly as a facing material, predominantly as a relatively thin veneer fixed to a solid background of other material.[2] In addition to the high cost of worked stone the cost of site labour in the erection of solid walling is considerable, especially in rubble work, so that the structural use of stone tends to be limited mainly to the outer leaf of cavity walling in areas where it can be supplied economically for this purpose from local quarries (see page 106). The necessary thickness of traditional stone walling results in relatively massive foundations and in greater extent of roofing to cover any given floor area, both of which constitute additional items of expenditure compared with construction in brick and blockwork. Plastering costs are high with rubble walling because of the uneven surface of the stonework and half-brick internal linings to solid rubble work have been used to produce a smoother face and reduce costs. However, tests have shown that even the traditional 406 mm stone wall will permit some rain penetration except in very sheltered positions so that solid wall construction is now generally avoided, for this reason as well as for reasons of economy in construction costs.

In addition to considering cavity walling a short review of the different types of solid stone walling is given here as some are still used occasionally in small-scale buildings and feature in maintenance and restoration work. The types of stones used for building and their characteristics are described in *MBS: Materials*.

Stone walling

Stones vary in their ease of working and, therefore, in economy of labour in cutting to shape, according to their hardness and the thickness of the beds.[3] For this reason stone walls are built in a variety of ways to take account of this, some of which affect not only the labour required in shaping the stones but also the labour in building the wall.

The two broad types of stone walling are (i) rubble walling (ii) ashlar walling. The first uses stones either as they come from the quarry or only roughly dressed to shape. The second uses stones very carefully dressed to plane faces, called ashlars, and laid with fine joints.

5.4.1 Rubble walls

These are built as *random rubble walling* using the stones of random size and shape as they come from the quarry, or as *squared rubble walling* using the stones after they have been roughly squared. In the latter laminated varieties of stone are used which split easily and require a minimum of labour to form reasonably straight faces.

Random rubble

In this walling as in all masonry longitudinal bond is achieved by overlapping stones in adjacent courses but the amount of lap varies because the stones vary in size. Spaces between stones are filled with small pieces called spalls. Since rubble walls are virtually built as two skins with the irregular space filled with rubble material, transverse bond or tie is ensured by the use of long header stones known as *bonders*. These extend not more than three-quarters through the wall thickness to avoid the passage of moisture to the inner face of the wall and at least one to each square metre of wall face is provided. Large stones, reasonably square in shape or roughly squared, are used for quoins and the jambs of openings to obtain increased strength and stability at these points.

Random rubble may be built as uncoursed walling as shown in figure 5.33 *A* in which no attempt is made to line the stones into horizontal courses or it may be *brought to courses* as in (*B*) in which the stones are roughly levelled at 300 mm to 450 mm intervals to form

[1] For further information on types of blocks and bonding patterns see *Design in Blockwork* by Michael Gage and Tom Kirkbride, Architectural Press.

[2] See *Stone Facings* Part 2.

[3] See *MBS: Materials* for reference to bedding planes of sedimentary stones and the need to take account of these in placing this type of stone in a wall.

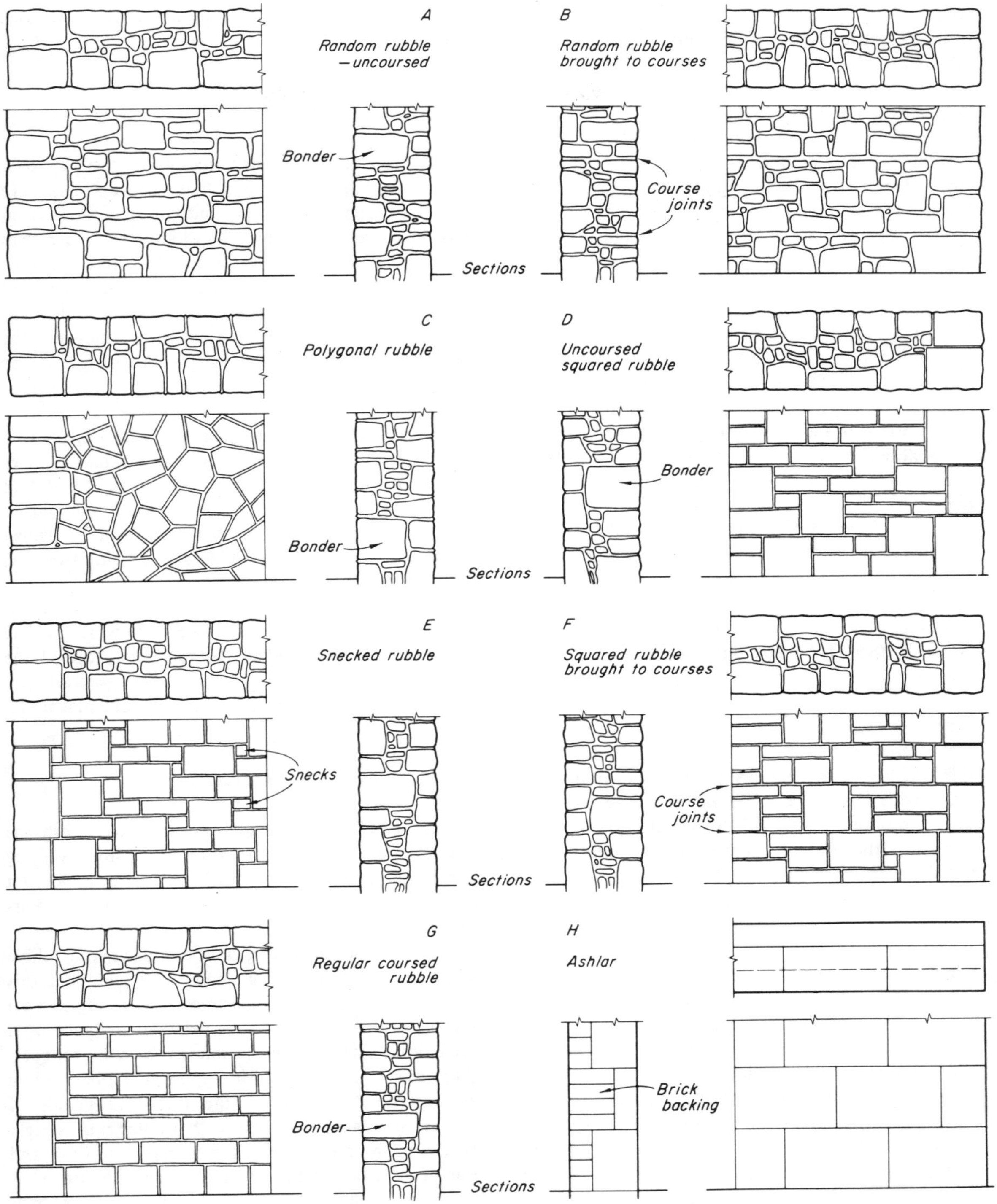

5.33 *Types of stone walling*

courses varying in depth with the quoin and jamb stones. Because the technique results in more care being taken in bedding and flushing with mortar and permits straight joints to be more easily avoided the walling is stronger than uncoursed work.

Variations of random rubble walling exist which are peculiar to certain areas. *Lakeland masonry* is constructed of slate with the stones laid to slope down towards the external face and only partially bedded in mortar near each face. *Flint walling* is constructed of flints with the wall face showing either (a) round, undressed flint stones, (b) snapped flints, which are stones broken transversely and laid to show the split face or (c) knapped flints, which are snapped stones with the split surface dressed to a square face. *Polygonal rubble* is adopted for hard, unstratified stones which are quarried in polygonal shapes. A minimum of dressing is applied to the face so that the stones fit together reasonably well as shown in (*C*), which shows a typical regional example from SE England known as *Kentish rag*.

The Building Regulations require solid walls of uncoursed stone or flints to have a minimum wall thickness of one and one-third times the thickness required for a similar wall constructed of bricks or blocks in order to attain adequate stability in this type of work. (See Table 5.1 with reference to weather resistance.)

Squared rubble

This is used in districts where stratified stone is available which may be split easily into appropriate thicknesses with straight bed faces and which requires little labour to obtain square stones. Rough squaring of the stones has the effect of increasing the stability of the wall and improving its weather resistance, since the stones bed together more closely, the joints are thinner and there is, therefore, less shrinkage in the joint mortar.

This walling may be built in four ways: (i) as *uncoursed squared rubble*, with stones of various depths laid in various face arrangements with no attempt to form courses as in (*D*); (ii) as *snecked rubble* in which long vertical joints are avoided by the incorporation of small stones called *snecks* which permit stones to overlap and thus break joint as in (*E*); (iii) as *squared rubble brought to courses*, with stones as used for (i) but brought up to level beds to form courses of varying depth as in (*F*); (iv) as *regular coursed rubble*, in which all stones in one course are the same depth, usually varying from 100 mm to 300 mm as in (*G*). Quoins and surrounds to openings in all these are formed in dressed stonework.

5.4.2 Ashlar walls

Ashlar is the name given to stones, usually over 300 mm and up to 450 mm in depth, dressed or sawn to blocks of given dimensions and carefully worked on face and beds to produce fine joints not more than 3 mm thick (*H*). To ensure sufficient stiffness against fracture in case of unequal settlement stones are limited in length and breadth relative to their depth: for soft stones the length not more than three times and the breadth not more than twice the depth; for harder stones not more than five times and three times respectively.

To economise in stone ashlar is usually a facing to a brick backing and the depths of the ashlar courses correspond with multiples of brick courses. The back of each stone in contact with the brickwork should be painted with bituminous paint as a precaution against damage to the stone by soluble salts passing from the backing into the facing.

Stonework may be face-finished in many ways which vary according to the tools used for the work and the amount of labour applied to the quarry face of the stones. These finishes and ashlar work generally are described and illustrated in works on stone masonry.

5.4.3 Cavity wall construction

It has already been pointed out that even with the considerable thicknesses of traditional stone walling some rain penetration is likely to occur in all but sheltered positions so that cavity construction is advisable (see table 5.1).

A cavity and inner skin can be applied to a rubble wall with a minimum thickness of 406 mm but this necessitates even larger foundations and results in even greater overall area to be roofed than the solid wall used alone. The economic problem here is to produce a thin outer skin of stone. Most types of rubble walling are unstable if less than 305 mm thick and the cost of preparing stones suitable for a thin outer skin is high. However, in certain areas, for example the Peak and Purbeck districts, where walling-stone beds are relatively thin and level so that stones require a minimum of shaping and preparation, it has been possible to develop outer skins from 150 mm to 230 mm thick making possible cavity walls with a reasonable overall thickness (figure 5.34). In other areas mechanical methods would be necessary to produce the narrow bed stones required. In fact, for many years limestone in the Bath area has been mechanically won and worked to produce blocks of fairly regular size with a 100 mm bed for use in the construction of cavity walls 255 mm to 280 mm in thickness. These mechanically produced blocks do, however, result in walling with an ashlar rather than rubble appearance.[1]

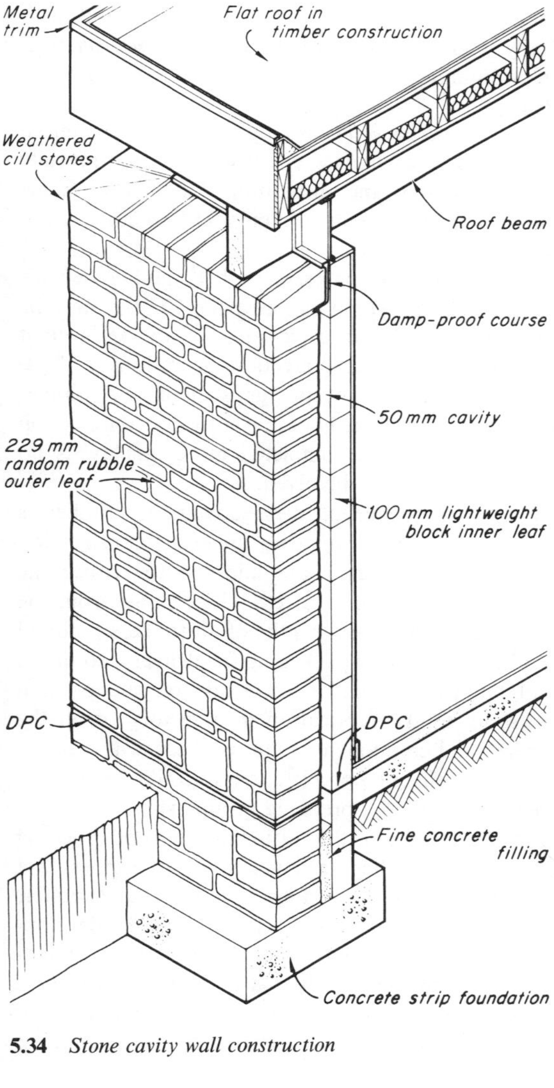

5.34 *Stone cavity wall construction*

Thermal insulation

The methods of thermal insulation described for brick walls may be applied to stone walls. When built at the present time stone walls are usually desired for their external face appearance so that external insulation is not suitable. This would also apply to existing stone walls retained for the same reason. The inner face of rough rubble walls would probably require a render coat before the application of internal insulation, especially that fixed adhesive.

Joints and connections

Adjacent stones in a course, or stones in adjacent courses, may be connected by means of *joggles* and *cramps* to prevent relative movement between the stones.

Forms of joggle joints are shown in figure 5.35. A joggle as at (*A*) prevents vertical displacement of one stone relative to the other. Joggles in bed joints prevent lateral displacement of stones in walls subjected to side pressure. These may be cut in the actual stones or formed as slate joggles (*B*) which is cheaper. Cement joggles (*C*), formed by filling with mortar grout the Y-shaped cavity resulting from the V-shaped sinkings in adjacent stones, are so shaped to prevent both vertical and lateral movement.

Cramps, also shown in figure 5.35, are used to prevent adjacent stones coming apart, particularly coping stones. These may be of metal which should be a non-corrosive metal to avoid spalling of the stone due to rusting of the metal (*D*), or of slate cut to a dovetail form (*E*).

Mortars

Mortars for stonework should be plastic enough to permit the stones to bed down evenly, have a permeability similar to that of the stone and have as low a content of soluble alkali as possible. Mixes fulfilling these requirements consist of finely crushed stone, lime and a very small proportion of Portland cement. Many stone suppliers recommend, or supply ready-mixed, particular mortar mixes for their stones.

[1] For further information on stone masonry generally reference should be made to BS 5390: 1976 *Code of Practice for Stone masonry*.

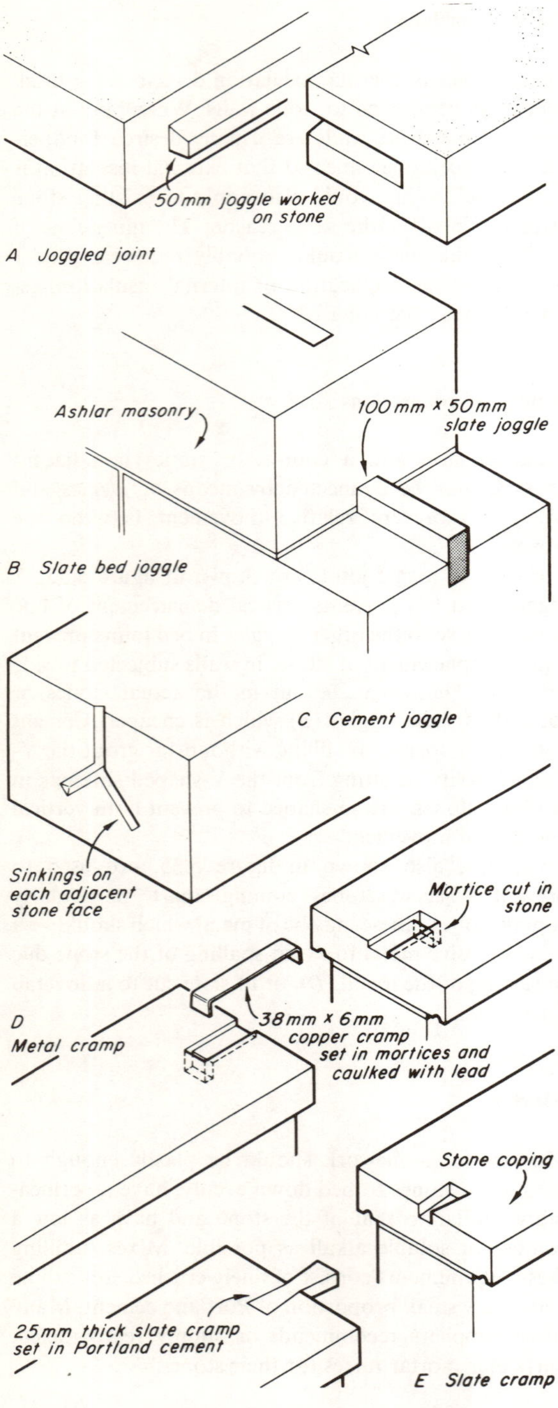

5.35 *Stone joints and connections*

5.5 TIMBER FRAME WALLS

In chapter 1 the distinction is made between a loadbearing wall structure and a framed structure, the latter being defined as one in which all loads are carried by the frame, the enclosing and dividing wall elements being non-loadbearing. Loadbearing walls for small scale buildings may be constructed by framing together relatively small timbers at close intervals so that a wall or partition panel forms a loadbearing system (figure 5.1). These are used in conjunction with small span timber floor and roof structures. While the whole of such a structure is in fact, a form of framed structure it is quite distinct from the type of framed structure defined above since there are no columns and beams acting as primary loadbearing members, and it is usually distinguished by being called *frame construction*, a building constructed in this way being a *frame building*.

The term *frame wall* is usually applied to an external timber wall involving in its construction appropriate external cladding and insulation. An internal frame wall is usually referred to as a partition, whether or not it is loadbearing.

As a structural material timber has favourable strength/weight/cost ratios (see page 127). It may be easily joined and fabricated and its low self-weight facilitates handling and erection operations and reduces the dead weight of the structure. A timber structure may be carried out in wholly dry construction and thus building is completed more quickly as no 'drying-out' period is involved.

As a consequence of these advantages timber is widely used in other countries for house construction where various methods are adopted to reduce the amount of site work involved. These range from the pre-cutting to specified lengths of all timbers required for a house and their packaging for delivery, so that a small builder has only to fabricate on site or in a workshop, to the fabrication in a factory of all the elements and components required for a whole house and their delivery complete on a trailer ready for speedy erection on a prepared ground slab.

In Great Britain the rationalisation of building methods has resulted in a wider use of timber as a structural material for houses, schools and similar small-scale buildings, and particularly in the development of building systems appropriate for such buildings, using timber as the basic material. In these building types the loading is light and the critical factor in the design of the structure is stiffness rather than strength. Timber is very stiff in relation to its weight.[1] It is, therefore, a suitable material and since the E value (see page 36) of both high and low grade timber is fairly constant it is structurally feasible, as well as economically desirable, to use the lower grades of timber.

Although timber is combustible its fire resistance is high in timbers of not less than 100 mm to 150 mm section. Smaller sections may be protected by non-combustible materials such as plasterboard to enable them to withstand the effect of fire for ½ to 1 hour or more (see Part 2 chapter 10). Current Building Regulations limit the use of structural timber by requiring non-combustibility in certain circumstances for a number of structural elements, but apart from these it is possible to use timber construction in many types of buildings up to 15·0 m high within the limitations laid down regarding the proximity of external walls to the nearest site boundary.

5.5.1 Frame wall construction

Basically a frame wall consists of vertical timber members called *studs*, framed between horizontal members of the same section at top and bottom, the top member being called a *top* or *head plate* and the bottom a *sole plate* or *piece*. The joints are simple butt and nailed joints and the frame is, therefore, non-rigid and requires bracing in order to provide adequate stiffness. Diagonal braces can be used for this purpose but the usual method, which is quicker and cheaper, is to use board or plywood external sheathing to stiffen the structure as well as to serve other purposes mentioned later. The studs are commonly spaced at 400 mm centres which is a normal spacing for timber floor joists and is related to the standard 1200 mm width of many types of sheet linings.

Four systems of frame construction may be used, the application of one or the other to a particular project depending on a number of factors which are referred to later. Two of these have been developed from the two traditional forms known as *platform frame* and *balloon frame* and all, in different ways, take into account the physical behaviour of the timber, the behaviour of the whole structure under load and the implications of the erection or assembly operations. In relation to these factors each, in certain circumstances, has advantages over the others.

The four types of frames are shown in diagram form in figure 5.36. In the *platform frame* the walls and partitions bear on the 'platforms' formed by the floor structures and the frames are single storey in height. In the *balloon frame* the wall studs and the ground floor joists[2] (if a suspended floor is used) bear on a common sole plate and the studs are continuous through two floors, the first floor joists being fixed individually to the studs. In the *modified frame* the wall frames are one storey high, the upper frame being erected directly on the lower. The ground and upper floor joists are fixed directly to the studs as in a balloon frame. The wall frames in the *independent frame* are constructed basically as in the modified frame but the upper floor structure is supported by a continuous bearer fixed to the inner faces of the studs. The floor, therefore, does not penetrate the thickness of the wall frame.

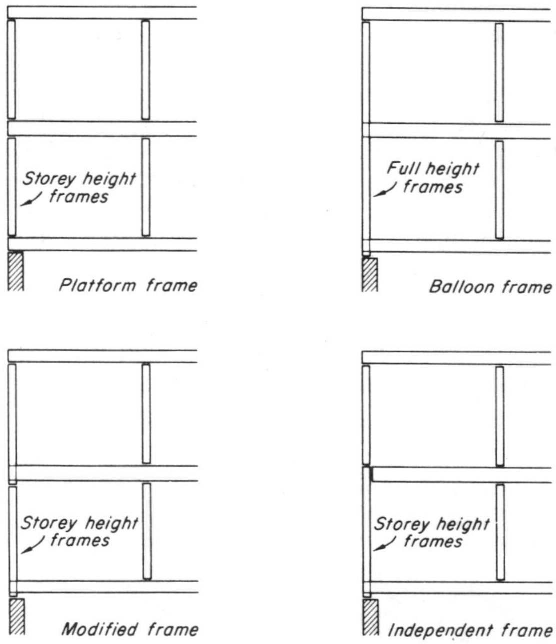

5.36 *Types of timber frame walls*

The advantages and disadvantages of these alternative methods of framing are discussed below.

In all types of frame the timber structure must be raised out of contact with ground moisture. This is accomplished by erecting it on a base wall or foundation beam rising to damp-proof course level or on the edge of a concrete raft floor, depending upon the type

[1] See also page 127 and Part 2 chapter 9.

[2] ie, the timber floor bearers. See chapter 8 for details of timber floor construction.

of foundation or floor structure employed (figure 5.37). As a base for the whole structure a *wall* or *cill plate* is set and carefully levelled on the damp-proof course and this must be securely anchored to the foundation by 13 mm diameter holding-down bolts at not more than 2·40 m centres, with at least two bolts to each wall or panel length. These should be built-in at least 380 mm in masonry walls or 150 mm in poured concrete. To maintain the effectiveness of the damp-proof course it must be sealed carefully at all bolt positions. The cill plate may be 100 mm by 50 mm when fixed to a concrete base but should be increased in width to 150 mm on a brick base wall so that the bolts may be set half a brick in from the face of the wall.

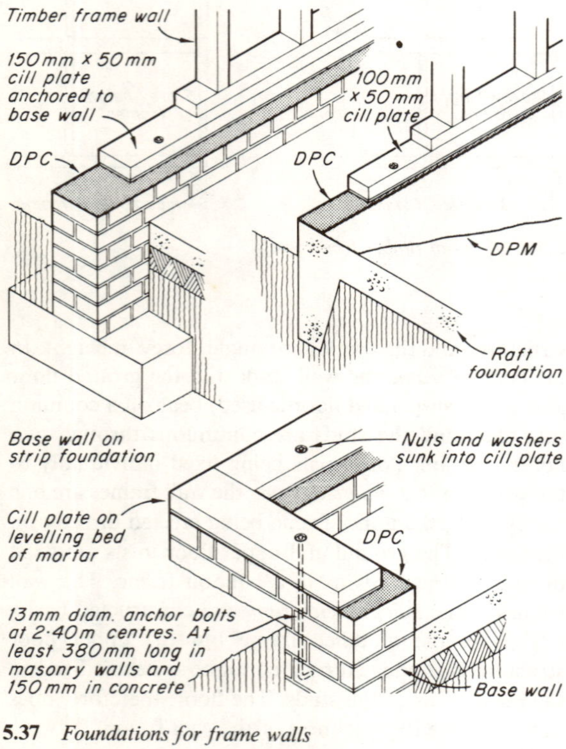

5.37 *Foundations for frame walls*

Platform frame (figure 5.38) The wall frames consist of panels of convenient size for handling and transport formed of 100 mm × 50 mm studs at 400 or 600 mm centres framed between a bottom sole plate and double head plates, the studs being end-nailed to the plates assuming that the usual method of fabricating horizontally off-site or on the floor platform is used. When erected the wall frames are fixed to the floor platforms by nailing through the sole plates. When a length of wall in any of the frame types is made up of a number of shorter panels, as is common when they are prefabricated off-site, the upper head plate is called a *head binder* and is nailed to the lower head plate on site to form a continuous member linking together and aligning the separate panels.

The floor structure consists of joists with a *header* joist at each end to close the floor cavities and flooring which completes the platform on which the wall frames are erected. The ground floor platform, if it is of timber construction, is built on the cill plate and the upper floor platform on the ground floor wall frame, in each case the joists being toe-nailed to the plates and end-nailed to the headers. The end joists on the return sides are nailed to the plates on which they bear.

As it is preferable for the finished flooring to be laid after the building is closed in a sub-floor of 19 mm boards or plywood may be used on which the finished flooring is subsequently laid. This is carried to the outer edge of the floor structure and serves as a useful working platform as well as enhancing the thermal insulation of the ground floor[1] and the sound insulation of the upper floor. If no sub-floor is used a joist must be placed 25 mm inside the end joist on the return sides to provide a nailing face for the flooring (figure 5.38). If vertical dimensions are related to finished floor levels and no sub-floor is used a filler-plate, the same thickness as the finished flooring, is fixed round the edges and on this the wall frames are erected before the flooring is laid.

The advantages of the platform frame method are threefold: relatively short timbers are required for the studs; the storey height panels facilitate transport when prefabricated construction is employed; the studs and floor joists align, a helpful factor when a grid layout is adopted; the cross grain moisture movement of the floor structure is the same throughout the building since both external wall frames and internal partitions bear on the floor platforms. A disadvantage is that since floor construction must precede the erection of the walls the covering in at roof level cannot occur at such an early stage as with the other methods and as a result temporary protection from the weather may at times be required for the floors until the structure is roofed in.

Balloon frame (figure 5.38) The wall frames consist of panels of convenient size framed up from the same size

[1] For thermal insulation of suspended timber ground floors see page 195.

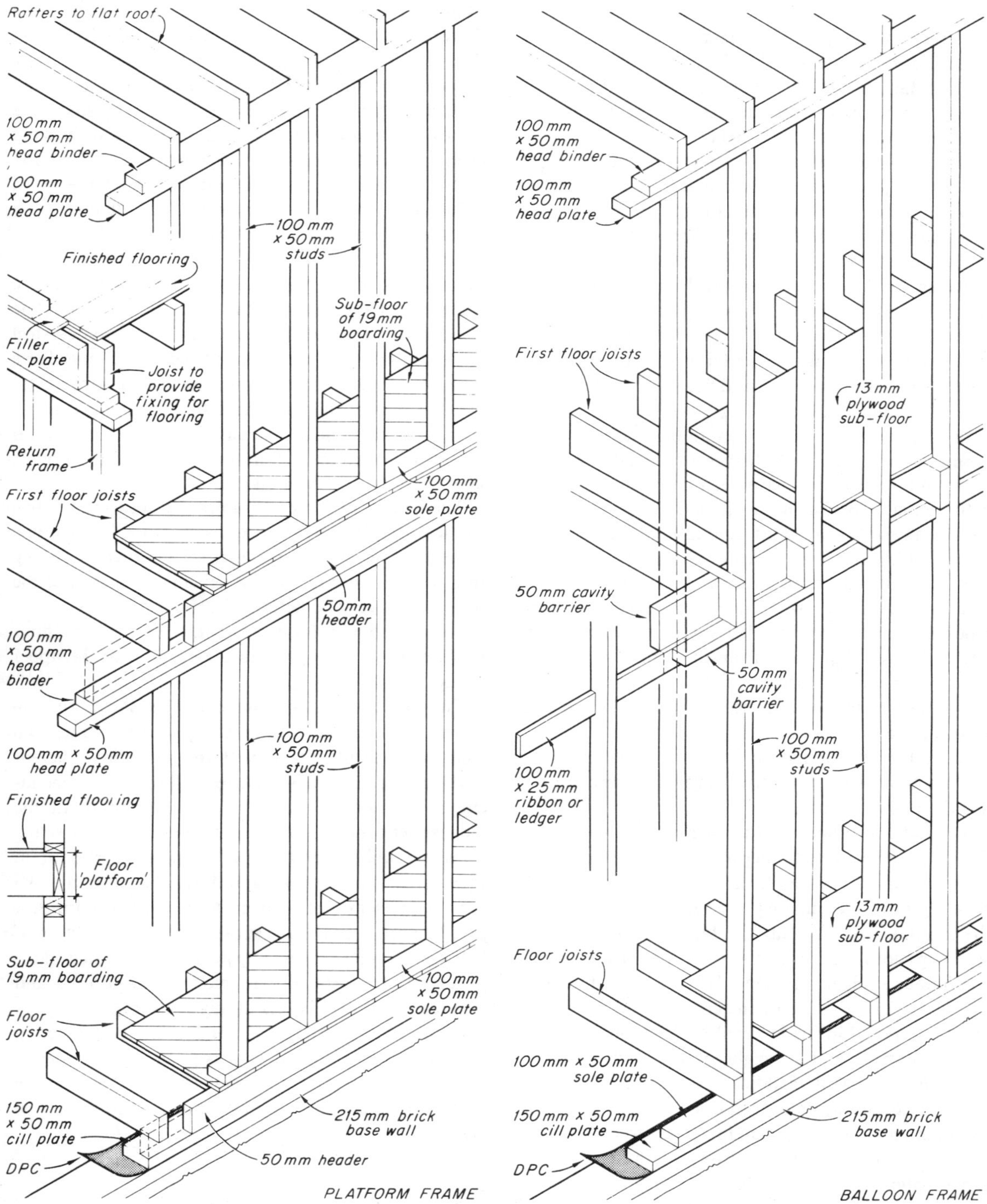

5.38 *Frame wall construction*

members and in basically the same way as for platform construction, but quite independent of the floor structures. The sole plate is nailed direct to the cill plate on the foundation wall or concrete raft floor and the studs extend from sole plate to double head plates at roof level unbroken by the floor structures.

Ground floor joists bear on the sole plate and the upper floor joists bear on a continuous member, called a *ribbon* or *ledger*, let into the studs. The ribbon serves as a bearing for the joists and provides lateral stiffening to the long studs, stiffening in the other direction being provided by the joists, all of which are face-nailed to the studs. Stiffening to studs in the return frames, parallel to the floor joists, is provided by fixing the studs to the outer joist or, preferably, to three or four joists by metal straps cut in and fixed to their top edges (see Part 2 figure 4.6 *E*).

In this type of frame continuous cavities from ground floor to roof exist between the studs when the inner and outer linings have been fixed, and these link with similar cavities in the floors. As a precaution against possible spread of smoke or flame through these cavities from the lower into the upper part of the frame and into the floor structure or from floor structures to the frame, cavity barriers must be built in to block the cavities between the studs a first floor level and to block off the floor from the wall cavities.[1] The cavity barriers are formed of 50 mm timbers cut-in between studs and joists as necessary. Details will vary according to the presence or otherwise of noggings required for fixing linings or skirtings, which may serve the function of cavity barriers provided they are not less than 38 mm thick.

The balloon frame has the advantage that the wall frames may be erected independent of the floor structure and up to roof level in one lift, which permits covering-in of the building at an early stage. Further, the continuity of the studs and face-nailing of joists to studs provides rigidity to the structure. However, the continuous studs necessitate relatively long timbers of small size. The absence of the cross grain of the floor joists interposed in the external walls minimises moisture movement in the height of the walls which is desirable when stiff claddings such as rendering or brick veneer are used; in such cases differential movement between frame and cladding needs to be kept to a minimum. At the same time cross grain movement still occurs internally since the floor is normally supported on storey height partitions and thus vertical movement is unequal across the whole structure. Joists and studs do not align as in the platform frame.

Modified frame (figure 5.39) As in the platform frame the wall frames are one storey high and are framed in the same way except that the lower frame has only a single head plate. The upper frames are erected on and fixed directly to the lower frames by nailing sole plate to head plate, and the ground and upper floor joists in each case bear on the sole plates and, as in the balloon frame, are face-nailed to the studs. As an alternative to nailing upper and lower panels may be bolted together before erection to permit assembly in one lift.

Independent frame (figure 5.39) The wall frames are framed up basically as in the modified frame but the upper floor structure is carried by a continuous steel angle or timber ledger which is fixed to the inner faces of the lower studs so that the floor does not pass into the thickness of the wall frame to gain support. However, some form of tie, usually a metal strap, between the floor structure and studs is required to provide lateral stability to the wall frames.

The heights of the wall frames are not fixed relative to the upper floor but the extension of the lower frame to the floor line as shown in figure 5.39 provides a fixing face for the inner lining and skirting by means of the upper sole piece and permits the supporting angle to be fixed to the lower frame so that, if desired, the floor may be constructed prior to the erection of the upper frame.

The last two systems use relatively short studs in storey height panels as in the platform frame and combine these advantages with the possibility inherent in the balloon frame of erecting wall frames independent of the floor to permit early covering-in. In the independent frame, however, the wall and floor structures are quite independent of each other both in erection and construction and thus, whereas in the two previous systems the floor must be constructed in situ joist by joist, even though subsequent to the erection of the wall frames, in this system prefabricated floor panels of any form of construction can be used and lowered on to the bearers at any appropriate stage of erection. This advantage can, of course, be obtained in a balloon frame if the ledger is fixed on the inner face of the studs. If required for reasons of a grid layout the joists and studs in an independent frame can be arranged to align.

[1] See Part 2, chapter 10, for the provisions of the Building Regulations in respect of cavity barriers in the building structure. See also BRE Digests 214, 215.

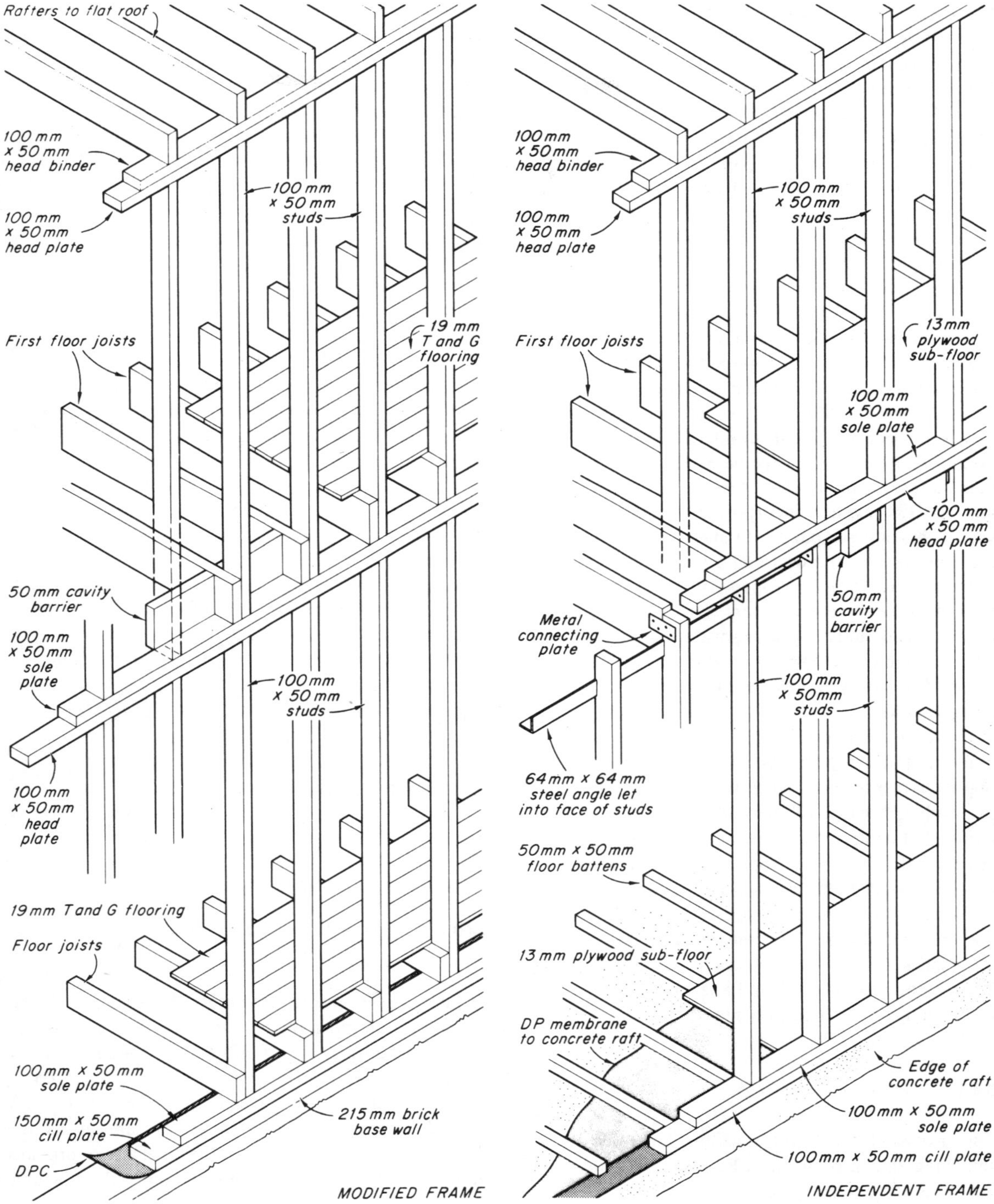

5.39 *Frame wall construction*

Prefabrication For buildings more than one storey high those systems employing single-storey height panels permit greater flexibility in elevational design for a given number of panel types than the balloon frame system. This is an important economic factor when components are to be factory produced and variety reduction is important (see page 19).

The smaller the panel, whatever the system, the greater the flexibility in design within a given range of panels and the more easily can they be handled, but the greater will be the number of joints between panels. Where joints between panels occur in traditionally constructed frame buildings they are covered by the external sheathing and internal linings applied after the erection of the panels (figure 5.43) but in a fully prefabricated system in which no layers are added to cover the joints a 'through-the-wall' joint is produced in which the main design problem is satisfactorily to exclude rain and wind while permitting easy assembly. Simple joint techniques have been developed in Canada and the USA which incorporate a seal to act as a vapour barrier (see page 116) and wind excluder near the inner face of the wall and a widening of the joint near the outer face. The 'open-joint' thus formed near the outer face serves to prevent rain penetration to the inside[1] and permits any warm water vapour that may pass through the seal and condense near the cold outer face to drain away.[2] Examples of these joints are shown in figure 5.40.

However, though simple in form, these joints cost money to produce properly in the factory and on site so that the larger the wall panels the less the overall cost of the walling because of the reduced number of joints required. Furthermore, it is usually cheaper to make large panels since every stop and start in a production process costs money and such panels are better suited to the incorporation of services, such as electric wiring, and the application of the complete interior and exterior finishes. Large panels such as these require mechanical handling for erection and adequate precautions to be taken against damage on site.

5.5.2 Junctions and openings

The provision of adequate nailing faces for linings, sheathing and skirtings must be ensured at angles and junctions and is usually accomplished by the introduction of extra studs or noggings as shown in figure 5.41.

When the layout of the structure is based on a system of co-ordinated dimensions with the faces of the linings

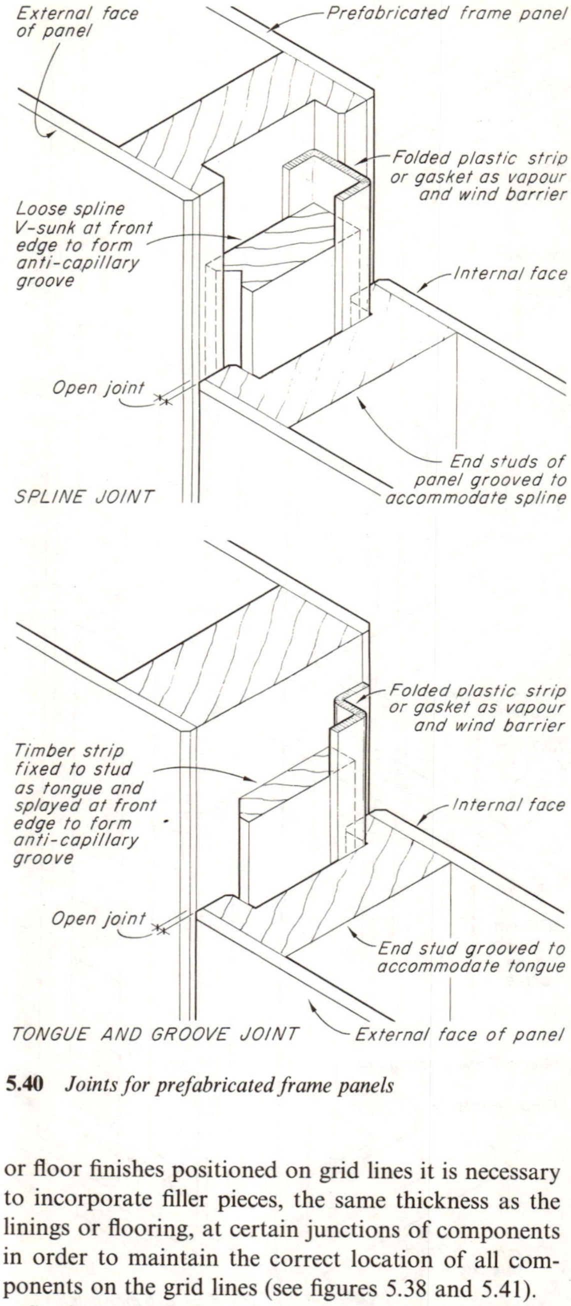

5.40 *Joints for prefabricated frame panels*

or floor finishes positioned on grid lines it is necessary to incorporate filler pieces, the same thickness as the linings or flooring, at certain junctions of components in order to maintain the correct location of all components on the grid lines (see figures 5.38 and 5.41).

Studs on each side of openings in frame walls are doubled to provide support for the lintel and to provide ample nailing surface for finishes round the opening

[1] By opening up the capillary path formed by the joint. See *Open Joints*, chapter 2.

[2] See *Condensation* chapter 2 *MBS: Environment and Services*.

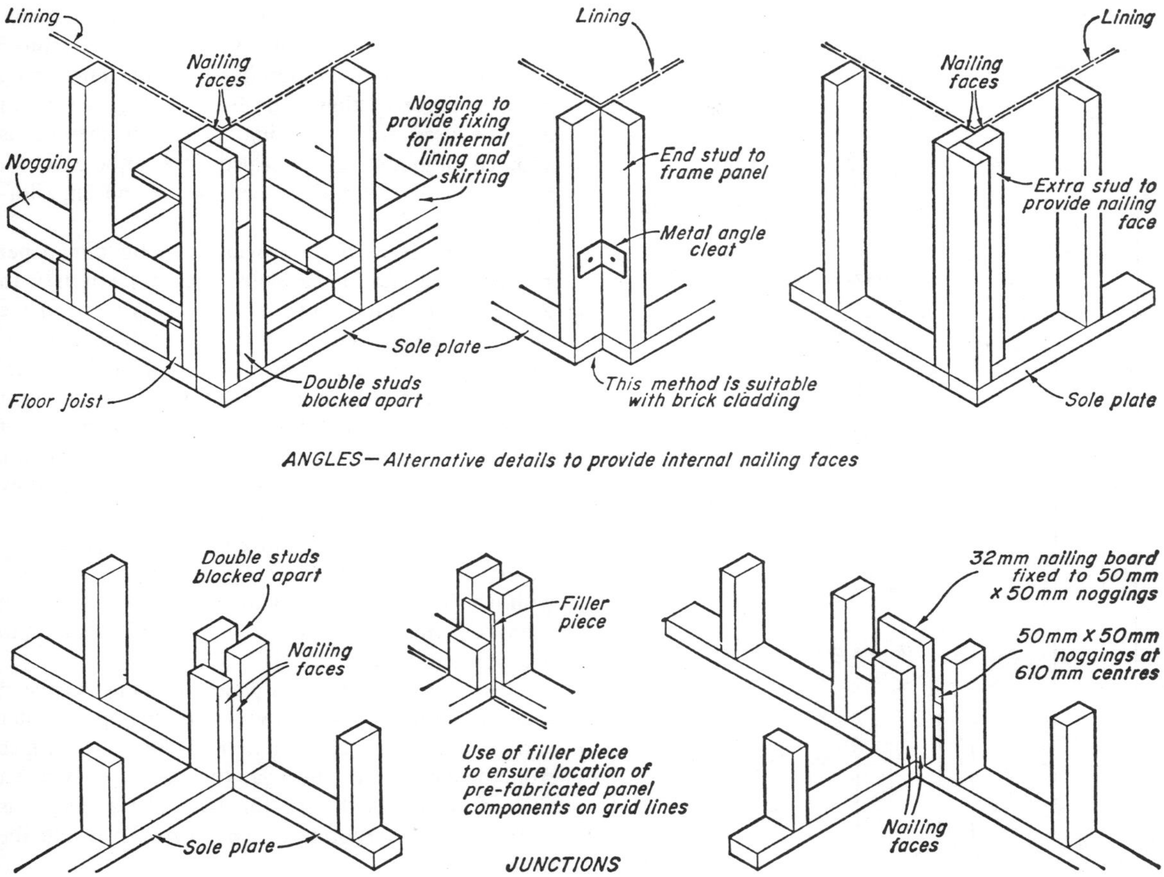

5.41 *Frame wall construction: junctions and angles*

(figure 5.42). In the case of door openings this also provides greater resistance against slamming. Lintels are formed of double solid timbers of a depth appropriate to the span or of plywood box construction using the wall frame and plywood sheathing as shown in figure 5.42. In the latter the head piece of the opening and the head plate over act as flanges and the sheathing takes the shear stresses with the studs acting as stiffeners. The lintel is thus an integral part of the wall panel. The traditional method of using diagonal struts in the framing over an opening to form an elementary truss involves a relatively large amount of labour in fabrication and it is cheaper and quicker to use one of these other methods. Cill members to window openings may be doubled if greater nailing surface is required.

5.5.3 Cladding to wall frames

A *sheathing* of 9·0 mm resin-bonded plywood sheeting or 25 mm T and G boarding is usually applied to the outside face of the frames (figure 5.43). This fulfils a number of useful functions. It provides rigidity to the total structure by bracing the framing panels and stiffens the individual studs in the direction of their smallest dimension. It produces stiffer individual panel components for handling and transport, forms an overall nailing surface for external claddings and improves the thermal insulation of the wall structure by sealing the spaces between the studs. Boarding is less frequently used now but where it is it should be laid diagonally across the studs to perform a bracing func-

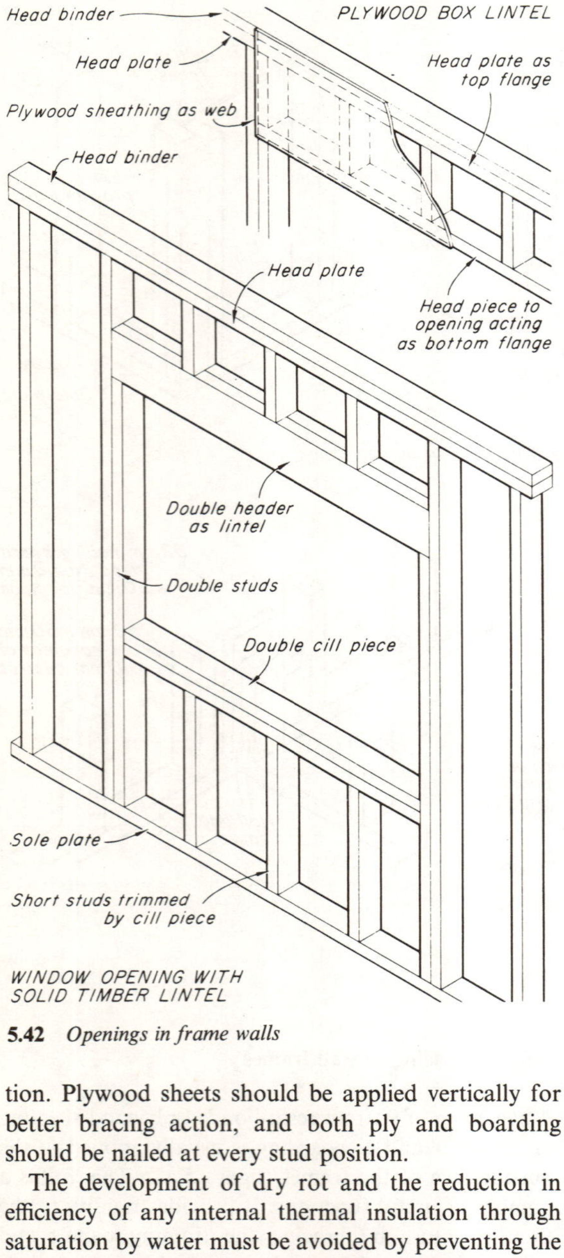

5.42 *Openings in frame walls*

tion. Plywood sheets should be applied vertically for better bracing action, and both ply and boarding should be nailed at every stud position.

The development of dry rot and the reduction in efficiency of any internal thermal insulation through saturation by water must be avoided by preventing the entry of water into the cavities of the wall frames. This is ensured by the use of a *moisture barrier* on the external face and a *vapour barrier* on the internal face (figure 5.44).

The function of the vapour barrier, placed on the inner, warm side of the wall frame, is to exclude from the wall cavities any water vapour passing through the inner lining from the interior of the building which might condense within the cavities.[1] Suitable materials are non-ferrous metal foil and polythene sheeting.[2] These are secured to the framing under the inner lining or, alternatively, aluminium-foil faced plasterboard may be used for the lining. Whatever method is adopted all joints must be sealed since, to be effective, the vapour barrier must be continuous and imperforate.

The moisture barrier is provided by a building paper stapled to the sheathing or outer faces of the studs, with good overlaps at the joints. It functions as a second line of defence against penetration of wind-driven rain or snow or of moisture from an external brick veneer. Since no vapour barrier at present is 100 per cent efficient it is essential that any small quantities of vapour which may pass into the wall cavities can pass to the outer air before it condenses. This is ensured by using a 'breather' type moisture barrier which excludes draught and external moisture but permits passage of water vapour. BS 4016: 1972 gives the requirements with which this type of building paper must comply in terms of permeability to water vapour and resistance to water penetration.

A frame wall resists water penetration in principle in the same manner as a solid wall with an applied outer surface, that is by the isolation of the outer surface from the inner surface by a gap (figure 5.2 *D*). Where a skin of brickwork is used for external cladding it is necessary to incorporate a cavity between it and the timber frame.

A wide variety of external claddings may be used on top of the sheathing, such as timber weather boarding as shown in figure 5.43 or WBP ply sheeting; hung shingles, tiles or slates; metal, plastic or asbestos cement sidings. For details of these reference should be made to Part 2 chapter 4.

A cladding of brick veneer should have a 40 mm space between the moisture barrier and the inner face of the brickwork (figure 5.43). The wall framing is set in from the face of the foundation bearing to form a base for the 102·5 mm brickwork. Bent galvanised light steel ties, which are flexible enough to avoid stressing the brickwork should the frame move vertically, are nailed to the framing and built into the brickwork at spacings of 1200 mm horizontally and 380 mm vertically (if studs are at 600 mm centres, then at spacings of

[1] See *MBS: Environment and Services* chapter 2 and *Structure and Fabric* Part 2, chapter 4.

[2] See *MBS: Materials* chapter 1.

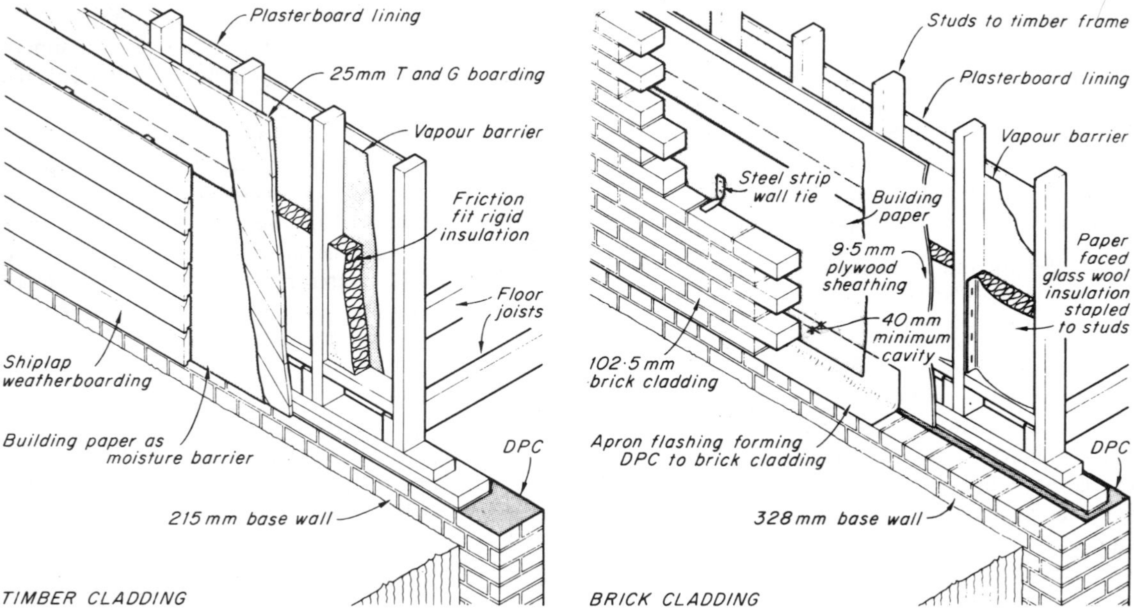

5.43 *Claddings to frame walls*

600 mm both horizontally and vertically). A flexible DPC is essential at the bottom of the cavity, extending to a minimum height of 200 mm up the face of the framing behind the moisture barrier. As in brick cavity walls open 'weep' joints 914 mm apart are formed in the bottom course.[1]

The cavity between the brick veneer and the moisture barrier must be closed at the top and around all openings by cavity barriers which may be of timber not less than 38 mm thick. The cavity must also be subdivided by similar cavity barriers placed at intervals of not more than 8 m apart. In terrace or semi-detached housing a cavity barrier must be placed in the plane of the separating walls if these are solid. If the separating walls are of cavity construction this cavity must be closed at the junctions with the external walls and that in the latter closed with cavity barriers on each side of the separating walls.[2]

The internal lining is commonly 13 mm plasterboard with taped and filled joints or with a skim coat of plaster. This provides fire protection to the timber structure and an added degree of rigidity to the wall frames. Thermal insulation may be 50 mm or more of mineral or glass fibre in the form of friction fit batts or paper-faced rolls of cavity width mats or quilt, with paper flanges on each side which are stapled to the studs, or it may be of rigid polystyrene or other type of insulation board secured by friction fit. The use of aluminium-foil faced plasterboard as the internal lining will give an increased insulation value.

[1] See Brick Development Association Design Note 6 *Brick cladding to timber frame construction* for provision of movement gaps in the cladding, and also for other aspects of this method of cladding.

[2] See Approved Document B Appendix G.

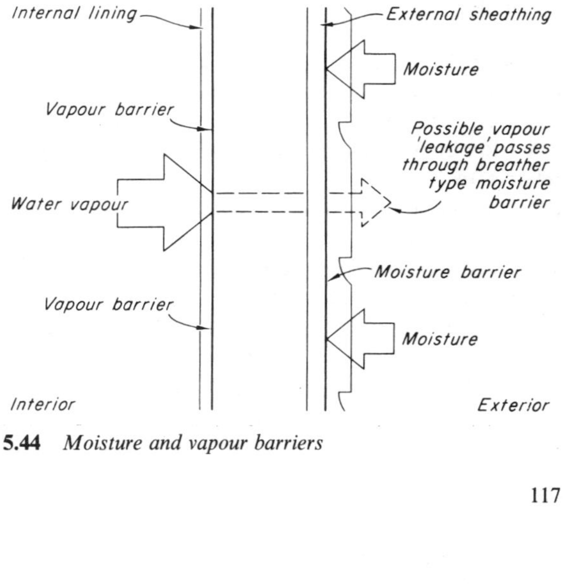

5.44 *Moisture and vapour barriers*

5.6 PARTITIONS

A partition is an internal wall other than a party, compartment or separating wall (see Part 2). Its primary function is to divide the space within a building into rooms. It may be loadbearing or non-loadbearing. The functional requirements which a normal partition should satisfy are the provision of adequate:

strength and stability
sound insulation
fire resistance.

A reasonable degree of sound insulation is usually required between the individual rooms in a building and in some circumstances a very high degree of insulation may be required. The main problem in such cases is to attain the required level of insulation with the lightest possible form of construction. Sound insulation is discussed generally in chapter 6 of *MBS: Environment and Services* and in respect of partitions in particular in chapter 11 of *MBS: Components*. Partitions which are to provide fire protection, such as those round escape stairs and along escape routes within buildings, must have a mimimum standard of fire resistance according to the class of building of which they form part. The standards required can usually be attained with the use of normal partition construction (see Part 2 chapter 10).

The thickness of a loadbearing brick or block partition is calculated in the same way as that for a loadbearing wall (see Part 2 chapter 4).[1] Loadbearing partitions are generally used only in one- or two-storey buildings of loadbearing wall construction. Bricks and concrete blocks are suitable for this type of partition.

In non-loadbearing partitions the compressive strength is not important provided the material is capable of bearing its own weight at the base. Thus they can be of lighter construction than loadbearing partitions. Transverse strength against lateral pressure, however, is important. Since this type of partition, in order to obtain a structure of minimum weight, is thin relative to its height, it must be considered, for the purpose of stability, as a slab spanning between supports which provide adequate restraint at the edges. Thickness as well as edge restraint is also important in relation to transverse strength. Generally accepted limits for height and length of non-loadbearing partitions in relation to the thickness are given for a number of materials in Table 5.3. Those for concrete block partitions are some typical figures based on the graph in BS 5628: Part 3: 1985 for partitions with restraint at top and both ends. Other graphs are also given in this specification for different degrees of restraint. 13 mm of plaster on one or both sides may be included in the thickness of the partition.

Edge isolation should be provided when structural movements are likely to produce cracks in the partition. Fibreboard, cork or other similar material is used for a resilient material and some degree of fixity may be obtained by sinking the edges into a chase sunk in the structure against which the partition butts, or in timber members fixed to the structure as shown in figure 5.45.

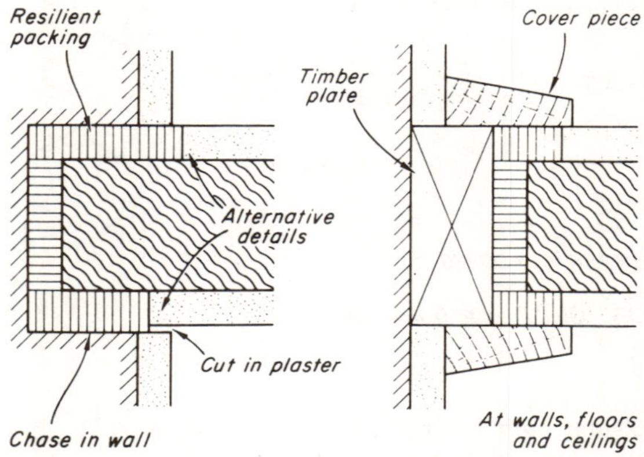

5.45 *Partitions — edge isolation*

This section is concerned with partitions constructed on the site with normal building materials. The numerous proprietary systems of demountable partitions which are used in certain types of building, such as offices, to provide an easy means of changing the room layout, are considered in chapter 11 *MBS: Components*.

Partitions may be constructed of bricks or blocks of various types, of slabs of different materials or as a framework covered with some form of facing. Bricks are now generally used only for loadbearing partitions, the design and detailing of which are similar to those for normal brick walls. Blocks as already explained, are larger and sometimes lighter than bricks and can, therefore, be more quickly laid. For this reason, and because they are cheap, they are widely used for partition work. Light materials can be made up into large units or slabs which although requiring two men to handle, need far less jointing.

The design and construction of partitions is covered

[1] The Building Regulations in AD A1/2, C10 provide means for determining the thickness when this is not calculated.

by BS 5628: Part 3: 1985 *Structural use of masonry* and BS 5234: 1975 *Internal non-loadbearing partitioning*, and much of the tabulated information given in Table 5.3, together with that in the remainder of the text, is based on these.

5.6.1 Block partitions

Partition blocks are normally of concrete. Hollow glass blocks are also used but their characteristics have little in common with concrete blocks and they are discussed separately. Concrete blocks are made with dense or lightweight aggregate concrete and with aerated concrete and are described in *MBS: Materials*. Reference is made to their use in wall construction under *Blockwork* earlier in this chapter.

Where the limits of height or length for a particular partition thickness would be exceeded, the partition must be divided into panels by rigid vertical or horizontal supports, so that the dimensions of each individual panel lie within these limits. Adequate lateral rigidity is provided by setting the edges of the partition at least 50 mm into a chase, groove or channel in the structure or intermediate supports, or by the use of metal wall ties.

Building blocks vary in the ease with which they may be chased and cut and this should be borne in mind in choosing the blocks and the thickness of partition. Vertical chases cannot readily be formed in solid dense concrete blocks or hollow blocks but special conduit blocks already grooved are available for bonding in with normal blocks. Lightweight aggregate blocks cut fairly easily. Horizontal chases should not exceed one-sixth of the thickness of the block in depth and vertical chases one-third.

5.6.2 Hollow glass block partitions

Hollow glass blocks are used where light transmission through a partition is required. They are hollow glass units manufactured by fusing together the rims of two glass coffers. Sizes range from 115 mm × 115 mm to 300 mm × 300 mm in thicknesses of 80 mm and 100 mm. Different surface patterns are produced by pressing flutes and other patterns on the inner and outer surfaces at various spacings, or on the inside face only to give a smooth external surface. The jointing edges are painted and sanded to form a key for mortar.

When the dimensions of a partition are greater than those given in table 5.3, vertical or horizontal intermediate supports of adequate strength and rigidity must generally be used. As at junctions with the main structure a 13 mm clearance must be provided at all junctions with intermediate supports. Lateral restraint at the edges can be provided by building the edges of the panel into a groove or channel formed in the structure or intermediate supports 10 mm wider than the block thickness and not less than 25 mm deep; this will allow 5 mm clearance at each face and 13 mm clearance at the end as shown in figure 5.46.

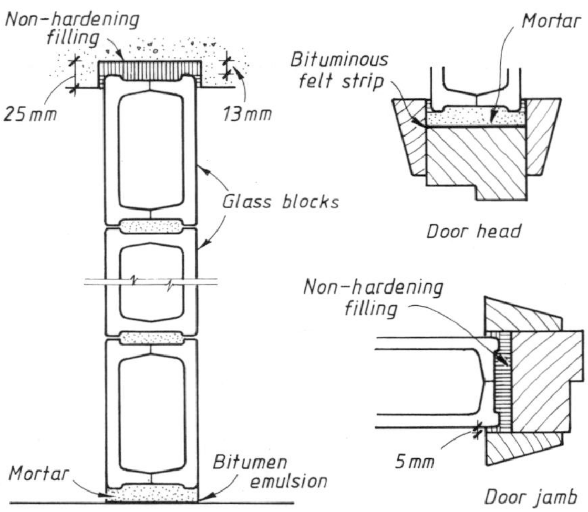

5.46 *Glass block partitions*

5.6.3 Slab partitions

Wood wool, compressed straw, plasterboard and cored plasterboard slabs can be used for the construction of non-loadbearing partitions.

Wood wool slabs These are described in *MBS: Materials*. Heavy duty slabs 50 mm or more in thickness are suitable for partitions. They are usually plastered.

Edge joints should usually be designed to allow longitudinal movement while giving support against lateral movement. This can be done by sinking the top and vertical edges into chases or grooves in the structure or intermediate supports. When the panels are short the edges may be restrained by expanded metal reinforcement on each face of the partition which is carried over the joint and securely fixed on each side. Plaster finish may be made discontinuous by a cut

Type of partition	*Strength and stability*			*Movements*
Concrete block	Thickness of block in mm	Length m	Height not to exceed m	Blocks to be fully matured to avoid shrinkage cracking and reasonably dry when used. Use weak mortars. Shrinkage joints not exceeding 6·0 m apart.
	50 75 90 100	3·00 4·00 4·00 5·00	2·25 3·50 4·40 4·70	
	Provide lateral rigidity — see text.			
Glass block	Maximum dimension in any direction—6·0 m. Maximum area—18·5 m^2. Provide lateral rigidity — see text.			Edge isolation at top and vertical edges of panels and openings with 13mm clearance. Support coated with bitumen emulsion before bedding bottom blocks. Build in every 3rd-5th course 64 mm wide open mesh non-rusting reinforcement with ends built into structure.
Slab	***Wood wool*** Heights not exceeding 3·66 m — 50 mm slabs. Heights not exceeding 4·90 m — 75 mm slabs. Heights greater than 4·90 m and lengths greater than 6·0 m provide intermediate vertical and horizontal support. ***Compressed straw*** Heights not exceeding 2·70 m — timber coupling strips at vertical joints fixed to slab edges. Heights greater than 2·70 m — full height timber studs at slab width spacing. Heights over standard maximum length of 3·66 m — intermediate horizontal supports required. ***Laminated Plasterboard*** Heights not exceeding 2.60 m—50 mm thick. Heights not exceeding 3·20 mm—80 mm thick. No limit on length. ***Cellular plasterboard*** Heights not exceeding 2·40 m—50 mm thick. Heights not exceeding 2·70 m—57 and 63 mm thick. No limit on length.			***Wood wool*** Maximum length of panel 6·0 m. Edge isolation or partial isolation by very weak mortar or mastic joint. Face reinforcement of expanded metal or wire netting at changes of longitudinal or transverse section (eg over door openings) to extend 305 mm on each side of point of change. Carry up door frames where possible. ***Compressed straw*** Thermal and moisture movement negligible in normal conditions. ***Laminated plasterboard and cellular plasterboard*** Negligible thermal and moisture movement. Edge isolation where movement of surrounding structure is likely to affect partition.

Table 5.3 *Block and slab partitions*

Mortar and jointing	*Openings*	*Fixings*
Cement, lime, sand 1:1:6 for 50 mm blocks, 1:2:9 for thicker blocks.	Thin partitions: carry up door frames and anchor to ceiling. Use grooved frames. Carry blocks on door frame. Thick partitions: tie frames to partition with metal ties or use grooved frames. Use reinforced blocks or RC lintels over all openings.	Lightweight fittings and fixtures—nail fixing to lightweight aggregate blocks; toggle bolts or similar fixing to hollow concrete blocks. Heavy fittings (WC tanks, LBs)—bolt through to steel or timber back plate for thinner partitions.
Fairly dry, fatty mortar. Suitable mix: cement, lime, sand 1:1:4. Joints about 6 mm thick.	Carry up door frames and secure to structure above or at sides, or secure frame by 5 mm perforated MS strips fixed to top of frame and running vertically or horizontally through joints to structure. Secure frames to partition by expanded metal ties in every 3rd or 5th course depending on size of opening.	Fixings cannot easily be arranged. Fittings and fixtures should be kept clear of this type of partition.
Wood wool Slabs laid horizontally; vertical joints broken bonded. Cement, sand 1:3 with up to ¼ part hydrated lime, or retarded hemi-hydrate plaster, sand 1:2. Mortar fairly wet. Joints as thin as possible. ***Compressed straw*** Built up dry.	***Wood wool*** Use grooved frames. For openings up to 1·50 m wide use single 610 mm deep slab as lintel with 100 mm min. bearings. Over 1·50 m wide use timber lintel with 150 — 230 mm bearings. For very wide openings use separate supporting frame to avoid excessive load on slabs. ***Compressed straw*** Carry up door frames and anchor to ceiling. Use grooved frames. ***Laminated plasterboard and Cellular plasterboard*** Carry up fixing battens and anchor to ceiling. Use rebated frames for laminated plasterboard.	***Wood wool*** Lightweight fittings and fixtures — use special nailing slabs with timber fillets let into centre or cement dovetailed timber blocks into slabs at suitable centres. Heavy fittings — fix pipes by bolting through to timber or steel back plate; use independent framing for WC flushing tanks and LB's. ***Compressed straw*** Nail fixing for light trim only. Nails driven at angle and adjacent nails opposed to form 'dovetail' grip. Lightweight fittings and fixtures — use 45 mm screws dipped in cold adhesive. Heavy fittings—WC tanks and LB's bolted through to timber or steel back plate. ***Laminated plasterboard and Cellular plasterboard*** Lightweight fittings and fixtures — screw fixing into plugged holes in laminated partitions; fix to cellular units by toggle or similar bolts. Heavy fittings — fix to timber plates inserted in core of laminated plasterboard and to battens screwed to plugs inserted in cellular units.

through the full depth of the plaster at the junction of adjacent surfaces or, if left continuous, the joints should be reinforced with scrim or metal mesh not less than 100 mm wide. Face reinforcement should be provided above all openings (figure 5.47).

Holes and chases can be made easily in wood wool slabs with normal woodworking tools. The depth of chases should not exceed one-third of the thickness of the slab and these should be covered over with light metal reinforcing strips before plastering is carried out.

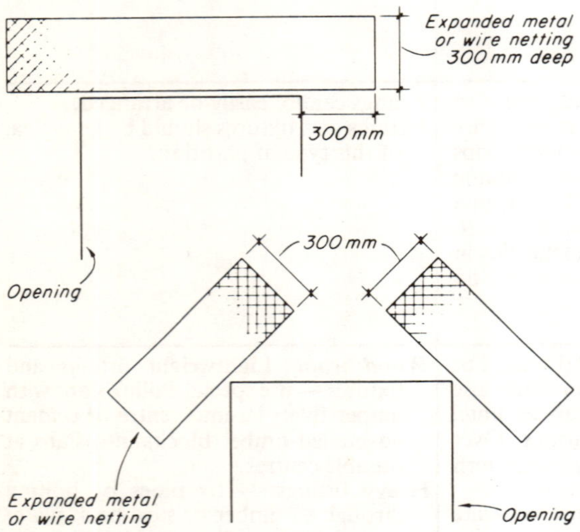

5.47 *Wood wool slab partitions – face reinforcement*

Compressed straw slabs These are made from straw by heat and pressure and are also described in *MBS: Materials*.

For heights up to 2·70 m the vertical butt joints between the slabs may be formed by timber coupling pieces slightly less in width than the thickness of the slab, nailed or screwed to the vertical edges of the slabs and grooved to key with each other (figure 5.48). The coupling pieces are secured at floor level to a 50 mm wide batten fixed to the floor and to the ceiling if this is of timber, or to a similar batten if not. The coupling pieces may be covered over by a cover strip or by a 125 mm wide Kraft paper scrim glued to the slabs on each side. This allows for any slight movement of the timber. Similar coupling members should be used at the junction of the partition with the structure. Alternatively, lateral restraint should be provided by sinking the edges of the slabs into grooves or channels in the structure, or formed in timber edge members. For heights over 2·70 mm, 100 mm timber studs should be used running from floor to ceiling and spaced to take the 1200 mm wide slabs. The minimum thickness of the studs should be 22 mm with the slabs held by fixing beads (figure 5.48). Alternatively, thicker posts grooved to take the slabs can be used.

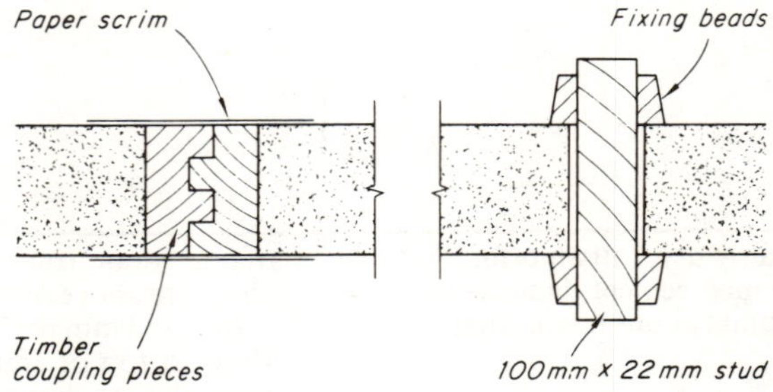

5.48 *Straw slab partitions*

Plasterboard This is described in *MBS: Finishes*, and is used in two ways in the construction of partitions: (i) as the layers in a laminated form built up on site, (ii) as a facing or skin on each side of a cellular core to form a self-supporting unit.

In the first method the partition is built up of three or four layers of plasterboard. 38 mm × 25 mm timber battens are fixed to walls, ceiling and floor and at corner angles and openings, to one face of which is nailed an outer layer of plasterboards. The plasterboards for the middle layer are then cut to fit between the floor and ceiling battens and are fixed by vertical bands of bonding compound to the outer layer. To complete the construction the final layer of boards is bonded to the middle layer in the same way and nailed to the timber battens as shown in figure 5.49. In the second method the partition is built up from 600 mm, 900 mm or 1200 mm wide panels consisting of two sheets of 9·5 mm or 12·5 mm plasterboard fixed to a square cellular core made of fibrous material, finishing 50 mm, 57 mm or 63 mm overall (figure 5.49). The panels are fixed by means of a timber batten fixed to the ceiling and walls and plugs let into the bottom of the panel as shown in figure 5.49. Fixing at door openings is by means of a storey height batten let into the vertical edge of the panel as shown and joints between panels are formed with a similar batten set half its depth in each core, the joint being finished with scrim and joint filler.

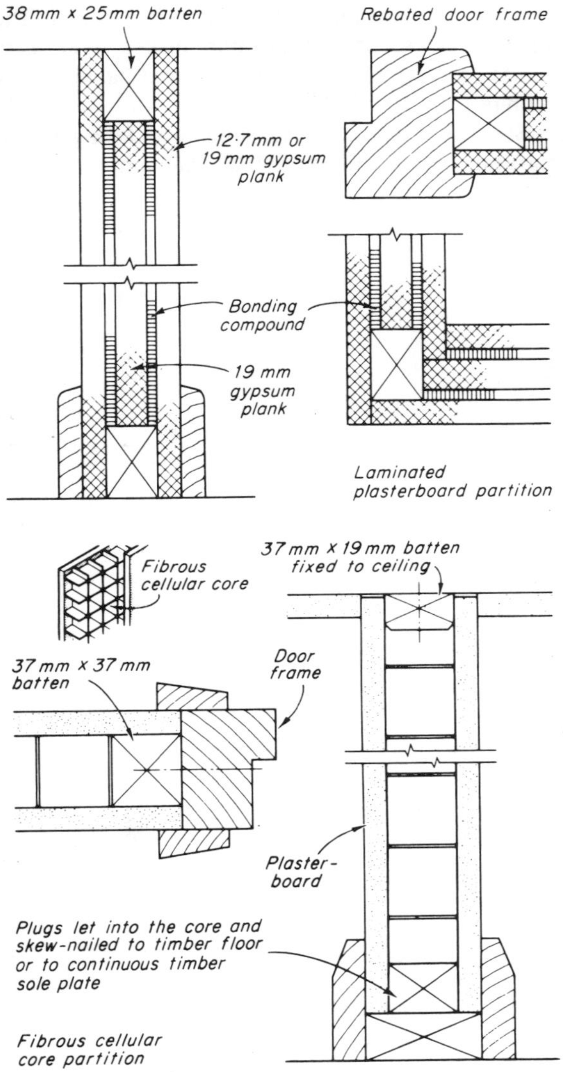

5.49 *Plasterboard partitions*

5.6.4 Timber stud partitions

Timber-framed partitions have largely been superseded for general use by the types of partition already described, but they are light and can, if required, support considerable loads which can be distributed directly to end bearings if the partition is trussed. It is a dry construction and a wide variety of facings can be applied, the spacing of the studs being varied to suit standard sizes of sheets.

The framing is essentially the same as for wall frames already described and consists of uprights or studs fixed between a timber sole plate and head, usually the same section as the studs. The studs are nailed to the plate and head and the latter are nailed direct to wood floors and ceilings or to fixing blocks let into concrete floors. Nailed joints at all points usually give sufficient rigidity together with the use of nogging pieces. These are short lengths of timber nailed tightly between the studs to stiffen them. They may be fixed in herringbone fashion but are usually fixed horizontally and spaced at centres at which they can act as cross fixings to any sheet coverings which may be used to face the partitions. The size of the studs will depend upon their height and spacing, and the latter depends upon the type of facing to be used. Table 5.4 indicates the appropriate sizes for the members of a framed partition based on a stud spacing of 457 mm centre to centre. The same table shows the centres at which studs and noggings should be fixed for a number of facings in general use. When 32 mm or 38 mm studs are used, those providing support to the meeting edges of facing sheets usually need to be wider to provide space for the fixing of two edges.

Openings may be simply formed by nailing head timbers between normal studs on each side of the opening, but a better way is to double-stud the sides of the opening with the inner studs cut short to form bearings for the head timber as in figure 5.42. Alternat-

Height up to metres		*Studs at 457 mm centres*	*Plate and head*	*Noggings*	
Lath and plaster	*Other facings*	*mm*	*mm*	*Main mm*	*Inter-mediate mm*
2·29	2·60	75 × 32	75 × 32 or 50	75 × 50	75 × 32
3·20	3·66	100 × 38	100 × 38 or 50	100 × 50	100 × 38
4·27	4·90	125 × 50	125 × 50	125 × 50	125 × 50
5·20	6·10	150 × 50	150 × 50 or 75	150 × 50	150 × 50

Facing	*Maximum centres of supports*	
	Studs, mm	*Noggings, m*
Plasterboard	305—610	1·22
Fibre-board	305—457	1·22
Hardboard	380—508	1·22
'Plastic' sheet	635	1·22
Plywood	305—1220	0·9—1·80
Wood wool slab	457—610	None

Table 5.4 *Sizes and spacing of members in timber stud partitions*

ively, the opening can be framed more rigidly if 75 mm or 100 mm thick posts are used to form the opening, with the same size timber for the head-piece which is jointed rigidly to the posts with wedged mortise and tenon joints. Fixings for skirtings, trims and fittings can be easily made by means of fixing studs and noggings framed into partition at the appropriate points. The open structure of the partition makes it an easy matter to accommodate conduits and cables freely within it.

Sheet and board coverings suitable for facing framed partitions are described in *MBS: Materials*.

6 Framed structures

The concept of the skeleton structure has been introduced in chapter 1 where the forms which it may take are briefly described.

Of these forms certain consist essentially of pairs of columns with members spanning between them, spaced apart to enclose the volume of the building. These are classified as (i) *building frame*, of columns and horizontal beams for single- and multi-storey buildings, (ii) *shed frame*, of columns and roof truss for single-storey buildings and (iii) portal or *rigid frame*, of columns and horizontal or pitched beam for single-storey buildings, the characteristic of this type of frame being the rigid connection between the columns and the spanning member.[1] (See figure 1.2 for illustrations of these.)

In a framed structure the loadbearing and the enclosing and dividing functions, which in solid and surface constructions are fulfilled by one element, are fulfilled by separate elements of construction — the former by the frame, the latter by the wall. In framed construction the wall, being relieved of the task of carrying loads from the rest of the structure may, therefore, be quite thin and light in weight, a factor which becomes increasingly significant with increasing height of structure and which has resulted in many developments in the field of external claddings and infill panels for framed buildings (see Part 2).

The advantages of the framed structure are (i) saving in floor space, particularly when internal structural supports must be provided, (ii) flexibility in plan and building operations, because of the absence of load-bearing walls at any level and (iii) reduction of dead weight, for reasons already given.

These advantages, however, do not necessarily make a framed structure economically advantageous in every circumstance, for example in the case of individual small-scale buildings[2] or of residential type buildings where the plan area is divided into rooms by walls and partitions (see Part 2 *Masonry Walls*). It can be said, broadly speaking, that framed construction becomes logical and is likely to be economical when the span of roof or floors becomes great enough to necessitate double construction (see page 150), involving beams or trusses applying heavy concentrated loads at certain points on the supporting structure which, in solid construction, would require the provision of piers.

In the case of industrialised system building, however, the framed structure can be economic even for small-scale building types. This is due to the economies deriving from large-scale production and to the reduction in erection time and of labour on site which should accompany the use of prefabricated components (see chapter 2).

Functional requirements

As already indicated, the primary function of a skeleton frame is to carry safely all the loads imposed on the building and this it must do without deforming excessively under load as a whole or in its parts. In order to fulfil this function efficiently it must provide in its design and construction adequate:

strength and stability
fire resistance.

Strength and stability are ensured by the use of appropriate materials in suitable forms applied with due regard to the manner in which a structure and its parts behave under load, as described in chapter 3.

Building frames may be classified according to the stiffness or rigidity of the joints between the members, especially between columns and beams. A *non-rigid*[3] frame is one in which the nature of the joints is such that the beams are assumed to be simply supported and the joints non-rigid. Rigidity in the framed structure as a whole is ensured by the inclusion of some stiffening elements in the structure, often in the form of triangulating members. (See page 51, and also *Wind bracing* Part 2.) Steel and timber frames are commonly jointed in this manner and sometimes precast concrete frames.

[1] The term *fully rigid* is applied to building frames in which all the connections between members are stiff or rigid. See part 2 chapter 5.

[2] The economic advantage of wall over frame construction for small-scale buildings is well illustrated in an analysis made by Stillman and Eastwick-Field of the costs of different ways of constructing a small office building. See *Architects Journal* 25 October 1956, 24 January and 25 April 1957. See also on this, Part 2 chapter 5.

[3] Often referred to as 'pin-jointed'.

A *semi-rigid* frame is one in which some or all joints are such that some rigididy is obtained, a technique usually limited to steel frames and which effects some saving in material. In a *fully-rigid*[1] frame all the joints are rigid. This results in considerable economies in material in the frame for reasons given on page 46. Depending upon the nature of the structure the joints alone may provide the stiffness necessary to prevent the frame as a whole deforming under lateral wind pressure, although additional stiffening elements are often required. This type of building frame can be constructed in steel and concrete.

An adequate degree of fire resistance in the frame is essential in order that its structural integrity may be maintained in the event of fire, either for the full period of a total burn-out or for a period at least long enough to permit any occupants of the building to escape. Concrete is highly fire-resistant but steel in many circumstances requires the provision of fire-protection, of which a number of forms exist such as encasure by concrete or by fire-resistant fibre board. Timber, although a combustible material which will easily burn in the form of thin boards, burns less readily when in thicknesses greater than about 150 mm. Its combustibility may also be reduced by the application of fire retardants. The subject of fire-resistance and fire protection generally is discussed in chapter 10 of Part 2 to which reference should be made.

Structural materials

The materials which are commonly used for framed structures are steel, concrete (reinforced or prestressed), timber and aluminium alloys, all of which have characteristics which make them suitable for this purpose in varying degrees according to the building type and the nature of the structure.

Materials for framed structures, particularly when these are tall or wide in span, need to be strong, stiff and light in weight.

The stronger a material the smaller the amount which will be required to resist a given force.

The stiffer the material the less will the structure and its members deform under load. Since the excessive deflection of beams and the buckling of columns must be avoided then in any given circumstances the use of stiffer materials will result in smaller members (see page 45). The relationship of the depth of a spanning member or structure, necessary to keep its deflection within acceptable limits, to its span is expressed as the *depth/span ratio* and is useful as a basis of comparison of the effects of using materials and forms of structure of differing degrees of stiffness. A small depth/span ratio indicates the achievement of adequate stiffness with minimum depth of spanning member.[2]

A material which is light in weight as well as adequate in strength results in structures of low self-weight. The self or dead weight of a structure, as well as the load which the structure is to carry, contributes to the stresses set up within it. Low dead weight is, therefore, an important economic factor especially in structures carrying light loads, such as roofs, for if the dead weight is considerably greater than the imposed load then the structure must be designed primarily to carry its own weight rather than the load imposed on it. Thus, the smaller the self or dead weight of a structure relative to the load to be carried the more efficient and, therefore, economic the structure. This relationship is expressed as the *dead/live load ratio* and, thus, a designer seeks to achieve the lowest dead/live ratio consistent with other factors in the design such as the effect of methods of achieving this upon the cost of the enclosing elements of the building, and the cost implications of fabrication methods which may be involved.

The relationship of the weight of a material to its strength provides an indication of its efficiency in terms of the weight required to fulfil the structural function. This is expressed as the *strength/weight ratio*, a high value indicating high strength with low self weight, resulting in a minimum weight of material to fulfil a particular structural purpose.

Steel is a material strong in both compression and tension and it is also a stiff material. A steel structure is, therefore, relatively economic in material because a small amount can carry a relatively large load and, because it is stiff, the structure and its members will not easily deform under load. It has a high strength/weight ratio. These characteristics make it suitable for both low-rise and high-rise building frames and roof structures of all spans.

Concrete varies in strength according to mix. The compressive strength of normal structural concrete is

[1] See footnote on page 125 regarding the term 'fully-rigid'. The term 'rigid frame' refers essentially to a single-storey roof frame as defined on page 125.

[2] See Part 2 for the implications of variations in the depth/span ratio.

about one-sixteenth that of steel, but its tensile strength is only about one-tenth of its compressive strength. Its stiffness is low compared with steel and its strength/weight ratio is low. To overcome these weaknesses structural members are *reinforced* in their tension zones with steel bars or *prestressed* in the same zones, usually by means of steel wires or cables. In reinforced form concrete is suitable for short-span low- and high-rise building frames and in prestressed form for wide span building and rigid frames. Prestressed concrete may also be applied to shed frames using precast roof trusses.

Timber varies in strength according to the species and to the presence or absence of knots and faults in the timber and the bending strength of structural softwoods varies from about one-thirtieth to one twenty-second that of mild steel. Compared with other materials its stiffness is low but in relation to its own weight which is quite light, it is relatively very stiff. Thus in structural applications compensation for its lack of stiffness can be made without excessive increase in weight of structure. It has a relatively high strength/weight ratio, and is suitable for lightly or moderately loaded low-rise building frames and for shed and rigid frames, particularly where the span and height of these are large.

Aluminium varies in strength according to the particular alloy, from about three-quarters that of mild steel to strengths somewhat greater than that of steel. Although stiffer than either concrete or timber aluminium alloys are only about one-third as stiff as steel but they are only about one-third its weight. They have, therefore, very high strength/weight ratios. These characteristics make them suitable for roof structures (which carry only light imposed loads), particularly those of long span, and less so for normal building frames. The high cost of aluminium usually precludes its structural use for other than very wide span roof structures.[1]

Layout of frames

As a general rule a layout with columns as closely spaced as the nature of the building will permit, thus resulting in short span beams or trusses, will be cheaper than one with widely spaced columns. The cost of the frame rises with increasing span of beams and falls with a decrease in span in spite of the greater number of columns, unless the latter are very tall. With smaller spans the beams reduce in size and cost and, similarly, the columns because of the reduced loads they carry.

The *spacing* of the frames is influenced largely by the economic span of the floor or roofing system which they support and this will vary with the imposed floor or roof loading and the type of floor or roofing system. It can be shown, however, in the case of building frames and rigid frames, that as the frame beams increase in span or frame columns increase in height there comes a point at which it may be more economic to increase the spacing between the frames, thus increasing the load on the beams but reducing the number of frames, rather than to maintain them at the most economic span of a particular floor or roofing system. In the case of shed frames, unless the trusses are of considerable span, a close spacing of the frames usually gives the cheapest structure.[2]

Wherever possible the layout of a skeleton structure should be based on a regular structural grid. The advantages of so doing are as follows:

(a) Loads on the structure are transmitted evenly to the foundations, thus minimising relative settlement and standardising the sizes of foundation slabs.

(b) It results in regularity in beam depths and column sizes and in the position of columns and beams relative to walls. This avoids the use of 'waste' material to bring beams and columns to similar dimensions either because they are exposed to view or, in the case of reinforced concrete, to standardise formwork. It also standardises the size of dividing and enclosing walls or panels.

(c) In reinforced concrete work the regular slab and beam spans minimise the variations in rod sizes.

(d) It permits greater re-use of formwork, both in precast and in situ concrete construction.[3]

6.1 BUILDING FRAMES

The circumstances in which shed frames and rigid

[1] See Part 2 generally for more detailed discussion of all these materials in this context.

[2] See Part 2 chapter 9 for a fuller discussion of the span and spacing of beams and trusses.

[5] For the significance of this see Part 2 chapter 5.

frames are used are those in which the primary structural problem is that of covering a single space with a roof and these are introduced in chapter 7. In this chapter building frames of limited height and span will be discussed.[1]

6.1.1 Steel frames

Small scale steel building frames are fabricated from sections, either rolled to shape from ingots of hot steel or cold-formed to shape from steel strip.

Hot-rolled-steel sections, known as Standard Structural Steel Sections, are standardised in shape and dimensions and are covered by BS 4: Part 1: 1982 and BS 4848. BS 4 covers beam, column, channel and tee sections and BS 4848: Part 4: 1972 covers angle sections (figure 6.1 *f, c, b, e, d* respectively). The majority of I-beam sections are known as Universal beams but BS 4 also covers a limited range of an earlier form of I-section having more tapered flanges than the standard Universal section (*a*). These are called *joist* sections to distinguish them from the latter. In addition there are I-sections with non-tapered flanges much wider relative to the depth of the web than the Universal beam, on account of which they are known as *broad flange sections* (*c*). Because of their shape characteristics these are particularly useful as columns and for this reason are called Universal Columns sections.[2] (In steel frames the columns are commonly referred to as *stanchions*).

Hot rolled hollow sections, both rectangular and circular, are covered by BS 4848: Part 2: 1975. The advantages of such sections for columns are described on page 48. Hot-rolled sections are used in all types of frame.

Cold-formed steel sections are pressed or rolled to shape from thin steel strip and are known respectively as pressed-steel or cold-rolled steel sections (BS 2994: 1976 covers cold-rolled sections). These sections can be formed to almost any shape (see figure 6.8 *A*) and strength can be varied by the use of different thicknesses of sheet steel.[3] They are most efficiently used in low-rise frames[4] of moderate span and loading where they may prove cheaper than hot-rolled sections particularly in circumstances where even the smallest hot-rolled section would not be stressed to its limit.

Construction with hot-rolled steel sections

Connections Connections between hot-rolled members of a steel frame are made by means of steel bolts and rivets with angles as cleats, or by means of welding.

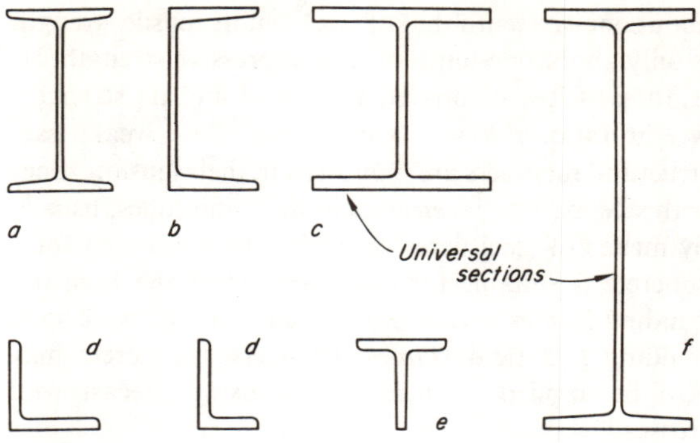

6.1 *Typical hot-rolled steel sections*

A welded joint is one in which the adjacent members are joined by the fusion of the steel at their point of contact. Additional metal at the joint is deposited in the process from steel welding rods. Welded connections are discussed under *Welded Construction* in Part 2 chapter 5.

Rivets with a snap head, which is almost semicircular in section, are used for structural work but as explained later, their use in modern steel fabricating is limited. They are driven hot so that on cooling they contract in length and form a tight connection.

Bolts are of three types: (i) 'turned and fitted' bolts, (ii) 'black' bolts and (iii) high strength friction grip bolts, each with hexagonal heads and nuts. Tapered washers are used to provide a flat bearing on bevelled surfaces. Turned bolts, as the name suggests, are turned to be parallel throughout the length of the barrel and must fit tightly into their holes. The working stresses permitted in turned and fitted bolts are the same as those for rivets but they are more expensive than rivets.

Black bolts are not turned to a precise diameter throughout and, therefore, cannot be a tight fit in their holes. As a consequence, the allowable stress in shear is not so high as for rivets and turned bolts and they may be used only for the end connections of secondary floor beams or for other connections where dead bearings are formed by seating brackets which resist the whole of the shear forces involved.

[1] See Part 2 for tall building frames, wide span roof structures and three dimensional frameworks.

[2] See chapter 3 page 49, and also Part 2 chapter 5.

[3] See Part 2 chapter 5 where methods of forming these sections are described.

[4] Up to three storeys although this can be exceeded if the lower stanchions are of hot-rolled sections.

Friction grip bolts, also called torque or, more correctly, torque-controlled bolts, are used instead of rivets and turned bolts. They are made from steel with a greater yield point than mild steel and are placed in holes large enough to permit a push-fit, making assembly easy. The nut is tightened to a predetermined amount by a torque-controlled spanner or a pneumatic impact wrench which presses together the surfaces in contact to such an extent that they become capable of transmitting a moment from one to the other by friction.

The introduction of friction grip bolts and the use of welding resulted in a reduction in the use of riveted connections and rivets are now used only in the fabricating shop and that to a limited extent. Shop fabrication is carried out primarily by means of welded connections. Site connections are normally bolted, the primary connections by means of high strength friction grip bolts, which permit the joints to be more easily made and with less expenditure of time and labour than with site riveting, and secondary connections by means of black bolts.

Connections between the different members of a steel frame can be classified as follows:

(i) compression connections, in which the load is transmitted directly from one member to another and the connection serves mainly to fix the two parts together (see figure 6.2 *C*)

(ii) shear connections in which the joint elements are stressed in shear (see figure 6.3, beam to stanchion connections at *C*, *D*)

(iii) tension connections, in which the joint elements are stressed in tension.

Compression connections are the most economical, but where these are not suitable shear connections are preferable to tension connections which, generally, are adopted in special cases only.

Connections are detailed so that fabrication is as simple as possible having regard to erection problems. For example, when the flange width of a beam is greater than the web depth of the supporting stanchion the beam flanges must be side notched in order to effect a web connection. Such notching could be avoided by increasing the depth of the stanchion section which might result in a reduction in fabricating costs such as to produce an overall economy in the frame, especially if a large number of similar beam connections were involved.

Building frames in hot-rolled steel Figure 6.2 *A* shows a small single-storey structure consisting of three frames made up of hot-rolled steel stanchions and beams set between end gable walls. To permit the enclosing elements of the building to be placed outside the line of the stanchions the beams cantilever a short distance beyond the stanchions and thus bear on top of them (*B*). A connection is made by forming a *stanchion cap* consisting of a cap plate welded to the stanchion, the top of the stanchion being machined true before the cap plate is fixed so that the load is transmitted directly through the plate to the end of the stanchion (*C*).

The foot of the stanchion must be expanded by means of a base plate large enough to reduce the pressure on the foundation to the safe bearing pressure of the concrete, and to provide a means of securing the stanchion to the foundation. The base plate itself will act as an inverted cantilever beam and this must be taken into account in establishing its thickness. The upper face of the base plate and the end of the stanchion are machined true for bearing and are welded together, either by means of angle cleats, or directly as shown in (*D*). The base is secured to its foundation by two holding-down bolts which are grouted into the concrete. Levelling up of the base to plumb the stanchion is done by means of steel wedges before grouting up with fairly dry cement and sand mix.

For low, short-span structures such as this example small section stanchions and a base such as that illustrated provide sufficient rigidity against side wind pressure on the building.[1]

Continuous angles are fixed across the ends of the beams and built into the gable walls. These provide a fixing for the fascia structure and the heads of the enclosing windows as well as serving to tie together the frames. The roof structure of such a building could be based on the use of wood wool slabs supported by timber rafters bearing on timber plates or MS angles bolted to the webs of the beams (see figure 8.14) or on the use of edge-reinforced wood wool slabs bearing on small secondary steel joists spanning between the main beams and spaced according to the slabs used.

Figure 6.3 *A* shows a steel skeleton frame structure for a small two-storey building. The stanchions are in one length with a base similar to that in the previous example but with a larger baseplate because of the greater load to be transferred to the concrete foundation (*B*). The roof beams to each frame are connected

[1] See Part 2 chapter 9 for further discussion on this.

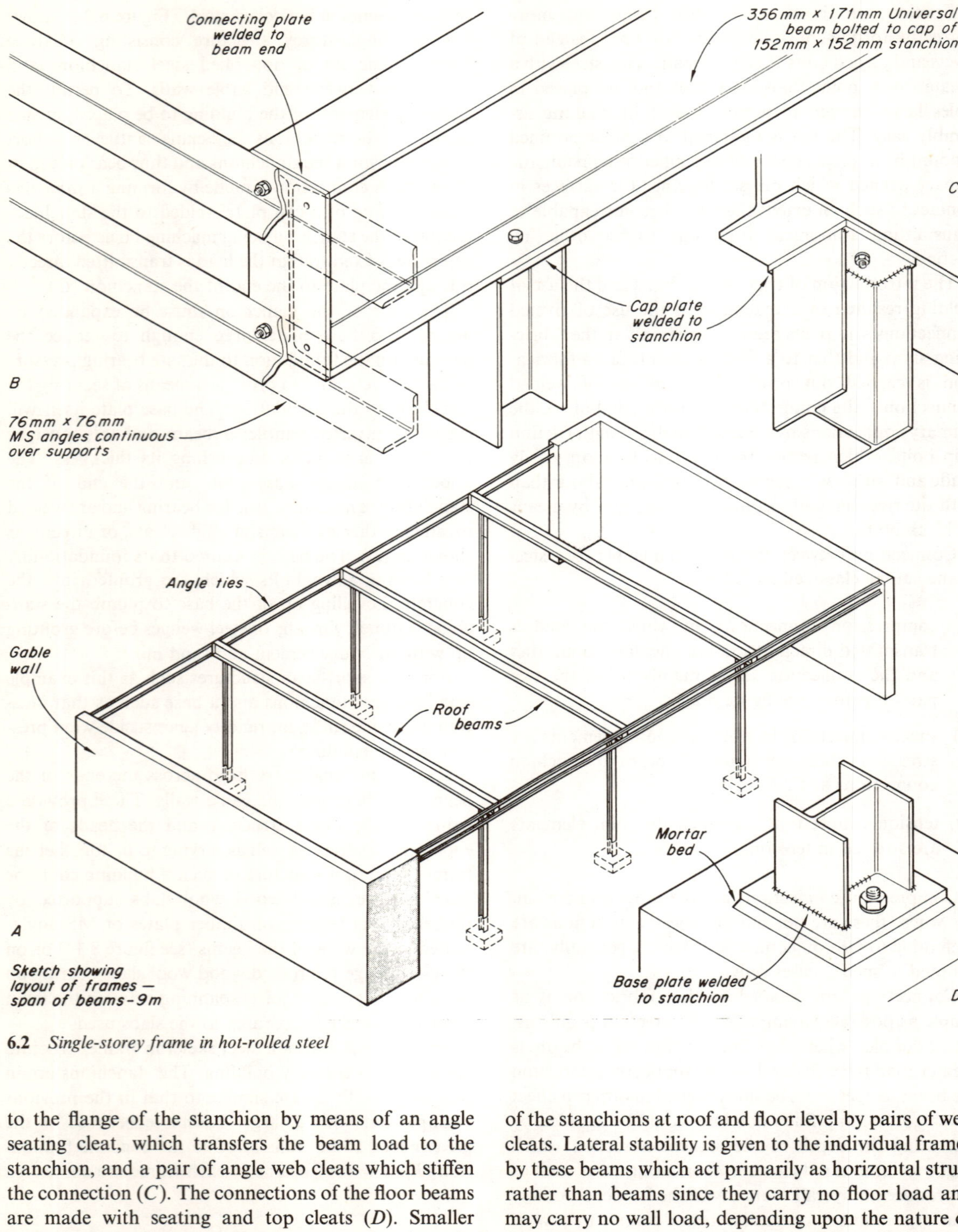

6.2 *Single-storey frame in hot-rolled steel*

to the flange of the stanchion by means of an angle seating cleat, which transfers the beam load to the stanchion, and a pair of angle web cleats which stiffen the connection (*C*). The connections of the floor beams are made with seating and top cleats (*D*). Smaller lateral beams called *tie beams* are connected to the webs of the stanchions at roof and floor level by pairs of web cleats. Lateral stability is given to the individual frames by these beams which act primarily as horizontal struts rather than beams since they carry no floor load and may carry no wall load, depending upon the nature of the enclosing elements of the building. Additional

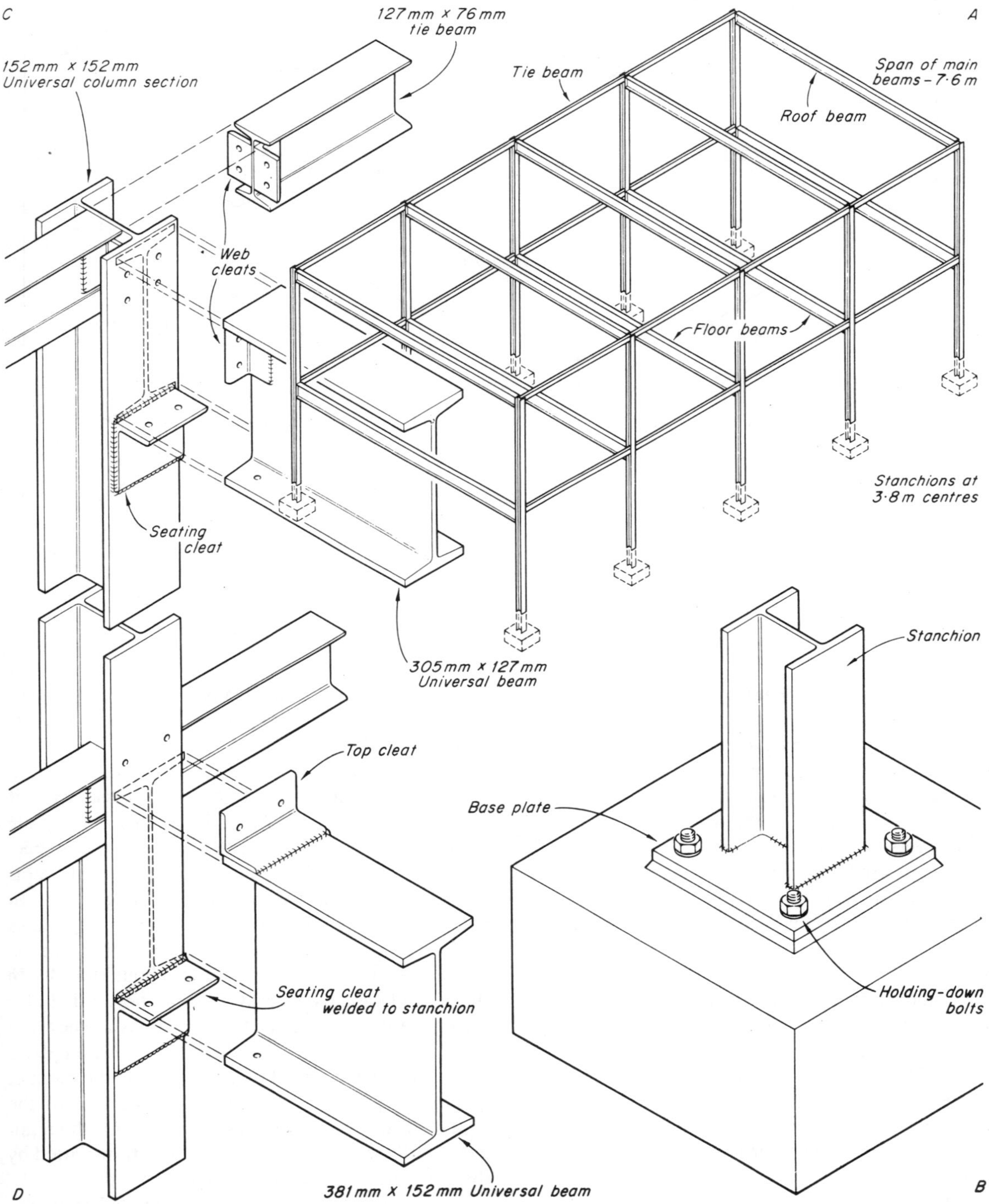

6.3 *Two-storey frame in hot-rolled steel*

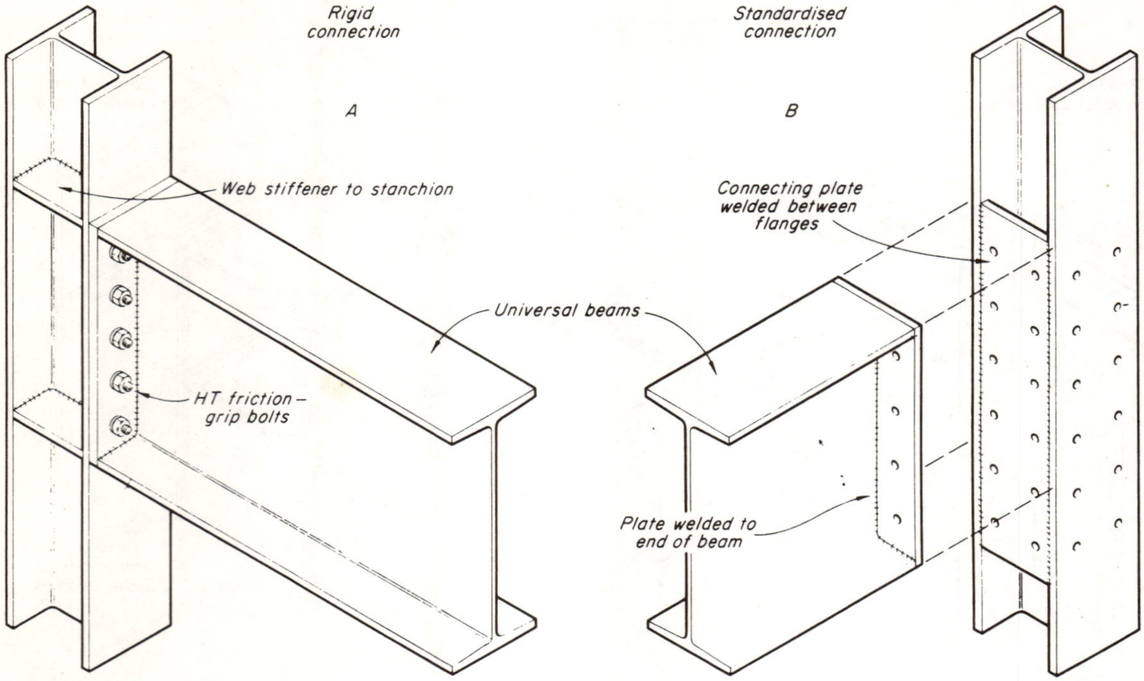

6.4 *Beam to stanchion connections*

rigidity in the direction parallel to the frames will be provided to the whole structure by gable walls at each end. These may be formed as brick or other masonry built into the upper and lower panels of the frame, thus exposing the frame or, alternatively, the frame may be recessed into the back of the wall the outer face of which covers the frame completely. The end frames in either case would be encased by concrete as fire-protection; the remainder of the skeleton would be protected in a similar manner or by one of the various forms of lighter, hollow casing (see Part 2 chapters 5 and 10).

The floor structure could be in situ cast concrete spanning between the floor beams although precast concrete construction could be quicker but probably dearer. (See chapter 8 for concrete floor structures.) A suitable flat roof construction would be timber joists spanning between the main beams and carrying wood wool slabs finished with screed and asphalt or other suitable covering.

It should be noted that Universal beam sections begin to be uneconomic at roof spans above about 10 m. The reason for this is related to the increasing self-weight of the solid-web beam with increasing span and has been referred to already on page 50. Alternative types of beam used for wider span roofs are described in Part 2 chapter 9.

Alternative methods of forming bolted beam to stanchion connections may be adopted (i) as a means of producing a semi-rigid or rigid joint other than by site welding or (ii) as a means of standardising the connection while allowing for variations in the sizes of beams and stanchions within a particular building system. In both cases this involves the use of connecting plates welded to the ends of the beams and in the first case the use of high tensile steel friction grip bolts (6.4 *A*). In the second case plates may possibly be welded also across the flange edges of the stanchions as shown in (*B*), to provide fixing on all sides, the stanchion plates and flanges being drilled to accommodate the number of bolts required for the largest connection in the system. The beam end-plates are drilled for the number of bolts required by the loads to be transmitted by the particular beam to which they are welded.

Fully welded frames are normally limited to single-storey structures because of the problems and time involved in site welding large numbers of joints in a multi-storey structure. Site joints in the former are few

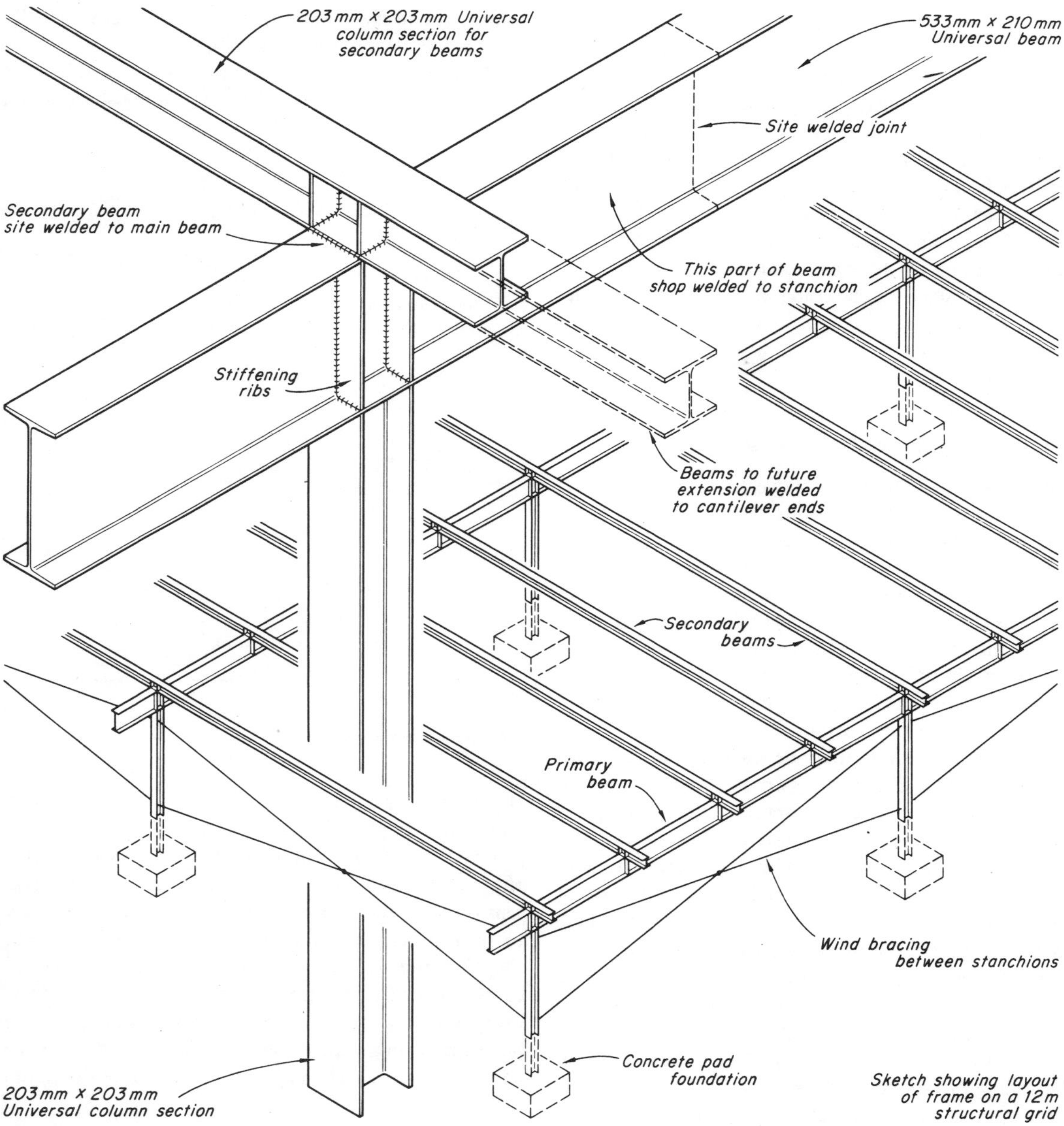

6.5 *Fully welded hot-rolled steel frame*

and can be so arranged relative to the necessary shop welding that they are simple to make as in the example in figure 6.5 which shows part of a large factory structure constructed on this basis. In addition to effecting economies in the skeleton frame by virtue of the continuity at all points[1] the welding of the joint produces neat connections free of all projections, thus enhancing the visual qualities of the structure which is wholly exposed to view. The stanchions in this example

[1] See chapter 3 page 46.

have portions of the beams shop welded to them to form cross-heads, between which the intermediate lengths of the beams are temporarily bolted in position while being site welded. This produces beams continuous over all stanchions and on these bear the secondary beams which are site welded to them at the supports. The function of the steel ribs welded across the beams at the junctions of the various members is to stiffen the webs of the beams.[1] Lateral rigidity is provided to the skeleton frame by diagonal *wind bracing*[2] in the perimeter panels since the enclosing wall and roof construction is of light-weight corrugated steel units and provides no stiffening.

Construction with cold-formed steel sections

Connections Connections between cold-formed steel members are made by various types of welds, bolts, cold-rivets and self-tapping screws.

Welded connections may be formed by means of gas or arc welding or by resistance welding. In the former methods the members are joined by the application of molten metal from a steel filler rod melted by a flame or electric arc; in the latter they are joined by the fusion of the metal at the interfaces of the members caused by the passage of an electric current through the members.

Resistance welding may be carried out as *spot-welding* in which the members pass between two thin copper electrodes which, at specified intervals, press them together and pass an electric current through them, the heat created causing the metal faces in contact to fuse together at the points of application of the pair of electrodes (figure 6.6), the spacing of these spots, the pressure and the electric current all being varied according to the nature and requirements of the joint; as *projection welding* in which groups of dimples pressed in one member fuse with the metal of the other (figure 6.6) or as *ridge welding* in which ridges formed in the rolling of the sections fuse together at their points of contact. Extensive tooling is required to form the ridges in the metal sections and to be economic a large quantity of the section must be required.

Welding is normally used in the fabricating shop to form composite structural sections and to join fixing plates and cleats to stanchions and beams which are then connected by means of bolts on site.

Bolted connections are normally used for site fixing of the component parts of the frame and these may require washers to be placed under both bolt head and nut to provide support to the thin walls of the members which, due to their thinness, are liable to buckle locally on the compressed side of the joint.

Cold rivets and self-tapping screws are used less for joining structural sections than for joining sheet metal.

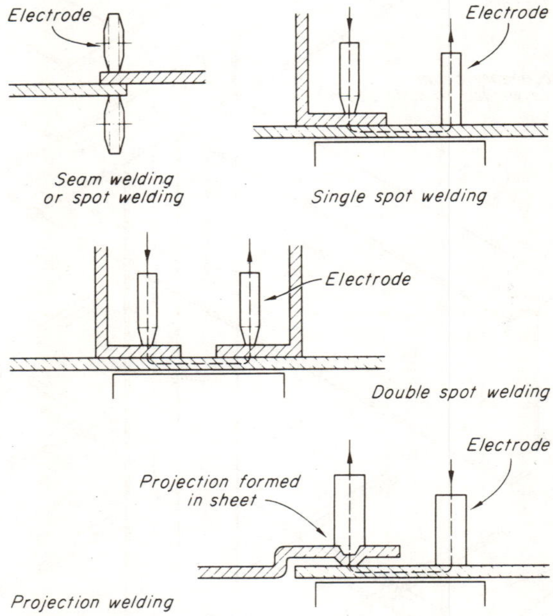

6.6 *Resistance welding*

Blind rivets, set from one side only, are commonly used, the simplest being the *spreading rivet* with a grooved dowel pin. The other form, known as a *pop rivet*, consists of a hollow rivet and pin. The pin, tensioned by the riveting tool, spreads the rivet shank to form a head and then breaks at its reduced neck, thus forming a closed rivet. Self-tapping screws are of case-hardened steel with threads designed to cut their own thread in the metal through which they pass, the holes drilled in the metal being the same diameter as the core of the screw. The methods are illustrated in figure 6.7.

Basic cold-formed sections are usually open shapes and mostly symmetrical about one axis only as shown in figure 6.8 (*A*), but closed and symmetrical sections may be formed by lock-seaming, bolting or welding basic sections together, resulting in sections having greater rigidity and torsional strength (*B*). So-called

[1] See Part 2 *Welded Construction* chapter 5.

[2] See Part 2 chapter 9.

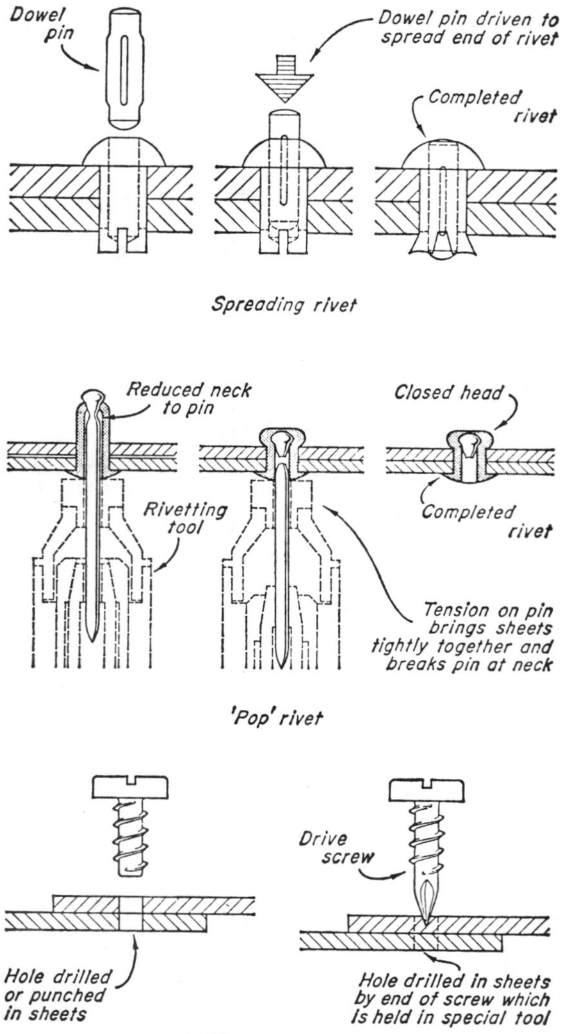

6.7 *Rivets and screws for sheet metal*

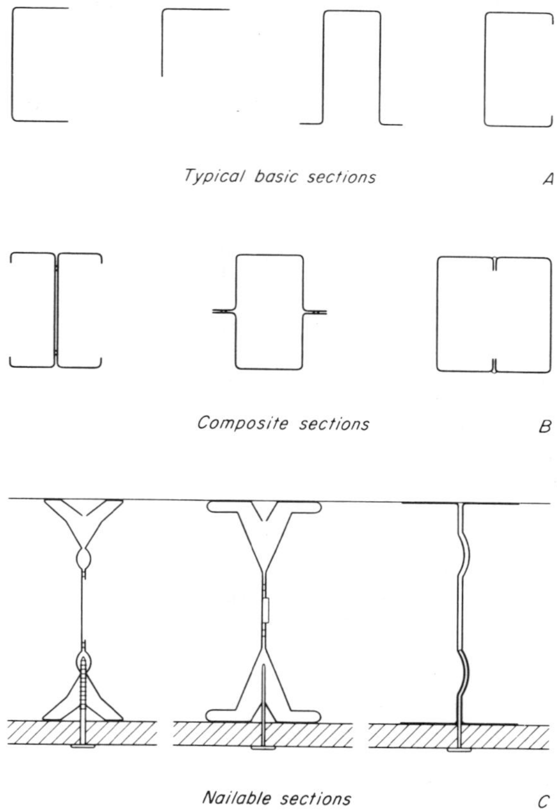

6.8 *Typical cold-formed steel sections*

nailable sections can be formed which provide a grip to fixing nails for deckings and linings without the use of separate fixing battens of timber. Similar sections are used for PK self-tapping screws to avoid site drilling (*C*).

Building frames in cold-formed steel Skeleton frames constructed with cold-formed sections invariably have composite symmetrical sections for the main members. Stanchions may be I-section formed of two lipped-channels resistance welded or bolted back to back or hollow-section formed of two similar channels welded lip to lip (figure 6.9 *A*, *B*, *D*). The latter, by virtue of the greater stiffness of the hollow form, is more suitable for tall stanchions. Main beams may be I-section for spans up to about 5 m to 6 m beyond which built-up lattice beams as shown in figure 6.11 are more economical. Tie beams are usually lipped channels or I-sections.

I-section members permit cleat and bolt connections similar to those for hot-rolled sections although, because of the thinness of the metal, stiffening plates welded on at the connections may sometimes be necessary (figure 6.9 *A*). Bolted connections cannot be made directly on to hollow stanchions. Fixing cleats and plates must, therefore, be welded on to these stanchions to provide for beams (figures 6.9 *B*, *C* and 6.10).

Stanchions may run in one length through the height of a two- or three-storey frame, but the adoption of various splicing techniques in industrialised systems designed for use in buildings of varying heights permits stanchions to be standardised in storey height lengths.

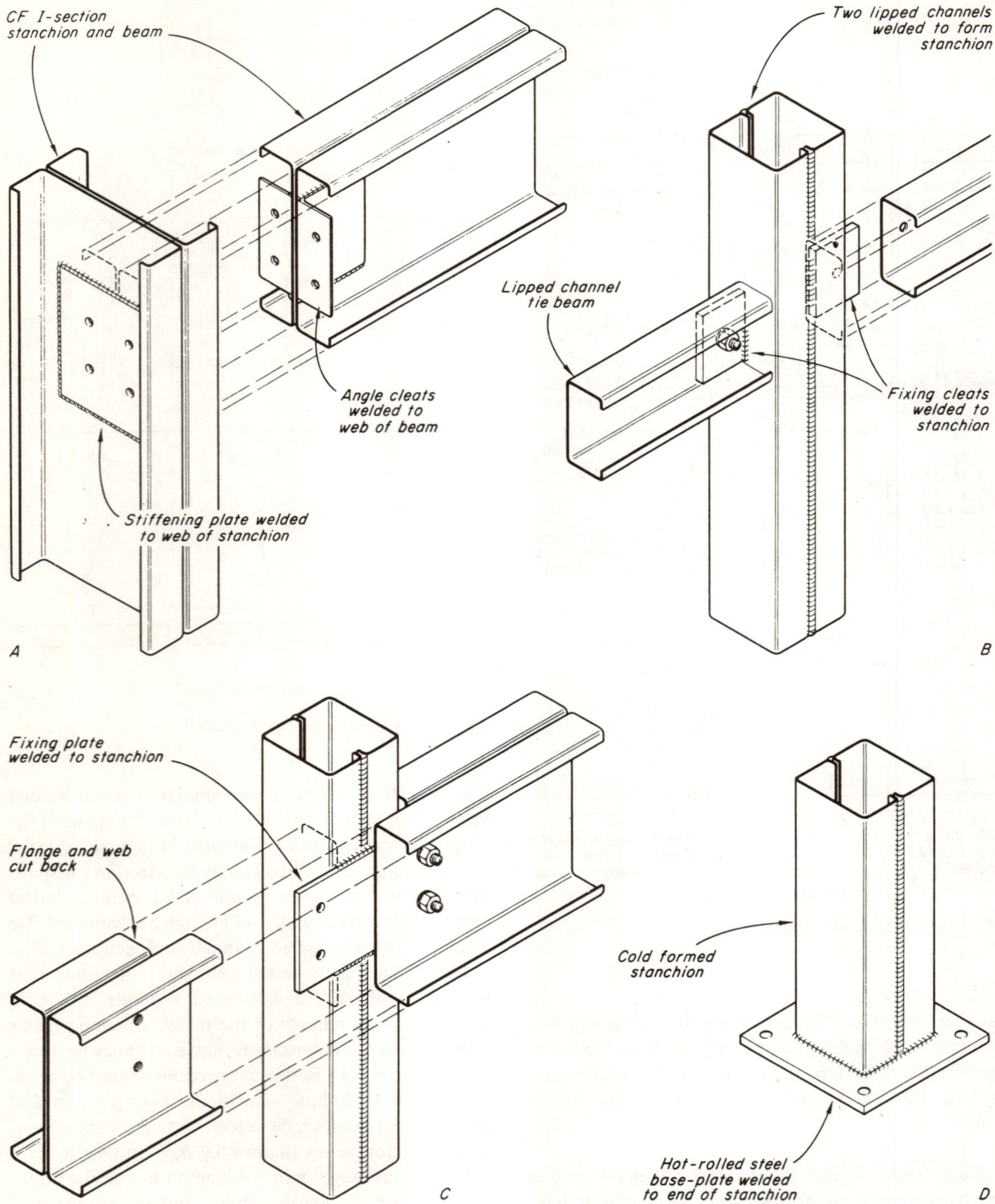

6.9 *Connections for cold-formed steel sections*

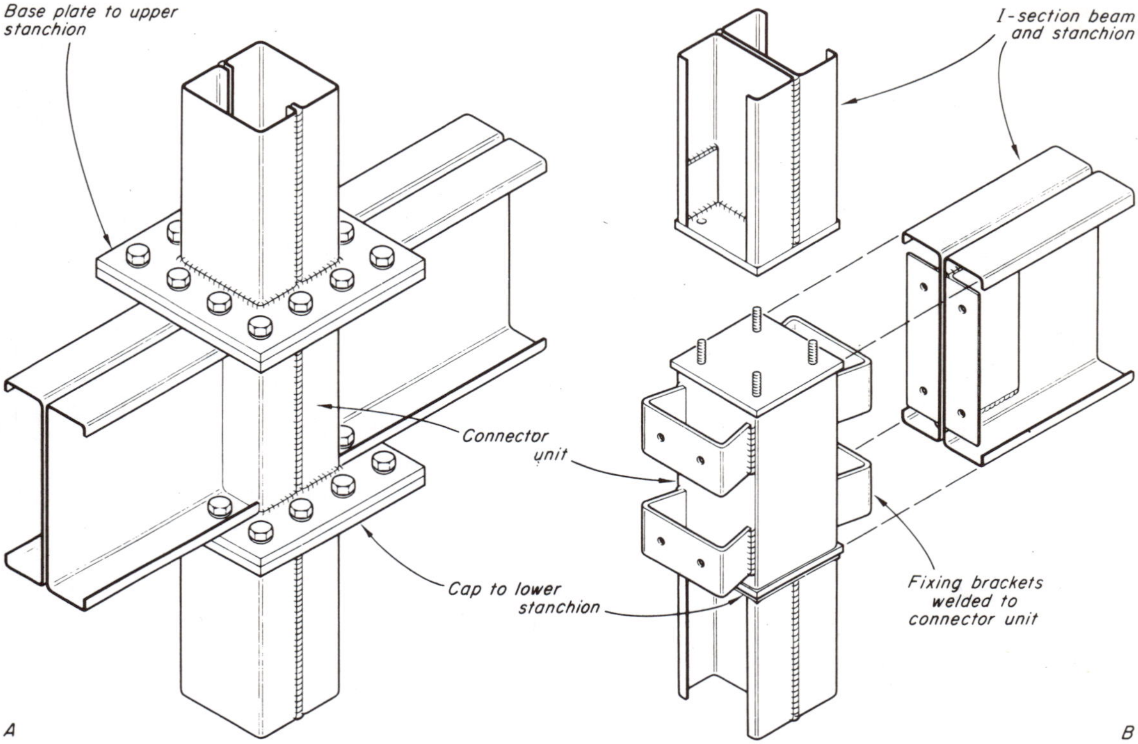

6.10 *Splices for cold-formed steel stanchions*

These techniques usually incorporate fixings for the beams. One method uses projecting cap and base plates welded to each stanchion length which form the stanchion splice and four-way fixing for beams. Another method interposes a component called a connector unit or 'loose header' between upper and lower stanchions. This joins the stanchions and provides fixing for the beam connections (figure 6.10 *A*, *B*). The open I-section in *B* permits the connection to be made with less projection of the plates than in *A*.

Steel lattice beams as shown in figure 6.11 are lighter in weight and, therefore, easier to handle and quicker to erect than comparable hot-rolled sections, although they will usually be rather greater in depth. The type of beam shown is suitable for small or moderate spans and light loads. For this reason they may be used with advantage in roof structures where a low dead/live load ratio is desirable[1] and are now widely used for the roof beams in structures such as those shown in figures 6.2 and 6.3 where the other members may be hot-rolled sections.

Two methods of connecting beam to stanchion are shown in figure 6.11, that incorporating 'key-hole' fixings requiring no site bolting and, therefore, saving time and labour on site.[2]

6.1.2 Timber frames

In small scale buildings with loadings such as those in houses a timber frame wall as described on page 109 is more than capable of supporting the load per metre run imposed by roof and floor. The spans are small and,

[1] See Part 2 *Choice of roof structure* chapter 9 for a discussion on this.

[2] For the use of cold-formed steel sections in building see BS 5950: Part 5: 1987.

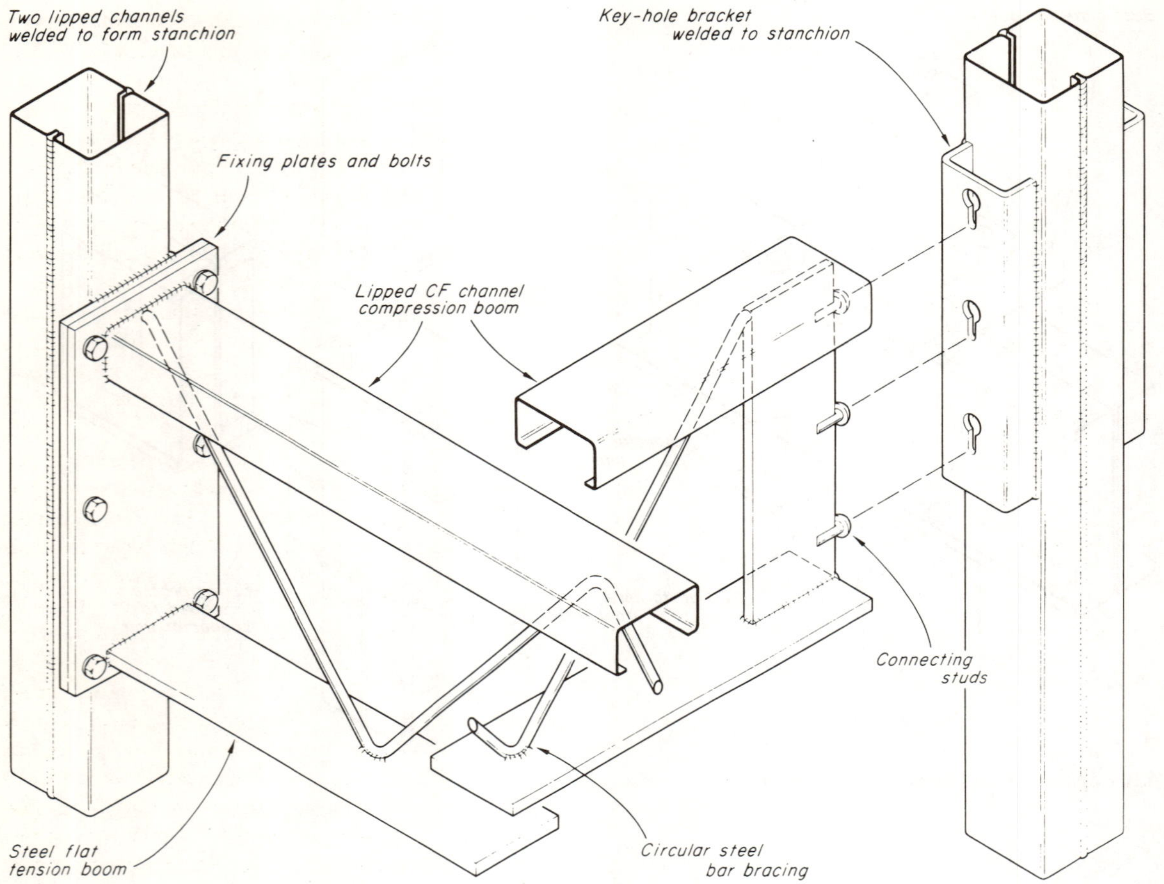

6.11 *Connections for cold-formed steel members*

therefore, loads may be evenly distributed by single roof and floor construction on to the wall panels and a framed structure is not essential (see page 125). The advantages of the framed structure in this context are that it permits the completion of the roof before that of the enclosing and dividing panels and any floors and provides flexibility in the planning of the building.

Skeleton frames constructed in timber may be fabricated from solid timber sections, built-up sections or glued and laminated sections.

Columns and beams Solid square or rectangular sections are generally the most economical in cost, but where members beyond the available sizes and lengths of solid timber are required it is necessary to form them by combining a number of smaller sections of timber. This may be accomplished by nailing or bolting together several pieces to form *built-up solid* sections, typical examples of which are shown in figure 6.12 *B*, *D*. Apart from obtaining the required sizes for large members there are advantages in building up solid sections from smaller pieces since these are easier to obtain and to season properly without checking and they may be built-up in ways that minimise warping and permit rigid connections between columns and beams.

In the case of built-up column sections involving butt joints in the length it is essential that the abutting faces be carefully machined and the joints staggered. Built-up box and I-sections used as columns are stiffer than a solid section for a given timber content and are particularly suitable for tall columns (see chapter 3). Another method of increasing the stiffness of lightly loaded columns is to provide the bearing area in two parallel

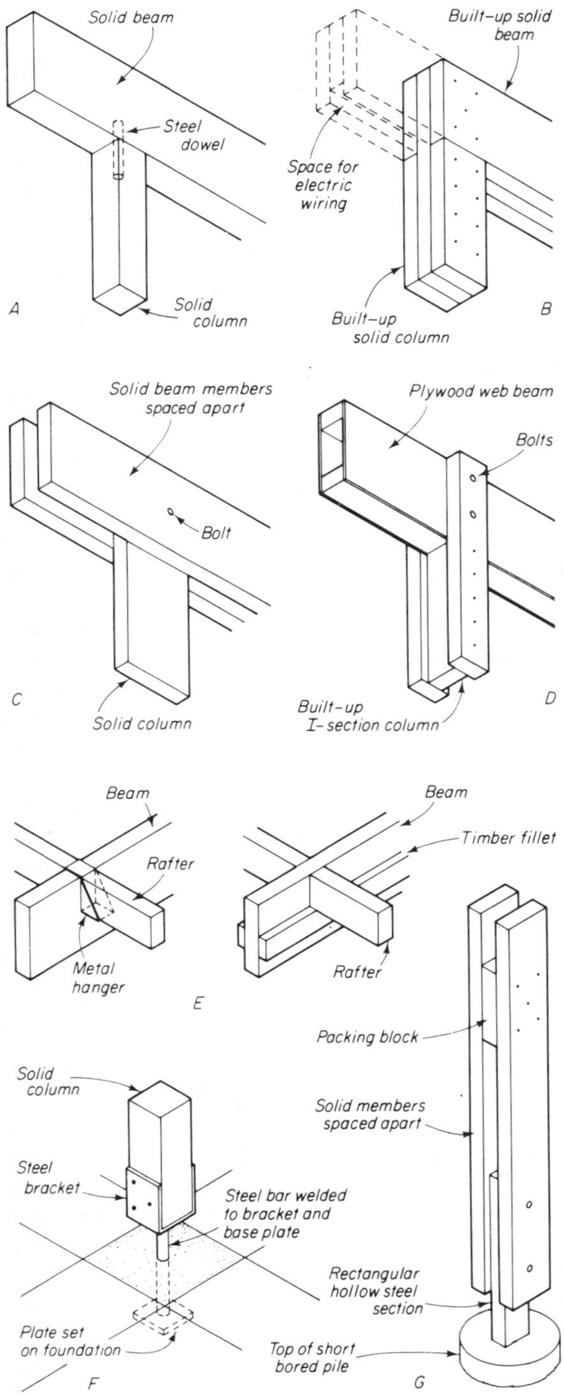

6.12 *Timber beams and columns*

members spaced apart by packing blocks at intervals and connected by nails, bolts or glue (figures 6.12 *G* and 6.17 *A*).

Built-up solid beams are normally built up of vertical pieces nailed or bolted together, nailing being satisfactory for beams up to about 250 mm in depth, although these may require the use of bolts at the ends if shear stresses are high. Where the imposed loading is light beams may be built up with solid flanges and plywood webs nailed or glued, or glued and nailed, together (figure 6.12 *D*). Such *web beams*, compared with solid beams, are very stiff relative to the amount of timber in them, especially those with two webs forming a box section as illustrated, and result in low dead/live load ratios. The thin webs necessitate stiffeners at intervals along the length of the beam. Increased shear resistance near the supports may be obtained by closer spacing of the stiffners at the ends of the beam (see Part 2 chapter 9). Standard proprietary beams of this type are available, one type of which obtains web stiffness by the use of a single corrugated plywood web rather than by separate web stiffeners (figure 7.16). The fire resistance of web beams such as these is very much lower than that of solid or glued and laminated timber beams by reason of the thinness of the parts (see page 126). If exposed to the weather the surface veneer of the plywood webs of this type of beam tends to check.

Glued and laminated sections, commonly called *glulam* sections, consist of timber laminations glued together to form square or rectangular sections (figure 6.17 *D*) and, for large span beams, I-sections. They are more expensive than solid or built-up sections but permit the use of higher permissible stresses in their design and are, therefore, suitable where loads are great or spans are large (see Part 2 chapter 9).

Connections Connections between beams and columns are made with nails, bolts, dowels and cleats according to the type of members. Those between solid and glulam beams and columns are made by metal dowels (figure 6.12 *A*) with or without side fixing plates, the latter providing a stiffer connection, or simply by side-nailing the beam to the column. Built-up members are connected by nails, bolts or bolts and timber connectors. Built-up solid sections or spaced solid members permit rigid connections to be made by passing one member, or part of it, through the other as in figure 6.12 *B*, *C* and *D*. If the beam in (*B*) continues over the column, as shown in broken lines, the outer pieces of the beam are made continuous and the centre piece is stopped short on each side of the column to allow its centre section to pass through the beam as shown. The deep junction formed by this method together with nailing or double bolting produces a relatively rigid connection. The construction shown in (*C*) is simple but the single bolt fixing does not produce a rigid connection and the thin column depends for resistance against buckling largely upon the infilling panels on each side.

Spaced beams as in figure 6.12 *C* permit the use of smaller sections than would be required for a single solid member and provide a space to conceal electrical wiring to lighting fittings. The solid built-up beam, however, has the advantage that one piece restrains the warping of the others. If accommodation for wiring is required this can be formed by making the centre piece less deep than its neighbours as shown in (*B*).

Beam to beam connections are made by means of metal hangers or by metal cleats bolted or screwed to the beams. Rafters and joists may be supported by hangers or timber fillets as shown in figure 6.12 *E*.

Column base connections are made in various ways depending on the relation of the column to the remainder of the building fabric. Free standing external columns normally are raised off the ground to isolate them from ground moisture. This may be done by means of a concrete stool or block with a damp-proof layer between the timber and concrete, the column being fixed in position by a steel dowel or by straps and bolts as shown in figure 6.16. Alternatives to this which isolate the timber from the ground and also fix the column in position are shown in figure 6.12 *F* and *G*. These three methods also hold the post against wind uplift, the effect of which can be considerable when the roof is flat and the structure is light. External perimeter columns normally bear on a continuous timber cill plate which is bolted to the concrete floor slab or perimeter dwarf wall and which also carries the infilling wall panels (figures 6.13, 6.14). If the loading on the column and the nature of the soil necessitates it a small pier and foundation slab or a short bored pile must be constructed under the column position (figure 6.15). Internal columns usually bear directly on the floor slab, thickened to form a base if necessary, and are secured in position by a metal dowel.

Building frames in timber Most small-scale timber framed structures take the form of *post and beam* construction in which resistance to racking distortion of the frame under working load is provided by the infill panels. A proportion of solid or near-solid panels is, therefore, necessary to ensure stability and can normally be provided. The choice of connection and form of junction between members of the frame in most cases thus depends largely on the degree of rigidity required for erection purposes before the panels are fixed.

Figure 6.13 shows post and beam frames constructed with built-up solid posts and single solid beams, the latter passing between the outer column pieces, to which they are secured by nailing, and bearing on the centre piece. The foot of each post bears on the timber cill plate bolted to the dwarf wall as for frame wall construction. Bearings for the rafters are provided by fillets nailed to the beam sides. The roof extends beyond the wall panels the heads of which are secured to noggings between two closely spaced rafters.

The frames in figure 6.14 are constructed with solid posts and spaced solid beams. The latter bear on shoulders formed at the head of the posts and are secured by two bolts to produce a rigid connection. In this example some lateral rigidity results from the provision of a deep tie beam immediately below the bearing of the main beams, this also being set into the face of the posts and bolted to them. This tie, together with the substantial member fixed to the paired beams above the glazing and some solid panels under some windows in the wall panels, would provide lateral rigidity to the structure. The beam ends are shown protruding beyond the eaves. This is common in the USA but, being exposed to the weather, the ends tend to check and may warp although precautions such as a flashing over the tops of the beams or soaking the ends with moisture resistant material are adopted to minimise this.

When solid beams are used and glazing is extended to their tops as shown precautions should be taken at the junction of glazing and beam to disassociate the two so that possible movement of the beam due to shrinkage will not cause breakage of the glass.

In figure 6.15 both posts and beams are of solid timber the former being dowelled to beams and floor slab. Since a dowelled connection does not produce a rigid post to beam junction rigidity must be provided by wall panels parallel to the beams. The header running over the tops of the posts provides some rigidity to the frames during erection and ultimate structural rigidity is provided by the top and bottom solid panels forming an opening for glazing. In this example the foundations are short bored piles under each post.

Figure 6.16 shows a double height house frame with the single floor raised above ground level. The posts are

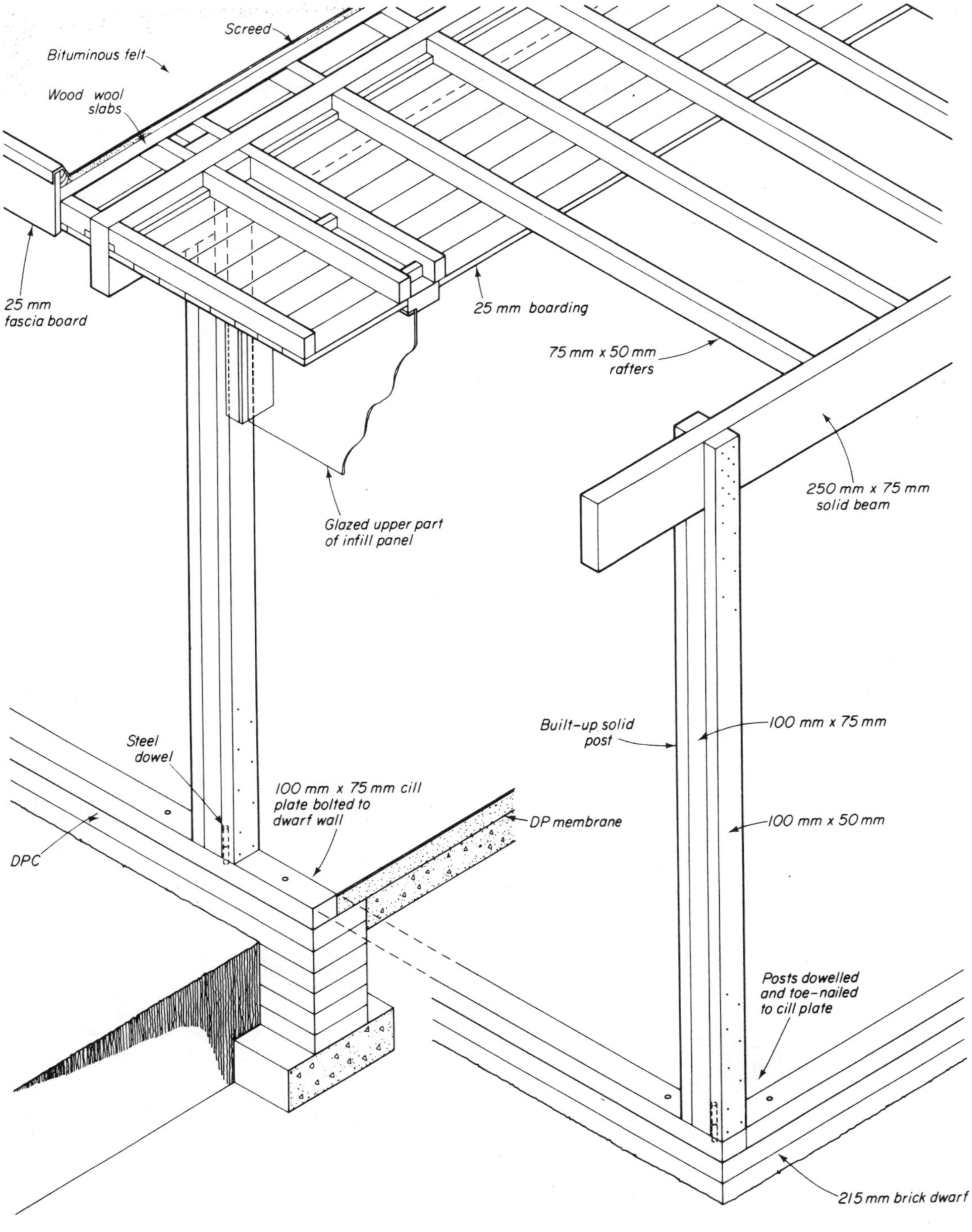

6.13 *Timber frame with built-up solid posts and solid beams*

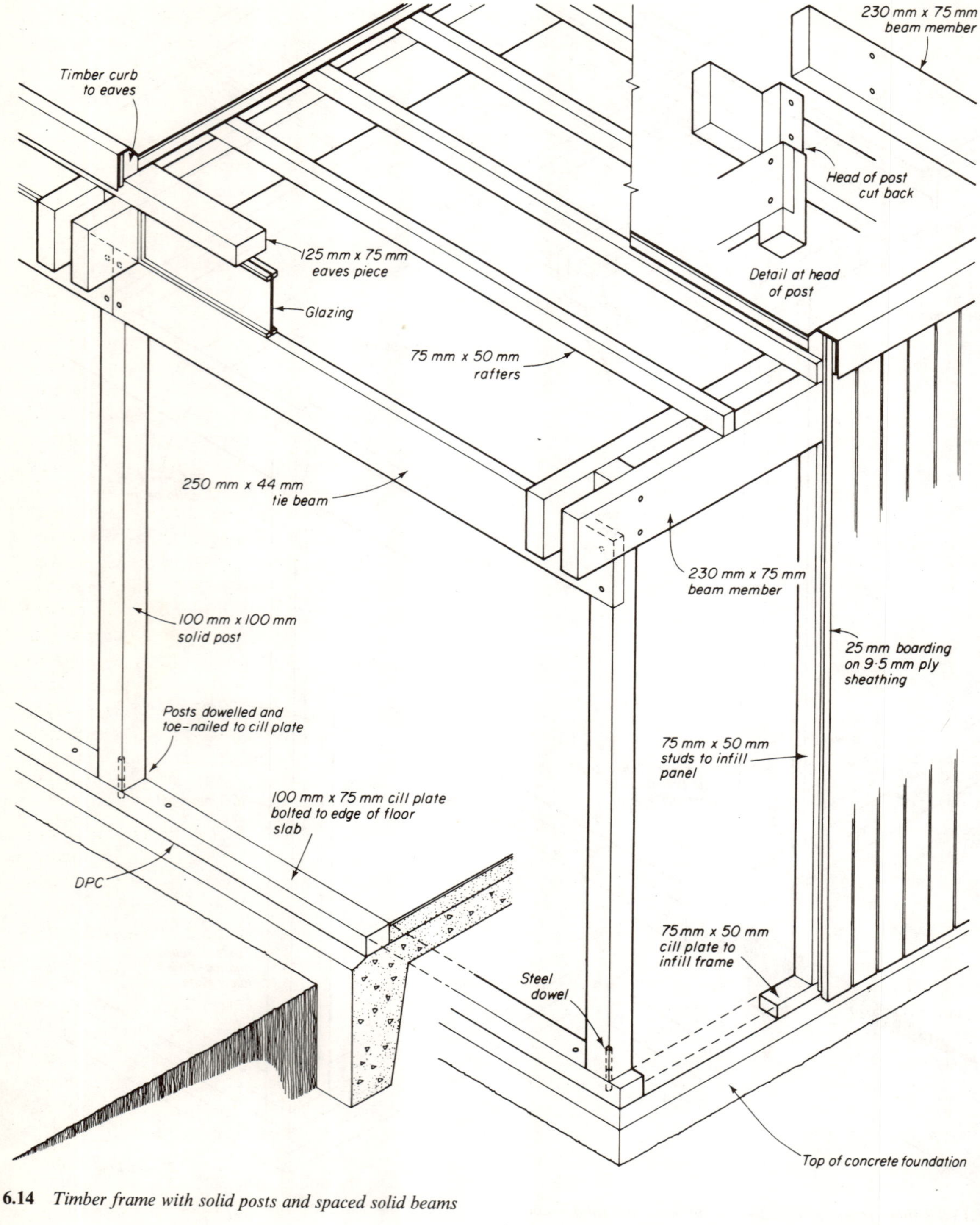

6.14 *Timber frame with solid posts and spaced solid beams*

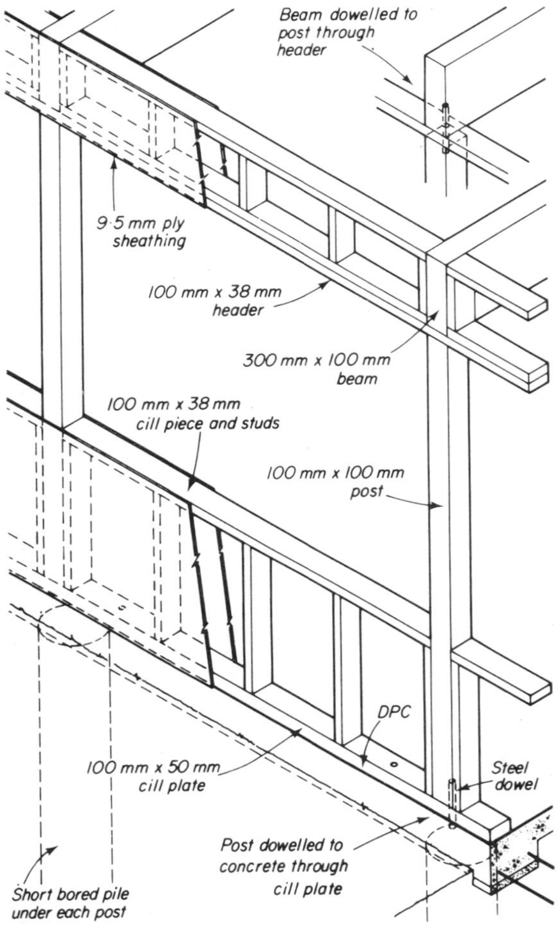

6.15 *Timber frame with solid posts and beams*

solid and the main bearing beams and lateral tie beams are formed as spaced beams with pairs of deep but relatively thin solid members. The depth of the junction between posts and beams and the use of three bolts at each connection produces stiff joints and rigidity in both directions. Packing blocks at the ends and centre of all beams provides the necessary stiffening against lateral buckling of the thin members. Further stiffening of the frame would be provided by enclosing one of the ground level bays on all four sides with solid panels, to form an entrance hall or other accommodation or, as in the steel frame on page 133, by diagonal braces of steel cable or rods placed in convenient panels at the same level.

The floor and roof consist of boarded timber joists and rafters made up into prefabricated panels bearing on top of the beams. Prefabricated wall panels are secured to the edges of roof and floor.

Prefabrication

The design of the last structure and its panel components permits full prefabrication of the parts and assembly by nut-and-bolt is simple and rapid. It is, however, a 'one-off' building. In any form of system building for a large market provision must be made for the variety of situations produced by varying spans and loading of beams and by single or two-storey buildings, by designing ranges of components with a minimum of variations in construction and dimensions which may be applied to a maximum number of building situations. Figure 6.17 shows solutions to some aspects of the problem which have been worked out in practice. (*A*, *B*, and *C*) illustrate column types in the same system, all of standard width. (*A*) is for the light loads and consists of two relatively thin but identical members of sufficient cross-sectional area to take the maximum design load for which they are intended, blocked apart to prevent them buckling. (*B*) is a built-up solid version with increased bearing area to take greater loads, the outer members and the overall dimensions being the same as in (*A*). (*C*) is an extension of (*B*) to provide further bearing area by the addition of a solid piece glued to (*B*), to receive which the rebate in the latter is formed. Thus by this means provision is made for a range of loading conditions by a minimum number of component parts with a standard width dimension.

Variations in span and loading of beams may be met by increased depth and by change of form while maintaining a standard width. Normal timber sizes set a limit to the depth of solid timber beams but deeper beams may be formed as plywood box beams as in (*E*) where imposed loads are light or by laminating thin boards in glulam construction as in (*D*).

The three-piece built-up solid column shown in (*E*) permits a splice connection to be formed for two-storey construction, by a simple variation of standard storey-height column components. The recess to accommodate the beam at the head of a standard column, formed by the omission of the centre piece at that point, is extended by the use of longer outer members. This receives the splice formed at the foot of the upper column by the use of shorter outer members. In this example the end of the ply box-beam is formed as a deep tenon which is accommodated by the column

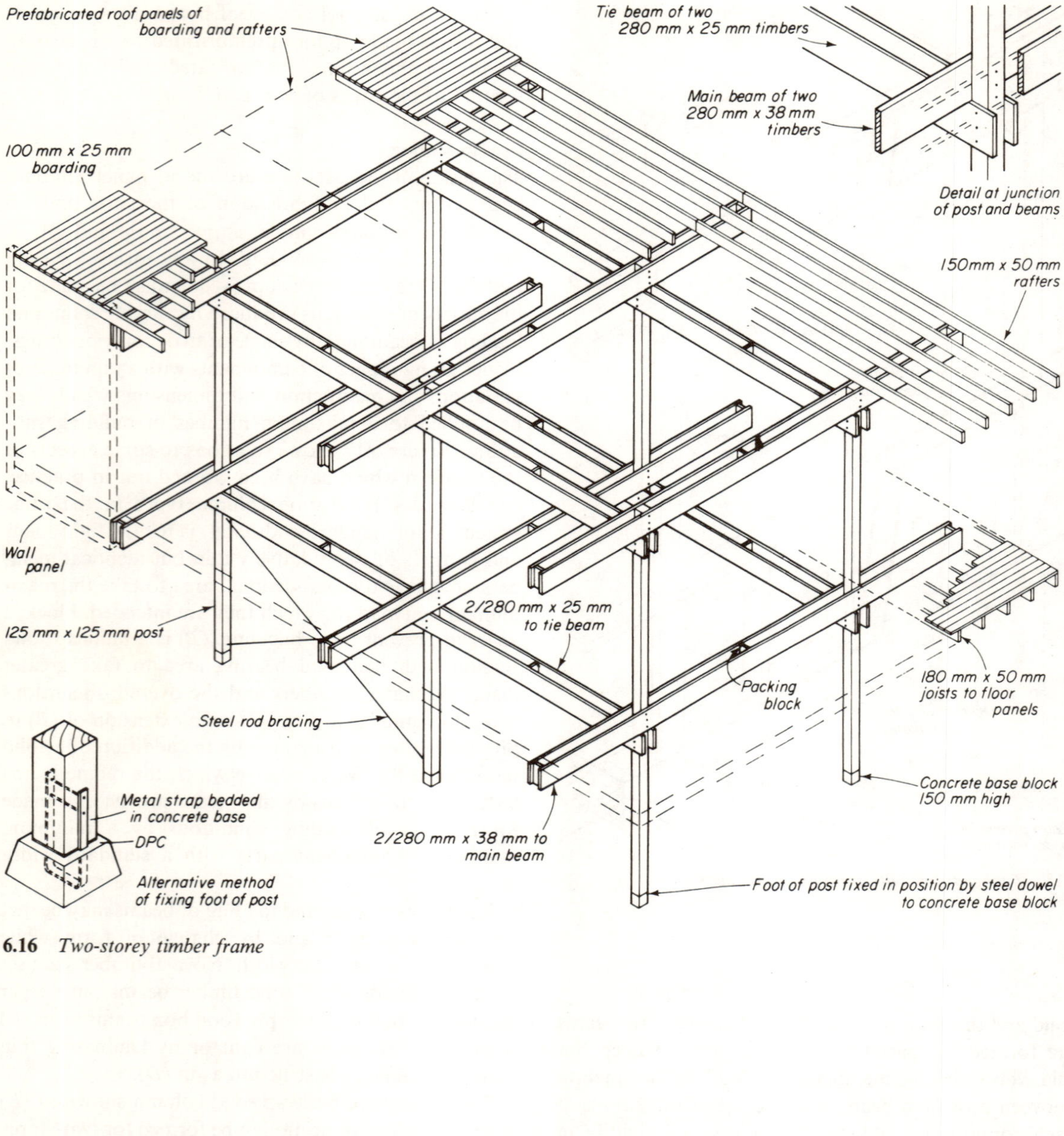

6.16 *Two-storey timber frame*

recess and bears on top of the centre piece. The connection is secured by bolts which pass through the edge members of the infill panels to hold them in position.

In single-storey framed systems the seating and fixing of beams may be accomplished by making the column shorter than the adjacent infill panels and using the latter to fix the beam in position as in (*G*) which requires the fixing of the panels prior to setting the beam in position. In (*F*) the built-up column has two

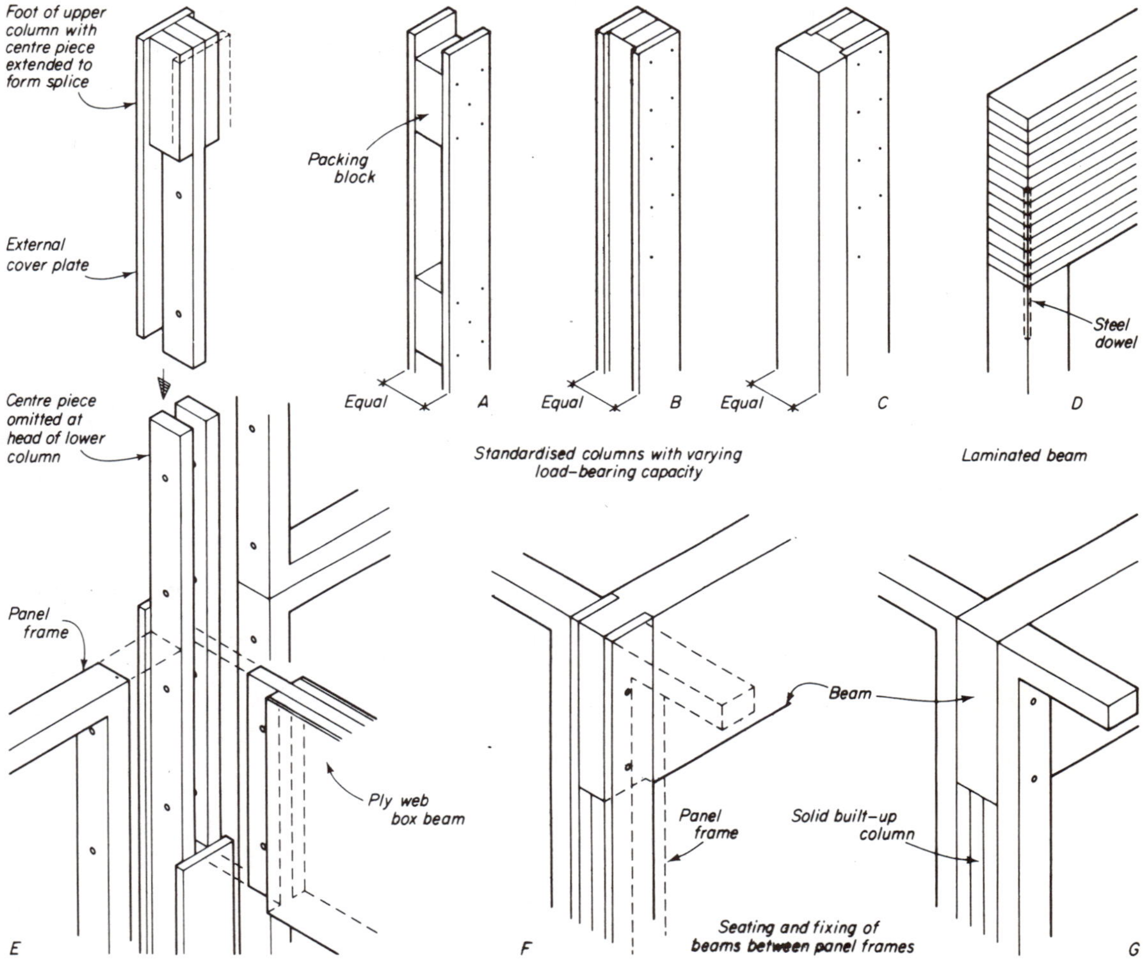

6.17 *Prefabricated timber frame techniques*

thinner outer pieces which carry up the full height and serve to hold the beam in position by nailing until the infill panels are offered up and bolted in position.

For speed and ease in assembling components relative to each other some way of positioning them, by jigs or other means, is desirable. Figure 6.18 illustrates a method of positioning panels and columns on a cill plate. The columns are notched at the foot in both directions to engage with blocks and a fillet fixed to the cill plate. The blocks fix the columns along the length of the cill piece and the fillet positions it laterally, as it does the panel components, the bottom rails of which are grooved to accommodate it.

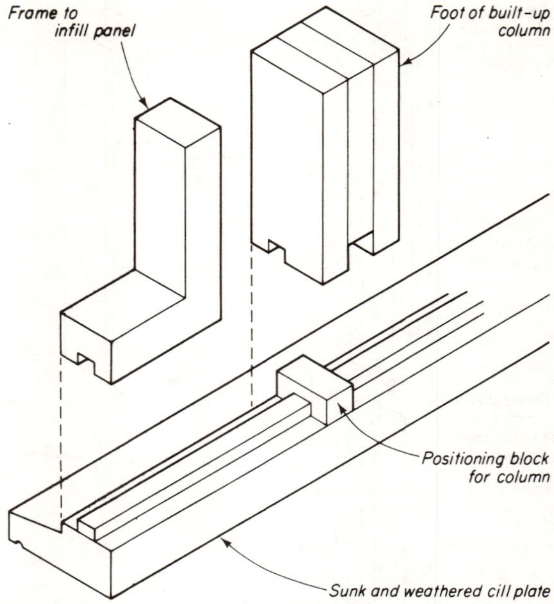

6.18 *Positioning of timber components*

7 Roof structures

A roof is an essential part of every building. Its most important function is to provide protection from the weather. In multi-storey buildings the span is usually not great and the roof is generally constructed in the same way as the floors. Domestic and small buildings of a similar nature are at present usually most economically covered by a flat roof of timber or reinforced concrete or a pitched roof of timber or light steel construction. The methods of constructing these are described in this chapter. In the design of single-storey buildings requiring roofs of medium and long span the structure of the roof is significant and is usually a critical factor. The reasons for this are discussed below and the types of roof structure which are used in these circumstances are described in Part 2.

Functional requirements

The main function of a roof is to enclose space and to protect from the elements the space it covers. In some types of roof, as will be seen later, the structure of the roof supports separate enclosing elements; in others the structural element is also the enclosing element. In either case to fulfil this function efficiently the roof normally must satisfy the same requirements as the walls. These are the provision of adequate:

- strength and stability
- weather resistance
- thermal insulation
- fire resistance
- sound insulation.

Strength and stability

Strength and stability are provided by the roof structure and a major consideration in the design and choice of the structure is that of span. The wide variety of roof types in different materials which have been developed is, in the main, the result of the search for the most economic means of carrying the roof structure and its load over spans of varying degrees. In all types of structure it is necessary to keep the dead weight to a minimum so that the imposed loads can be carried with the greatest economy of material. Where spans are large this factor is of greatest importance. In the case of small buildings and in those divided into small areas, or in which columns in relatively large areas are not objectionable, the problem is simple because the roof may then be supported at reasonably close intervals and a light, economic roof structure used. As already mentioned, in the majority of multi-storey buildings a flat roof similar in construction to the floors is normal and in multi-cell single-storey and small-scale buildings a flat or pitched roof of simple construction. In these buildings the problem is easily and economically solved; it is in wide span single-storey buildings that the problem becomes difficult. Structures of this type involve problems peculiar to themselves due to the absence of intermediate support for the roof.

In the previous chapter on page 126 reference is made to the economic significance of keeping the dead weight of a structure to a minimum. By so doing the efficiency of maximum load carried by a minimum of self-weight may be achieved. This is particularly important in roof structures where the loads to be carried are relatively light[1] and especially so when spans increase and the dead weight becomes increasingly significant. The degree of efficiency in this respect, as shown in the last chapter, is indicated by the *dead/live load ratio*, expressed in terms of loads per square metre of area covered or per metre run of roof structure. The structural problem in the design of wide span roof structures is, therefore, primarily that of achieving a dead/live load ratio as low as possible while having regard to all other factors relating to the design of the building as a whole.

In solving this problem two factors are important: the characteristics of the materials to be used and the form or shape of the roof. Reference is made on page 126 to the strength, stiffness and weight of materials and it is shown that if they are strong less material is required to resist given forces, if they are stiff they will deform little under load and the structure may be of minimum depth; if they are light in weight the self-weight of the structure will be small. All of these contribute to a structure of small dead weight. All materials, however, do not satisfy these criteria to the same degree but a deficiency in one can often be

[1] For roof loadings see Part 2 chapter 9.

countered by the use of an appropriate form of structure. For example, the stiffness of glass fibre reinforced plastics is not as great as that of some other materials although they are very strong, but economic roof structures can result by using them in curved or bent forms. Steel, concrete, aluminium, timber and plastics are all used for roof construction and economic structures may be produced with all of them if their characteristics are carefully considered and an appropriate type of structure is chosen accordingly. This subject is discussed in more detail in Part 2 chapter 9.

In addition to the dead load and the superimposed loads of snow and, possibly, foot traffic the roof must resist the effects of wind. The pressure of wind varies with its velocity, the height of the building and the locality of the building, that is its exposure to the wind. Wind may exert pressure on some parts of a roof and suction on others, both in varying degrees at different points according to the pitch of the roof as shown in figure 7.1 which indicates patterns of distribution. Higher suctions and pressures occur at the edges of the roof than at other parts and on flat roofs and those of low pitch the suction over the windward side can be considerable, with proportionately greater suction at the windward edge. An extreme case is the monopitch[1] roof with a wind blowing diagonally across it from its higher edge, when the suction exerted on the high windward corner can be considerably greater than the value laid down in the normal regulations for roof loadings. The effect of this can be increased if the roof incorporates a wide eaves overhang resulting in upward pressure on the underside in addition to the suction on top.

When very light roof coverings are used, such as some forms of aluminium sheeting, the supporting structure as a consequence tends to be light and the weight of the cladding and roof structure as a whole may not be heavy enough to withstand the uplift of excessive suction occurring during short periods of very high wind. In such circumstances the fastenings to the cladding and the fixing of the roof structure to frame or walls must be so designed as to prevent them being stripped off.[2]

The problem of wind resistance is often aggravated by large areas of lightly clad wall and roof exposed to the wind with only a comparatively small amount of structural framing by which the effects of wind can be resisted. This is typical of many large industrial buildings and often necessitates heavy foundations as a means of holding down the building structure as a whole when this is light and the building is extensive in area (see Part 2).

Weather resistance

Adequate weather resistance is provided by the roof coverings and the nature of these will affect the form and some details of the roof structure. Roof coverings are discussed in *MBS: Components*, to which reference should be made.

Thermal insulation

In most buildings, because of its position, the provision of thermal insulation in the roof is essential, particularly in the case of single-storey buildings where the roof area may exceed that of the walls, with a relatively greater heat loss. Thermal insulation, however, is rarely a factor affecting the choice of the roof type since the normal methods of providing it are generally applicable to all forms of roof. These methods vary and involve the incorporation of flexible or stiff insulating material in or under the roof cladding or structure or the use of self-supporting insulating materials such as wood wool slabs which are strong enough to act as substructure to the covering. As with insulated frame walls a vapour barrier is essential in some methods of roof insulation,

[1] A roof sloping in one direction only. See figure 7.9.

[2] See page 159 also BRE Digest 119 *Assessment of wind loads.*

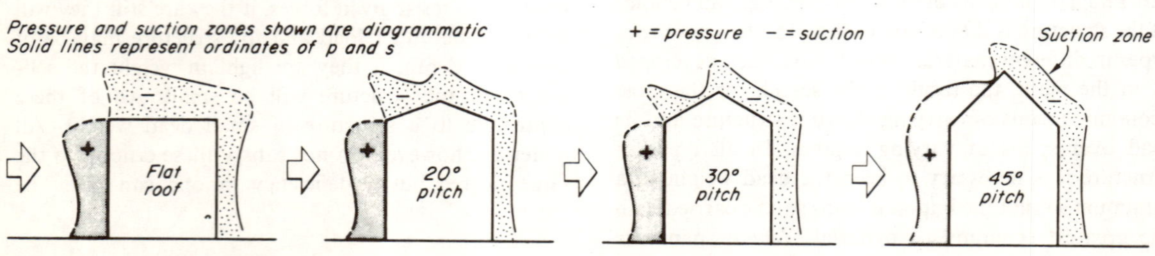

7.1 *Effect of wind on buildings*

together with adequate ventilation of the roof space, in order to prevent condensation in the roof space, resulting in possible dry rot and deterioration of the structure (see page 116).[1] In the case of concrete surface structures it is possible to use lightweight aggregate concrete for the structure to provide, either fully or partially, the required insulation.[2]

Fire resistance

The degree of fire resistance which a roof should provide depends upon the proximity of other buildings and the nature of the building which the roof covers. Adequate fire resistance is necessary in order to give protection against the spread of fire from and to any adjacent buildings and to prevent early collapse of the roof. The form of construction should also be such that the spread of fire from its source to other parts of the building by way of the roof cannot occur. These matters are discussed fully in Part 2 chapter 10.

Sound insulation

Most forms of roof construction provide for the majority of buildings an adequate degree of insulation against sound from external sources. Only in the case of buildings such as concert halls in noisy localities might special precautions be necessary and only in such cases is it likely to be a factor affecting the choice and design of the roof structure. The fact that weight and discontinuity of structure are important factors in sound insulating construction makes this problem peculiarly difficult in the case of roofs.

Types of roof structure

Roofs may be broadly classified in three ways (i) according to the plane of the outer surface, whether this be horizontal or sloping (ii) according to the structural principles on which their design is based, that is the manner in which the forces set up by external loads are resolved within the structure of the roof (iii) according to their span.

Flat and pitched roofs

A roof is called a *flat roof* when the outer surface is horizontal or is inclined at an angle not exceeding 10 degrees and a *pitched roof* when the outer surface is sloping in one or more directions at an inclination greater than this.

Climate and covering materials affect the choice between a flat or pitched roof. The effect of climate is less marked architecturally in temperate areas than in those with extremes of climate. In hot, dry areas the flat roof is common because it is not exposed to heavy rainfall and it forms a useful out-of-doors living room. In areas of heavy rainfall a steeply pitched roof quickly throws off rain, while in areas of heavy snowfall a less steeply pitched roof, say not more than 35 to 40 degrees, preserves a useful 'insulating blanket' of snow during the cold season, but permits thaw water to run off freely.

Coverings for roofs consist of *unit* materials, such as tiles and slates laid close to and overlapping each other, and *membrane* or sheet materials, such as asphalt, bituminous felt or metal sheeting, with sealed or specially formed watertight joints. With the former the open joints necessitate the use of a pitched roof so that water may run off quickly, without passing through the covering. The actual pitch or gradient of the roof slope varies with the length and width of the units and with the form of the units, many of which are made to minimise the entry of water at the joints and thus permit a low pitch. Membrane materials can be used on pitched or flat roofs. Sheet metals must be laid at a slight slope or *fall*, and some require the provision of steps or *drips* at intervals down a flat roof. Other membrane materials may be laid quite flat although there are sound arguments for giving a slight fall to all flat roofs whatever the type of covering used.[3]

Two- and three-dimensional roof structures

From a structural point of view roof structures may be considered broadly as two- or three-dimensional forms. Two-dimensional structures for practical purposes have length and depth only and all forces are resolved in two-dimensions within a single vertical plane. They can fulfil only a spanning function. Three-dimensional structures have length, depth and also breadth, and forces are resolved in three dimensions within the structure. These forms can fulfil a covering and enclosing function as well as that of spanning and are commonly referred to under the general term of 'space structures'. Two-dimensional structures include beams, trusses and rigid frames of all types, including arch ribs. Volume is created by the use of a number of such two-

[1] See also BRE Digest 180.

[2] See *MBS: Materials* for insulating materials.

[3] These considerations are discussed fully in *MBS: Components*.

dimensional members carrying secondary two-dimensional members in order to cover the required space (figure 1.2). Three-dimensional, or space structures, include cylindrical and parabolic shells and shell domes; doubly-curved slabs, such as hyperbolic paraboloids and hyperboloids of revolution; folded slabs and prismatic shells; grid structures such as space frames, space grids and grid domes and barrel vaults; suspended or tension roof structures (figure 1.2). All these forms cover space and in the case of domes are capable of completely enclosing.

As indicated earlier in this chapter, roofs may be constructed of a number of different materials, each of which in different circumstances may prove more suited than the others to the requirements of a particular building or the particular type of roof structure selected. Some types of structure may be constructed satisfactorily in more than one of these materials, but the basic principles of the structure are the same in each case.

Roofs constructed of two-dimensional members are classified as single, double and triple roofs according to the number of horizontal stages necessary economically to transfer the loads to the supports. In *single* roof construction the roofing system[1] is carried directly by one set of primary members spanning between the main supports and spaced apart at the economic span of the particular roofing system being used (figure 7.2 *A*). As the span of the primary members increases a point is reached at which it becomes more economical to use larger members spaced further apart using these to support secondary members to carry the roofing system (figure 7.2 *B*). This is referred to as *double* roof construction. In some circumstances spans are such that three sets of members are required to produce an economic structure, resulting in three stages of support. This is called *triple* construction (figure 7.2 *C*). This classification is applied to both flat and pitched roof and to floor construction.

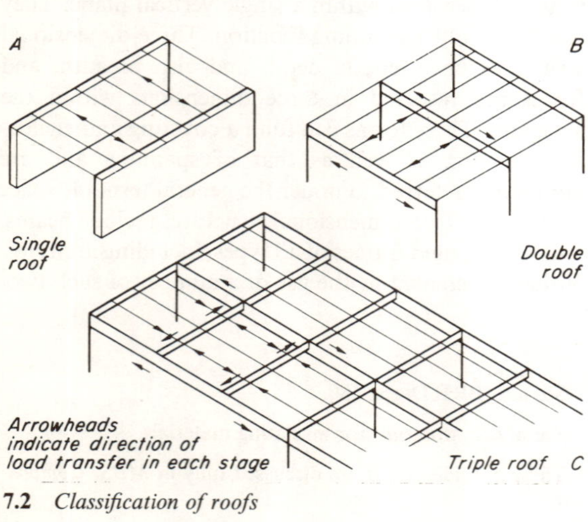

7.2 *Classification of roofs*

Long and short span roofs

Reference has already been made to the fact that span is a major consideration in the design and choice of a roof structure although other factors, including functional requirements and considerations of speed and economy in erection, have an influence as well and these are discussed in Part 2 chapter 9.

Roof structures are classified in terms of span as:

short span	up to 7·60 m
medium span	7·60 m to 24·40 m
long span	over 24·40 m

Short span construction will usually be cheapest as far as structure is concerned. The span of the structure is usually fixed by the proposed use of the building, as this will dictate the minimum areas of unobstructed floor space required. In some types of building a considerable number of internal supports will be permissible, whereas in others very large unobstructed areas will be essential. As an increase in the distance between supports usually results in an increase in the cost of the structure (page 127), the minimum spans compatible with requirements of clear floor area should always be adopted in design.

Three dimensional structures are normally not economic over short spans although certain circumstances referred to later can make them so. In this volume structures appropriate to short span construction are considered. Reference should be made to Part 2 for a consideration of three-dimensional forms and other two-dimensional forms appropriate to medium and long span construction.

7.1 FLAT ROOFS

7.1.1 Single flat roofs in timber

The roof structure consists of timber bearers called *joists* spanning between supports (figure 7.3 *A*) and is basically the same as that for suspended timber floors

[1] The term *roofing system* implies the weather-resisting covering material together with the sub-structure to which it is attached as distinct from the roof structure which supports it.

described in chapter 8. Thickness of joists and factors affecting their size are the same (see page 193). It should be noted that a greater imposed loading must be assumed for a roof with access not limited to that necessary merely for purposes of maintenance and repair than for one in which access is limited to these purposes.[1] This obviously results in smaller joists in the latter case for the same spans and spacings of joists.

The spacing of the joists depends on the material used for the base or substructure to the roof finish. 400 mm to 450 mm is normal for 19 mm tongued and grooved boarding, 400 mm for 18 mm chipboard and 600 mm for 22 mm chipboard or 50 mm wood wool slabs.

The maximum economic span of joists to flat roofs accessible for purposes other than maintenance and repair is the same as for domestic floors, that is about 4·90 m requiring 225 mm × 50 mm or 63 mm joists at 400 mm centres. In roofs accessible only for maintenance joists of this size can span to about 6·00 m. Nevertheless, shorter spans are likely to be more economic since shorter lengths of timber are relatively cheaper than long lengths and the required sections will be smaller. As with all spanning members the joists should therefore, wherever possible, span the shortest distance over any area. Deep joists will require strutting at intervals as described for floors.

The roof joists bear on 100 mm × 75 mm timber members called *wall plates* (figure 7.3 *B*, see also page 194), or, where these would have to be built-in to the inner leaf of a cavity wall, on MS *bearing bars* (figure 7.5 *A* and page 197). The function of these is to provide a bearing for the joists and to distribute their loads uniformly to the wall.

Where supporting walls carry up as parapets above the roof the joists may be supported at the bearings on wall plates, bearing bars or metal hangers in any of the ways described for floor joists and the same considerations will apply (see page 197 and figures 8.8 and 8.9). When the roof is not accessible to traffic there is less objection to the use of metal hangers on the inner leaf of a cavity wall, since the load at each bearing will be much less than in a floor and the consequent moment of eccentricity will be smaller.

The slope or *fall* of the roof, essential with some roof coverings and commonly provided with all[2], may be obtained by laying the joists to fall in the required direction. If, however, a level ceiling is required battens of timber varying in depth are laid on top of horizontal joists to form a sloping top surface. These are called *firrings* or *firring* pieces (see figure 7.3). When the joists run in the direction of the fall the firrings may be tapered strips nailed to the top of each joist (*B*) or they may be battens of varying depth fixed across the joists at appropriate centres (*C*). When the joists are at right angles to the fall similar firrings varying in depth are nailed to the top of each joist (*A*). Fixing to the top of individual joists permits a minimum thickness of firring of 13 mm. Fixing at right angles necessitates at least 38 mm to 50 mm minimum thickness to give enough stiffness; the firrings throughout will, therefore, be this much deeper. Timber boards may tend to curl slightly and should be laid parallel to the fall so that the slight corrugations which form will not impede the flow of water. This necessitates joists or firrings at right-angles to the fall. If the direction of joists or firrings prevents this the boarding should be laid diagonally to the fall (*B*). This problem does not arise with deckings such as chipboard, plywood or wood wool slabs which are now more commonly used than boarding because of the speed in laying them and of the insulating value of wood wool. The use of wood wool slabs, which require a top screed, permits the necessary fall to be formed in the screed and eliminates the need for firrings.

As already mentioned steps or drips down the roof are required when sheet lead or zinc covering is used and the direction of joists and firrings must be arranged with this in mind. The simplest way is to lay the joists at right-angles to the fall, the difference in levels in the decking required to form the drip being achieved by the user of deeper joists and firring pieces above the drip positions (figure 7.3 *D*). If this type of roof with sheet metal covering is set within parapets a *parapet gutter* must be constructed at the lower end of the roof by means of which the water is conducted to rainwater outlets (figure 7.3 *A*). The necessary falls in the gutter to the outlets may be achieved by fixing the gutter bearers to fall (*A*) or by the use of cross bearers of varying depth resting on horizontal bearers as shown in (*E*). If the gutter is long drips will be required along its length and sufficient depth must be provided between the edge of the roof and the base of the gutter to allow for these. This type of gutter is known as a *parallel* or *box* gutter.

[1] For example, in respect of houses not exceeding three storeys in height, the Building Regulations 1985, require 1·50 kN/m² in the former and 0·75 kN/m² for pitches up to 30 degrees in the latter case.

[2] Minimum falls appropriate to various flat roof coverings which should avoid the 'ponding' of water on the roof are given in BS 6229: 1982. These vary from 1:80 to 1:60.

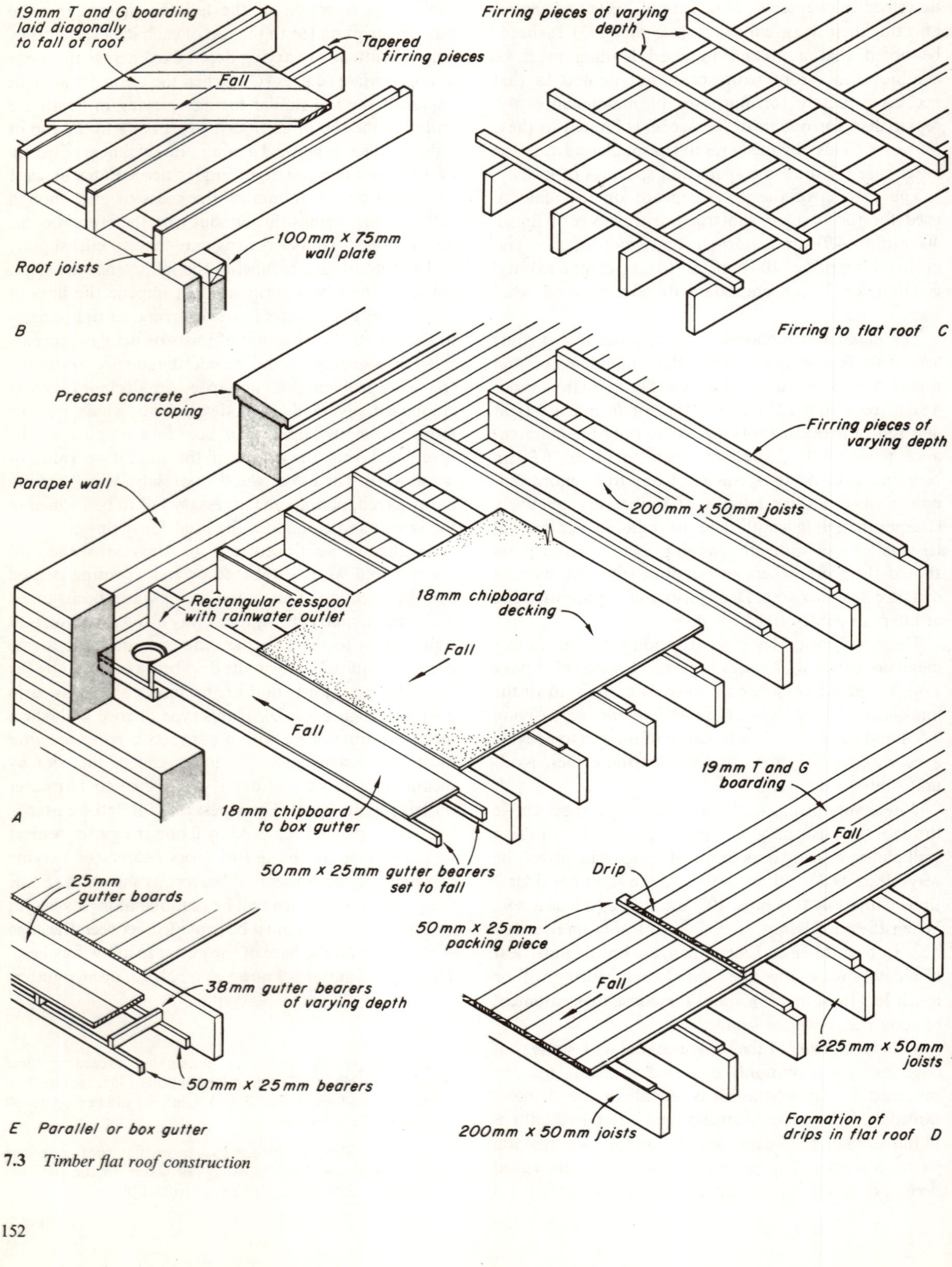

7.3 *Timber flat roof construction*

Thermal insulation
The basic timber flat roof, illustrated in figure 7.3, may be insulated in two ways: by placing the insulating material on the ceiling lining between the roof joists or by placing it above the roof decking, resulting in what is known as *cold deck* and *warm deck* construction respectively. There are two forms of the latter.

As with timber wall frames (see page 116) the possibility of condensation occurring within the roof cavities must be avoided since this may reduce the efficiency of the insulating material and cause damage to the roof members. The continuous and impervious nature of flat roof coverings, however, necessitates a somewhat different approach to the problem of preventing such condensation.

Cold deck construction In this method (figure 7.4 *A*), with the insulating material placed immediately above the ceiling, the roof deck and roof space are at or near the external air temperature and, for much of the year, substantially below that inside the building. There is, therefore, considerable danger of warm humid air from inside the building penetrating and condensing within the cold roof space. Two precautions must be taken to avoid this.

Firstly, a vapour barrier or check should be incorporated on the underside, that is the warm side, of the insulation to prevent the passage of water vapour into the roof. This involves not only the provision of a layer of material of high vapour resistance such as non-ferrous metal foil or polythene sheeting,[1] but also avoiding leakage paths for air round services penetrating the barrier or through joints in the barrier material itself and at its perimeter.

Secondly, because in practice some leakage of water vapour is inevitable, the roof space must be adequately ventilated in order to disperse such vapour and thus prevent it condensing on the underside of the cold roof deck. The Building Regulations require the space between the insulation and roof deck to be at least 50 mm deep and cross ventilation to be provided on two opposite sides of the roof by permanent vents of an area at least equal to that of a continuous gap 25 mm wide along the full length of those sides. The vents must be situated at each end of each space between the joists. Where the joists must be at right-angles to the flow of air counter-battens should be used to form a ventilation space across the roof above the joists. Adequate cross-ventilation may be difficult to achieve in flat roofs with a span in excess of 5 m or with a complex plan shape and in such cases the ventilation openings and the space above the insulation should be substantially increased. For roofs exceeding 10 m span and for those of more complex plan shape than a simple rectangle the Building Regulations require the area of ventilation to be 0·6 per cent of the roof area.

In flat roofs large enough to require sub-dividing by cavity barriers it may be necessary to incorporate ventilating cowls at certain points which penetrate the roof deck and covering in order to achieve adequate ventilation within each subdivision of the roof.

The insulating material may be in the form of mat, quilt or loose fill and may need to be secured against displacement by the flow of air through the roof spaces. The thickness required will depend upon the material used and the nature of the ceiling lining.

It will be seen that the provision of adequate cross ventilation is a primary factor in the successful functioning of this type of construction. During periods of windless conditions this will be lacking and the risk of condensation will be high, especially during cold weather. Warm deck construction, described in the following paragraphs, avoids this risk.

Warm deck construction In this method (figure 7.4 *B*)

[1] See *MBS: Materials* chapter 1.

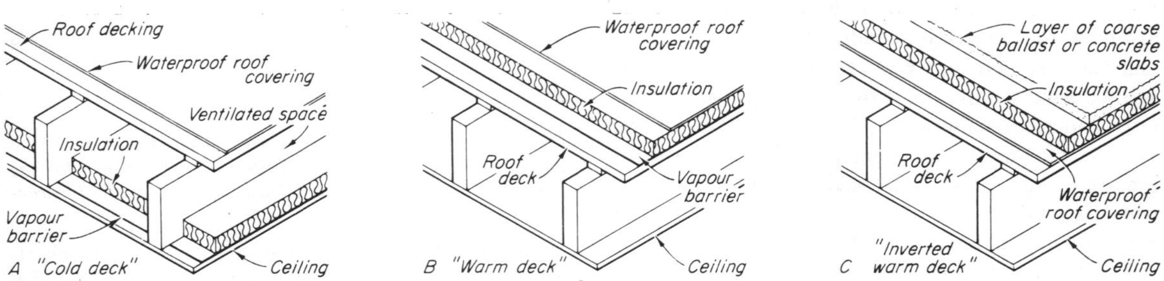

7.4 *Thermal insulation of flat roofs*

the insulating material is placed between the waterproof covering and the roof deck, with a vapour barrier on the warm side of the insulation. The material must be capable of withstanding normal roof loads and materials such as resin-bonded mineral fibre slabs and expanded polystyrene or polyurethane slabs meet this requirement.

The thickness of the insulation must be enough to ensure that the vapour barrier and roof members below it are warm enough at all times to prevent condensation of any water vapour which can pass freely into the unventilated roof space.[1]

In this form of construction precautions must be taken to prevent the thermal movement of the insulation rupturing the waterproof roof covering. In the case of expanded polystyrene, which exhibits considerable movement, an isolating layer of bitumen impregnated fibre building board is placed between the two. It is necessary to avoid wetting the insulation during construction because it is sandwiched between the impermeable layers of vapour barrier and roof covering and could not dry out. It would thus be less effective and some types might rot.

Insulation above the roof structure protects the latter from extremes of temperature and from the consequent excessive thermal movement.

Inverted warm deck construction This form of construction (figure 7.4 *C*), in which the insulation is placed above the waterproof roof covering, avoids some of the problems associated with the previous form. The insulating material must be unaffected by water and capable of withstanding normal roof loads — expanded polystyrene is a typical material. Sufficient insulation must be provided to ensure that the underside of the waterproof covering is warm enough at all times to prevent condensation of any water vapour passing into the roof space which, as in the previous form, is unventilated.

In calculating the thermal insulating value of the material used a 20 per cent loss in effectiveness is assumed to cover wet conditions. Since insulating materials are light in weight the insulation should be secured against wind uplift by a layer of coarse ballast or by concrete paving slabs. Alternatively this, together with a reduction in dead weight, could be achieved by the use of insulating slabs faced on the upper surface with glass fibre reinforced cement and bonded to the roof covering underneath with bitumen.

This form of construction has the advantage that the waterproof roof covering as well as the structure is protected from extremes of temperature and thus from excessive thermal movement.[2]

Eaves treatment

When the supporting walls carry up as parapets the roof will be enclosed by the walls. If, however, the roof is carried over the top of the walls its edge will be exposed as an *eaves* which can be finished (i) close to the outer wall face to form a *flush eaves* or (ii) beyond the outer wall face to form a *projecting eaves*.

Flush eaves The ends of the joists terminate at the outer wall face and are covered, or finished, with a 25 mm or 32 mm wrot board called a *fascia* or *fascia board* nailed to the joist ends tight against the wall (figure 7.5 *A*). If there is no eaves gutter the fascia should extend above the roof level as described below and as shown in (*B*) and its bottom edge should be splayed to form a drip. An alternative, and better detail, is to project the joists slightly and introduce a wrot timber fillet 13 mm above the bottom of the fascia to form a slightly larger drip (*B*). The outer leaf of the wall is carried up between the joists as *wind* or *beamfilling*. The return eaves parallel to the joists is formed against the face of the return wall which is carried up to the same height as the beamfilling as shown.

The thickness of the fascia board will vary with its depth. If a deep fascia is required this can be formed with tongued and grooved boards fixed to timber *grounds* if too deep for a single board. This will minimise movement and the boards can be thinner than necessary with a single board (*C*). Alternatively, exterior grade plywood can be used as in (*B*). If a narrow fascia is desired the joists may be reduced in depth beyond the joist bearing as for the projecting eaves shown in figure 7.6 *B*.

When the roof discharges into an external eaves gutter fixed to the fascia (often then called a 'gutter board') the roof finish or a flashing will turn over the board into the gutter (figure 7.5 *A*).

If rainwater drainage is required on one side only, or if drainage is entirely internal, discharge over the edges

[1] For methods of assessing the risk of condensation in the building fabric see BS 5250: 1989 *Code of practice for the control of condensation in buildings.* See also *MBS: Environment and Services.*

[2] For comprehensive lists of the advantages and disadvantages of these methods of insulation see BRE Digest 312 *Flat roof design: the technical options.*

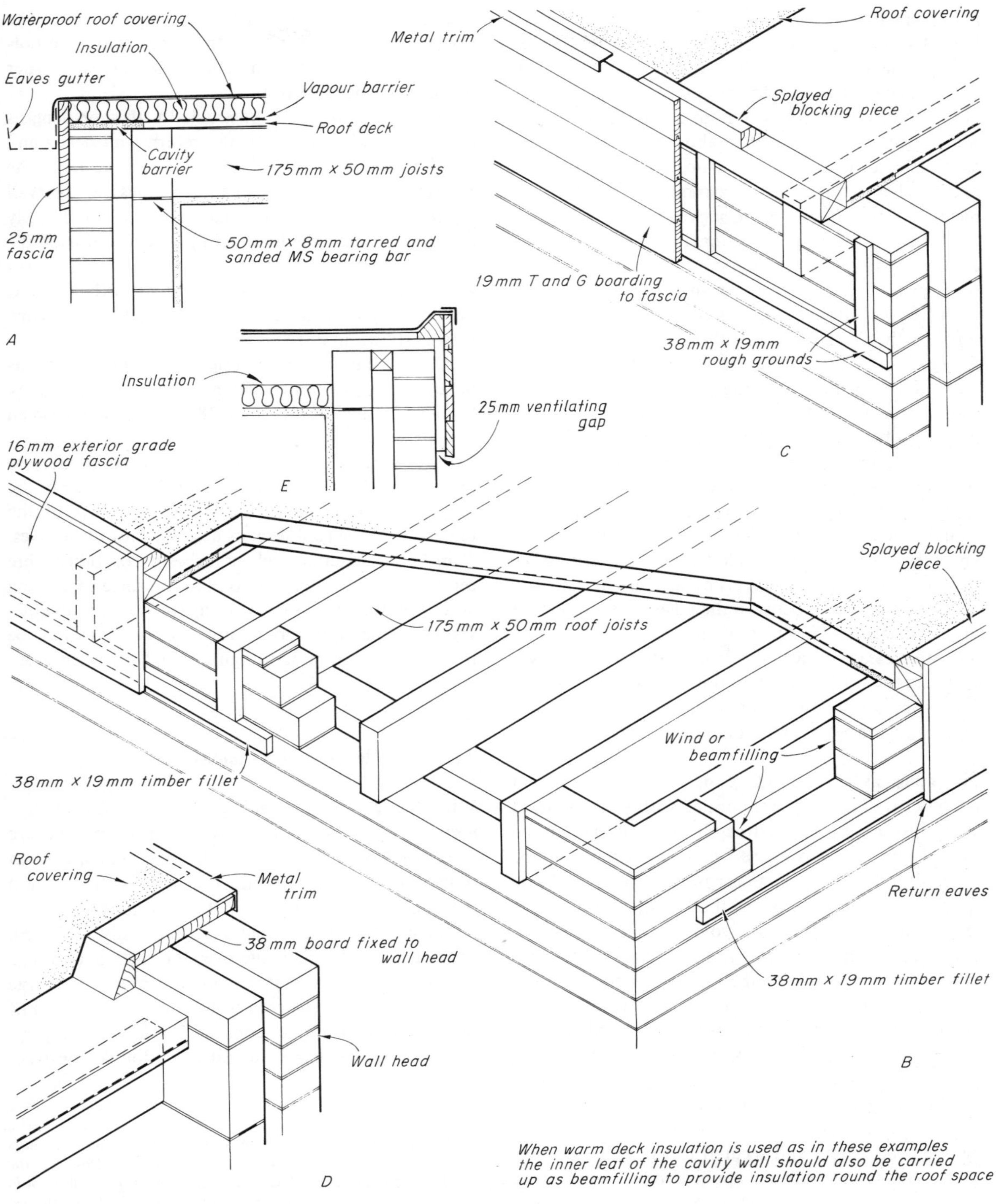

7.5 *Timber flat roofs — flush eaves*

is prevented by extending the fascia above the roof level. The upstand is stiffened by a triangular or splayed timber blocking piece which, in the case of bituminous felt coverings, also provides the splayed angle necessary at the upturn of the felt (*C*). If no fascia is required the wall may be carried up slightly above finished roof level as in (*D*) and the roofing be carried over the wall head and finished with a metal flashing or some form of metal roof trim (see *MBS: Components*).

When cold deck construction is adopted the flush eaves presents difficulties in providing ventilation to the roof space. The details shown in figure 7.5 *A* and *D* would be unsuitable unless tubular vents were taken through the fascia and wall thickness into each roof cavity. In the case of *C* the bottom fillet could be omitted to provide ventilating gaps on each side of the roof and the vertical grounds be extended to give fixing for the bottom of the fascia as shown at *E*. All such vents and gaps must be equivalent in area to a 25 mm continuous gap running the full length of the eaves as explained on page 153. To achieve this the grounds would need to be thicker than those shown in *C*. Vents and gaps of such widths should be protected against the entry of birds and large insects by wire mesh.[1] The wall head or beam filling between the joists must stop short of the underside of the roof deck to give a 25 mm minimum clearance and the wall cavity be closed by a cavity barrier of fire-resistant fibre board equivalent to 9 mm asbestos building board or of timber not less than 38 mm thick fixed between the joists.

Projecting eaves The ends of the joists project beyond the outer wall face and are finished with a fascia board as in flush eaves. The full depth of the joists may be carried through as in figure 7.6 *A* or the ends may be reduced in depth to give a narrower fascia as in (*B*). A splay cut as shown is better than a square cut at the reduction in depth as this avoids a concentration of stress at the cut. A *closed eaves* is formed by fixing a *soffit* to the underside of the joists, formed of 19 mm tongued and grooved boarding (*A*, *B*), 13 mm exterior grade plywood or 6·5 mm asbestos or other fibre cement sheet (*C*). The back of the fascia should be grooved to take the front edge or tongue of the soffit, the back edge of which in the case of plywood or fibre cement sheet should be fixed to a 25 mm fillet cut between the rafters and secured to the wall. An *open eaves*, that is one with no applied soffit, requires the planing of the joist ends where exposed to produce a finished surface and somewhat more care in building up the beamfilling between the joists as this then becomes a continuation of the main, exposed wall face. The fascia should project 13 mm to 16 mm below the bottom of the joists (or the soffit, in a closed eaves) in order to form an adequate drip.

The return eaves are constructed with short lengths of joists cantilevering over the return wall heads and fixed back to the first joist inside the wall, if the projection is no more than the thickness of the wall (figure 7.6 *A*). If greater than this it is structurally advisable to omit the first main joist and tail back the cantilever joists to the next one to give a longer lever arm. In either case support to the angle of the fascia is provided by a diagonal joist tailed back in the same way as shown in (*A*).

Where the inner leaf is stopped at wall plate level as in (*B*) and (*C*) the wall cavity may be closed by a cavity barrier of timber not less than 38 mm thick cut between the joists.

Eaves ventilation for cold deck construction may be provided by keeping the eaves soffit 25 mm clear of the wall face and stopping the wall head short of the underside of the roof deck as described for flush eaves. Thin soffit material such as asbestos or other fibre cement sheet does not lend itself to this since edge fixing at all points is necessary as in figure 7.6 *C*. Individual vents of the appropriate area could, however, be formed in the soffit as in figure 7.28 *C* between every pair of joists.

7.1.2 Double flat roofs in timber

When the span of the roof is greater than the economic limits of timber joists the span and, therefore, the size of roof joists can be kept within these limits, as with timber floors, by the introduction of cross beams, and the same methods may be used (see page 201).

Three-dimensional structures, as already indicated, are not generally economic over the short span range although structural systems prefabricated on a large scale can sometimes prove to be so. A folded form of proprietary light timber construction in this category[2] which is suitable as an alternative to double construc-

[1] BS 5250: 1989 on the control of condensation in dwellings recommends a mesh size of 4 mm but BS 6229: 1982 *CP for flat roofs* recommends that ventilation openings to flat roofs should be protected by mesh the apertures of which 'should not be less than 8 mm to avoid excessive air flow resistance'.

[1] Produced under the trade name of *BTC-Dek*.

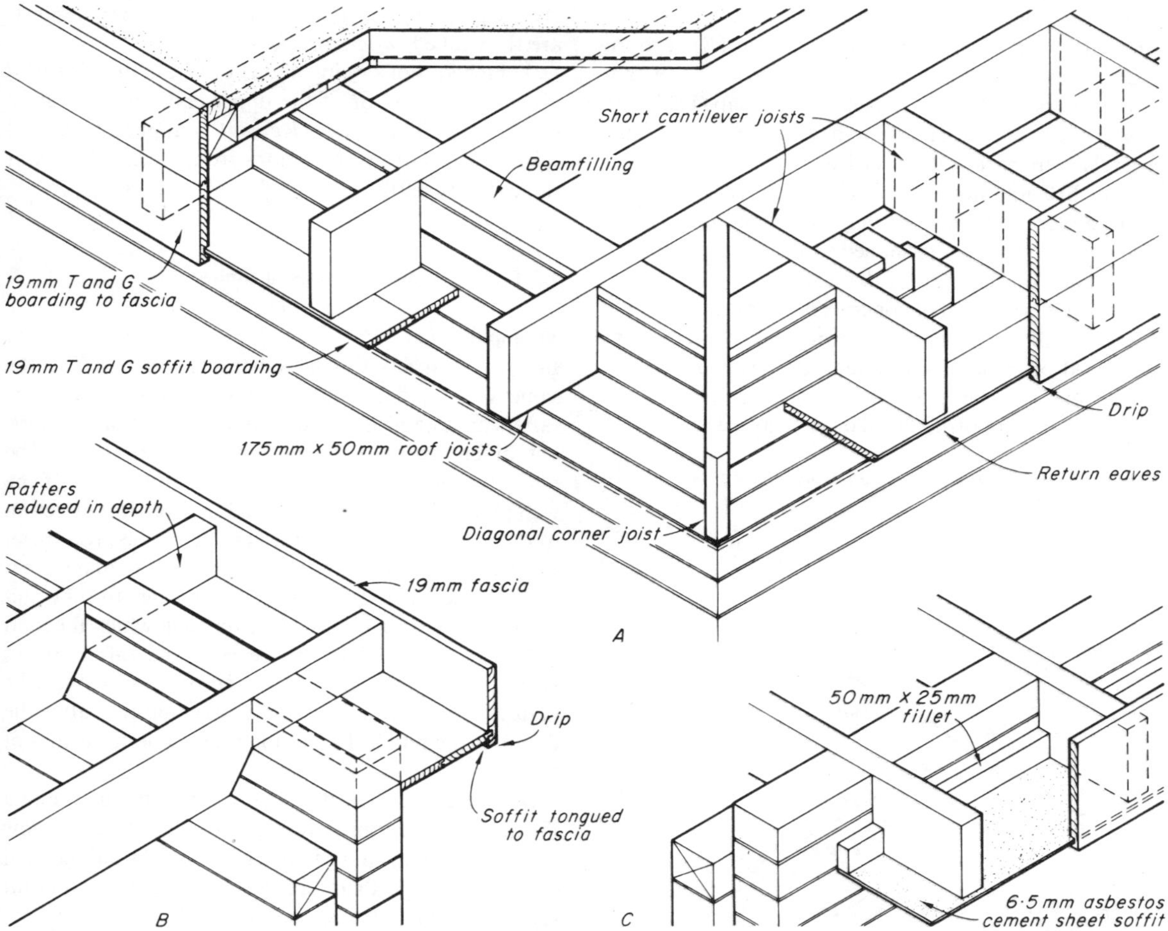

7.6 *Timber flat roofs — projecting eaves*

tion for spans in the upper part of this range is shown in figure 7.7. It will be seen that this consists basically of folded or trough units with top and bottom 'flanges' of softwood connected by plywood webs to produce a stiff form. These are constructed in a range of depths in panel widths up to three troughs wide. A number of special components are supplied for trimming openings and for forming projecting eaves on return walls.

7.1.3 Reinforced concrete flat roofs

The construction of one-way spanning roof slabs of in situ cast concrete is the same as for floors (see page 201). For the reasons given under 'Floors' precast concrete may be used in the construction of flat roofs, but it is not likely to be so economic as in situ construction for very small spans.

A flush eaves may be formed by fixing the fascia to fillets cast in the edge of the concrete slab as shown in figure 7.8 *A* or to the face of the wall if the outer leaf is carried up as in (*B*). The latter method avoids the use of edge shuttering to the slab when it is cast. A deep, flush concrete fascia may be formed by turning up the edge of the slab to form a low parapet as shown in (*C*).

A projecting eaves may be finished with a timber fascia board which should project 13 mm to 16 mm below the underside of the slab (*D*). Alternatively, if no eaves gutter is required, the edge of the slab may be formed with a weather groove on the underside and a small upstand at the top to contain the screed and prevent water discharging over the edge (*E*). Proprietary precast units[1] may be used to provide a fair-face concrete fascia. These are basically L-shaped, bed on top of the roof slab and hang over its edge. The units are relatively narrow and vertical joints occur at intervals of 610 mm along the fascia (*F*). Precast concrete eaves gutters to a flat roof may be formed in a similar manner with edge units shaped to form a gutter which are bedded on the head of the wall.

Thermal insulation of concrete flat roofs is usually in the form of either of the warm deck constructions described on page 154, although the insulation might be on the slab soffit together with a vapour barrier where the nature of the insulation required it.

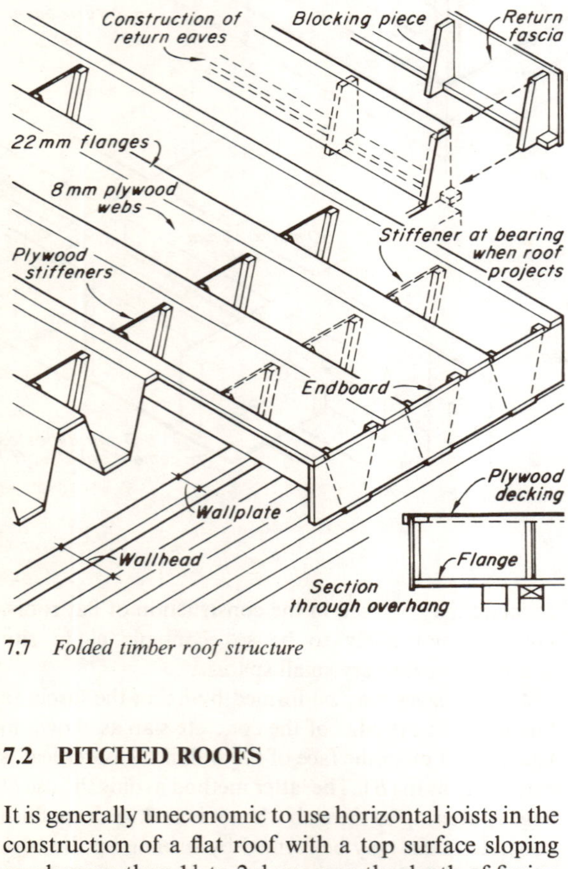

7.7 *Folded timber roof structure*

7.2 PITCHED ROOFS

It is generally uneconomic to use horizontal joists in the construction of a flat roof with a top surface sloping much more than 1½ to 2 degrees as the depth of firring becomes excessive (the minimum slope recommended for sheet metal roofing is normally 38 mm in 3·00 m which is rather less than 1 degree). Considerably greater inclinations than these are required for unit roof coverings. With these steeper slopes it is cheaper to lay or 'pitch' the joists to the required slope: the joists are then called *rafters* or *spars*.

A pitched roof sloping in one direction only is called a *monopitch roof*, or, if the upper end abuts a wall, a *lean-to roof*. The term *ridge roof* is applied to a roof sloping down in two directions from a central apex or ridge (figure 7.9).

7.2.1 Single pitched roofs in timber

Monopitch roof

In many respects this is constructed in basically the same way as a flat roof and the size and spacing of the rafters are established in the same way, but one important difference of detail occurs at the bearings. The horizontal joists of a flat roof bear on the top of the wall plate and transfer their load in a vertical direction to the plate. The inclined rafters of a pitched roof meet the plates at an angle and their load tends to make them slide off the plate. To reduce this tendency to slide and to provide a horizontal surface through which the load may be transferred to the bearing, the rafters in all pitched roofs are notched or *birdsmouthed* over the plates (figure 7.10). To avoid weakening the rafter the depth of the notch should not exceed one-third that of the rafter.

Should timber boarding be used as substructure to a membrane type roof covering, although its use is now less common as indicated earlier, it should be laid diagonally, since there are no cross firrings to permit laying with the fall (see page 151).

Eaves details are similar to those described for flat or ridge roofs depending on the pitch. If the roof is 'contained' within the walls the latter may terminate just above the tops of the rafters and be finished with the roof covering and a trim in a similar manner to figure 7.5 *D* or they may rise above the roof as a parapet in which case it will be necessary to form a parapet gutter at the lower end of the roof as for a ridge roof (see figure 7.26). More commonly, the roof carries over the walls to finish as a flush or projecting eaves as in figure 7.10 *A*. At the return walls the roof will finish as a flush or projecting *verge* as described for ridge roofs on page 177.

An external eaves gutter, although cheaper, can be avoided by forming an internal gutter behind the lower eaves fascia as shown in figure 7.10 *B*. It is essential that

[2] Produced under the trade name of *Finlock Flat Roof Fascias*.

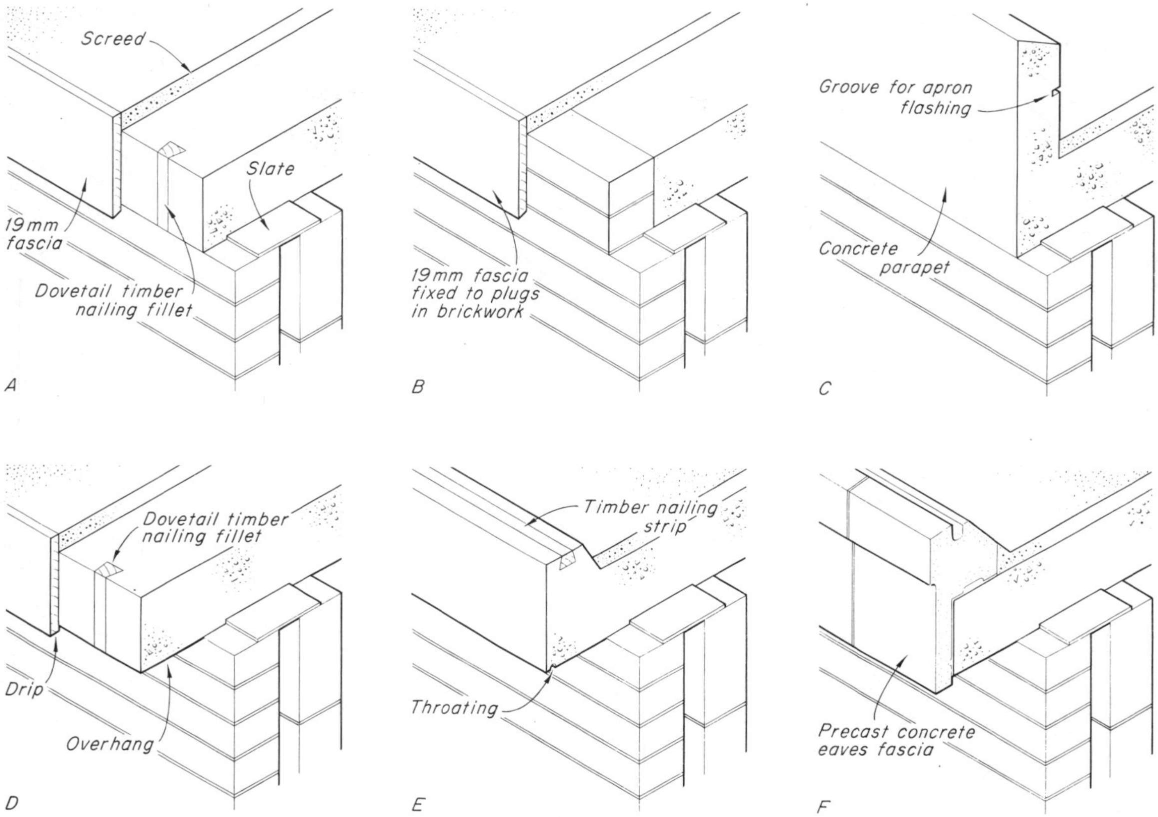

7.8 *Concrete flat roofs — eaves treatment*

the front of the gutter be at a lower level than the back so that in the event of a pipe blockage water will drain over the front rather then seep back into the roof structure and possibly into the building. Such a gutter formed behind a flush eaves occurs over the wall head and involves an internal down-pipe and gutter connection.

In the details shown in figure 7.10 the wall cavity would require closing by a timber cavity barrier as described on page 156.

Reference has been made to the need for adequate achorage against wind suction on this type of roof (see page 148), particularly if the roof covering is very light, resulting in a roof structure of small members and, therefore, of light weight. If the tendency of a roof to be sucked off its bearings cannot be resisted by its own

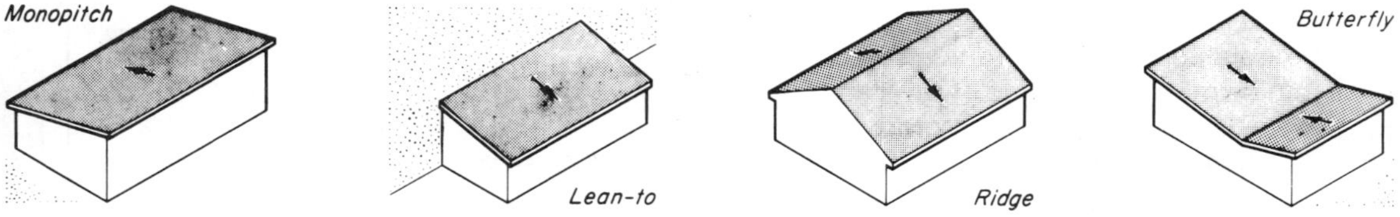

7.9 *Types of pitched roof*

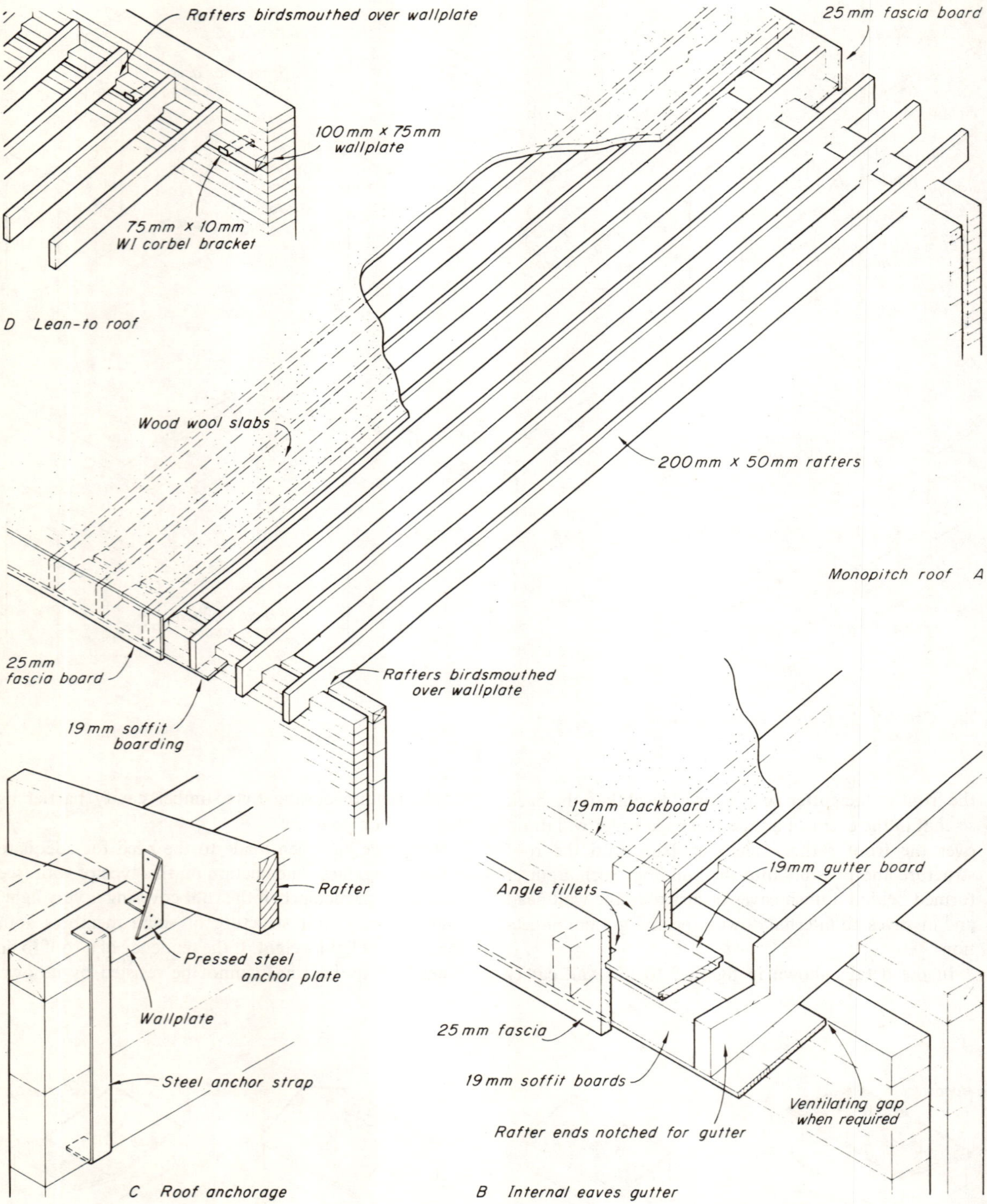

7.10 *Monopitch roof*

dead weight, including that of the roof covering, then the weight of the wall structure must also be employed to provide resistance and this is effected by means of anchorages which tie together roof and walls.[1] The rafters are anchored to the wall plate by timber fillets well nailed to both or by galvanised steel anchor plates (figure 7.10 *C*). If the wall is of masonry construction the wall plates must be secured to the wall head by galvanised bolts or straps as shown in (*C*), up to 2 m apart carried down far enough to bring in to play the required weight of wall above the point of anchorage. For small span roofs this will usually need to be five or six brick courses down since the tensile strength of the joints is generally discounted in masonry walls.[2] The Building Regulations require this type of anchorage to be provided for all roofs except a normal tiled or slated roof with a pitch of 15 degrees or more and with its main timber members spanning on to the wall head at not more than 1200 mm centres.

If the wall is a timber or light-metal frame, in addition to the roof anchorage special anchorage of the wall frame to the floor or foundation slab may be necessary.

A roof formed of two monopitch roofs falling to a *valley* (see page 178) is given the colloquial name of *butterfly roof* (figure 7.9). A cross wall or beam is required under the valley to provide support for both sets of rafters. With felt and similar coverings a triangular fillet in the valley may be all that is required to form a gutter; with low-pitch tiling or slating it is necessary to form a wide gutter by means of gutter bearers framed in a similar manner to that shown in figure 7.26 *A*.

Monopitch roofs can be constructed with the rafters laid at right-angles to the pitch when this produces the shorter span and permits smaller sections to be used.

Lean-to-roof This is a monopitch roof of which the tops of the rafters are pitched against a wall (figure 7.9). The feet of the rafters are birdsmouthed over a wall plate as for a monopitch roof, and the upper ends over a plate supported on the wall by corbel brackets (figure 7.10 *D*) or by any of the means of supporting floor joists described on page 197. All the relevant details of construction are the same as in a monopitch roof.

Couple roof

This is the simplest, but not necessarily the most economic, form of ridge roof sloping down in two directions from a central apex or *ridge* as it is technically termed. It consists of pairs, or couples, of rafters pitched against each other at their heads with their feet bearing on opposite walls (figure 7.11 *A*).

When two spanning members are arranged in this way the junction at the ridge forms a mutual support so that the span of each is the distance between this point and its lower support. The depth of the rafters in a couple roof may, therefore, be considerably less than that of those in a flat or monopitch roof of the same overall span. This is an advantage from the point of view of economy of rafter material, but the arrangement of rafters results in a tendency for the ridge to drop under the roof load with a resultant outward spread of the rafter feet.[3] In order to keep the roof stable this outer spread or thrust must be resisted by sufficiently heavy supporting walls. If the walls are tall they will, therefore, be thick and expensive. For 215 mm solid or 250 mm cavity walls of normal height the roof must be limited to a maximum clear span of about 3·00 m to keep the thrust within acceptable limits. The clear roof space given by this roof can, however, be used with advantage over wider spans than this if the roof pitch is steep and the eaves are low (figure 7.11 *B*). This has the effect (i) of reducing the outward thrust of the rafters and (ii) of reducing the height of any supporting walls and, therefore, their tendency to overturn, so that their thickness may be kept to a minimum (see pages 38 and 42).

The feet of the rafters are birdsmouthed over wall-plates and the upper ends butt against a flat board called a *ridge piece* or *board*, to which they are nailed. This board facilitates fixing of the rafters and keeps them in position laterally.

Close couple roof

This roof results from the introduction of horizontal members to tie together the feet of each pair of rafters and prevent their outward spread (figure 7.11 *C*). This forms a simple triangulated structure and produces vertical loads on the supports with no tendency to overturn the walls so that their thickness need take no account of this. These members, known as *ties*, are spiked to the feet of the rafters at plate level and if they are used to support a ceiling, as commonly is the case, they are called *ceiling joists*.

[1] The total weight required may be found by calculating the maximum wind suction by reference to CP3: chapter V: Part 2.

[2] See Part 2, beginning of chapter 4 on this.

[3] See chapter 3 figure 3.11.

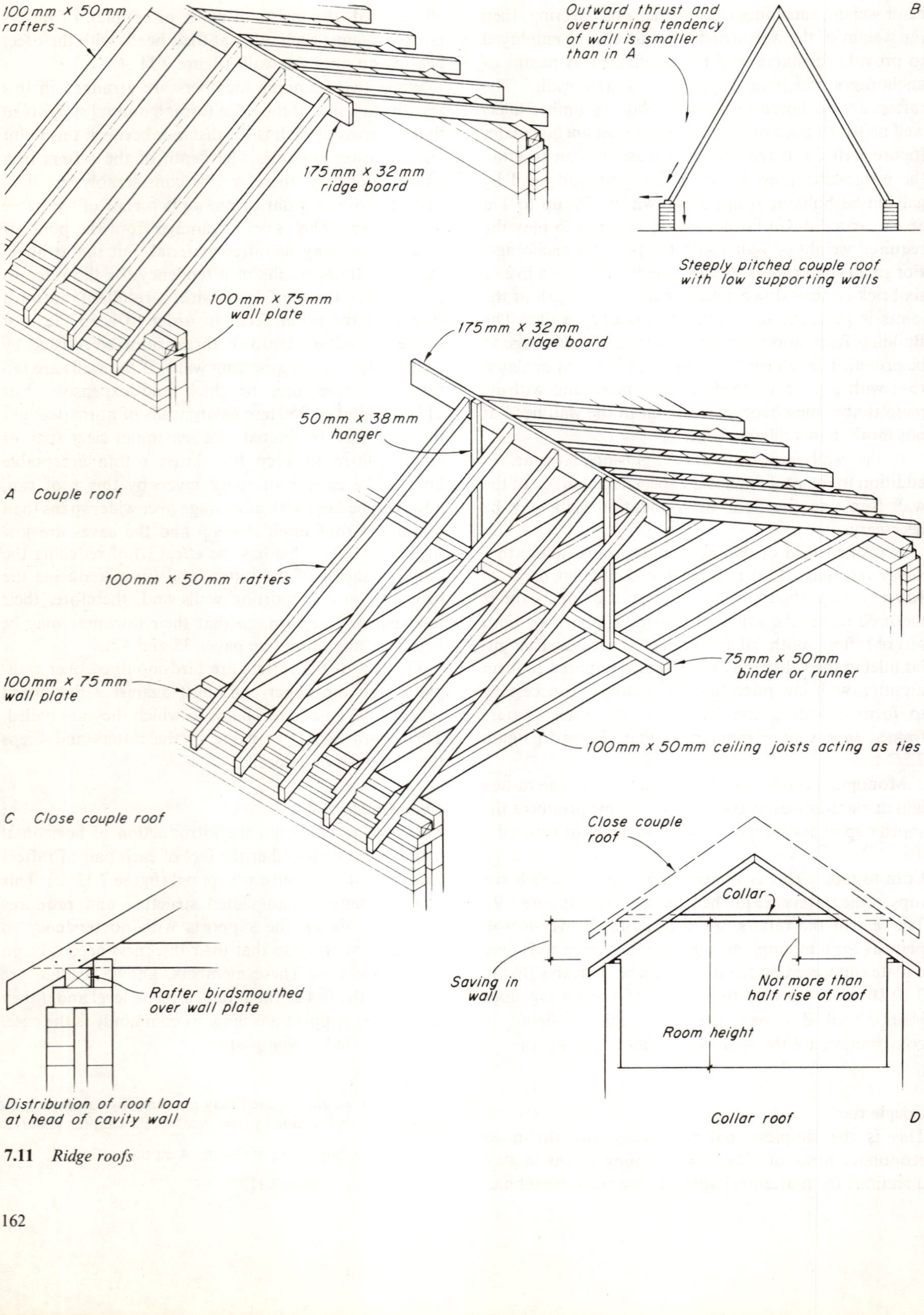

7.11 *Ridge roofs*

The maximum economic span of this roof is about 6·10 m, this being limited not by the spread of the rafters but by the economic sizes of the roof members. It is generally found most economic to restrict the depths of rafters to about 100 mm and, depending on the weight of the roof covering and the pitch and spacing of the rafters, this depth can be used over spans of about 4·60 m to 5·20 m.

The function of ceiling joists as ties can be fulfilled by quite small sections but, as they act also as beams supporting their own weight and that of the ceiling, they tend to sag or deflect and they must be large enough to keep this within acceptable limits.[1] For spans of the order given above quite large ceiling joists would be necessary and it is found more economic to reduce their effective span by suspending them from the ridge as shown in (*C*). The longitudinal 75 mm × 50 mm *binder* or *runner* skew nailed to the joists permits the *hangers* to be fixed to it at every third or fourth joist spacing rather than to each joist, thus economising in timber. Fixing of hangers to runners should be deferred until the roof covering has been laid in order to avoid deflection of the ceiling joists due to the transfer through the hangers of any slight movement of the roof structure as it takes up the load.

Collar roof

In this roof tie members are used but at a higher level than the feet of the rafters and they are called *collars* (figure 7.11 *D*). It can be used for short spans not exceeding 4·90 m when it is desired to economise in walling, since the ceiling will be raised and the roof may, therefore, be lowered on the walls to the same extent for a given height of room. The influence of the collar on the spread of the rafters is less marked the higher it is placed and half the rise of the roof is the maximum height at which it should be fixed. The size of the collars is the same as for close couple ties of an equivalent span. In the past a dovetail halved joint at the junction of collars and rafters was normal but this involves considerable labour and it is cheaper and stronger to use a bolt and timber connector (see page 170).

Thermal insulation

Pitched roofs may be insulated either in the plane of the rafters or, if a horizontal ceiling exists, in the plane of the ceiling (figure 7.12). Because of the inclination of the rafters the first position requires a greater area of insulation than in the ceiling plane but it must be adopted for couple and collar roofs and for mono-pitch or lean-to roofs without horizontal ceilings. Problems with regard to condensation arise in both methods as in the case of cold deck flat roof construction and are approached in precisely the same way. Condensation may occur on the underside of the cold, impervious underlay to slating or tiling or on the underside of any decking provided to take a waterproof membrane covering on, for example, a monopitch roof.

In ceiling insulation the material is commonly laid between the joists as a quilt although slabs, mats or loose fill can be used (figure 7.12 *B*). A vapour barrier must be placed below the insulation and the precautions in respect of this outlined under cold deck flat roofs must be taken (page 153). The roof space above must be adequately cross ventilated by means of permanent vents in two opposite eaves.

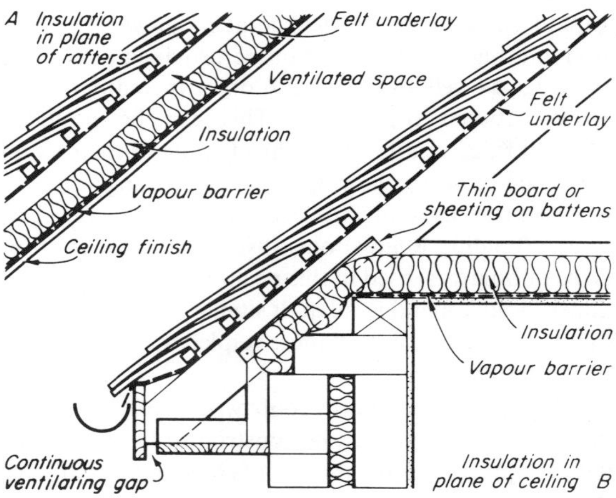

7.12 *Thermal insulation of pitched roofs*

The ventilation area at each eaves for roofs with a pitch less than 15 degrees must be at least equal to a continuous gap along the eaves 25 mm wide, with the same provision for increasing span and complex plan shape as for cold deck flat roofs. This may be reduced to 10 mm wide where the pitch is 15 degrees or more. In addition, gable or ridge vents may be incorporated with advantage.

The roof insulation must be so arranged at all the edges as to avoid such gaps between it and the wall insulation as would form potential cold bridges. In ensuring this at the eaves there exists the possibility of the roof insulation blocking the ventilation paths from

[1] See Part 2, 6.1.2 regarding these limits.

eaves to roof space and means of avoiding this must be adopted, such as thin boards fixed between the rafters over the wall head (figure 7.12 *B*) or proprietary eaves ventilator panels or tubes placed in the same position. The gap so formed, separating the felt underlay from the insulation, should be at least 25 mm irrespective of the eaves opening size.

When insulation is placed in the plane of the rafters (figure 7.12 *A*) the space above the insulation should be at least 50 mm deep as for cold deck flat roofs and the ventilation area at the eaves should always be equivalent to a continuous gap 25 mm wide, irrespective of the pitch of the roof. Where the insulation is much thicker than 25 mm rafters deeper than 100 mm, which is normal for domestic ridge roofs, will be required in order to provide for the necessary sag in the felt underlay and the 50 mm space for ventilation. A vapour barrier must be incorporated below the insulation and high level ventilation be provided at or near the ridge as well as at eaves, in order to ventilate each space between the rafters. This can be provided through ridge ventilators or through tile ventilators on each side close to the ridge, giving openings at least equivalent to a 5 mm wide continuous gap as required by the Building Regulations. Insulation in the form of slabs or boards is used in this position or, alternatively, paper faced mineral wool quilt similar to that shown in figure 5.43, but with the flanges stapled to the underside of the rafters.

Mono-pitch and lean-to roofs with insulation in the plane of the rafters must be constructed in accordance with the provisions for ridge roofs having insulation in the same position. The high-level ventilation to a mono-pitch roof which carries over the walls as in figure 7.10 *A* will be through the top eaves; that to a lean-to roof may be through the roof covering as near as practicable to the wall face or may be arranged at the actual junction of roof and wall.

As elsewhere, the application of these methods of insulation has been omitted from the illustrations of pitched roof construction in order to maintain the clarity of the structural detailing.

7.2.2 Double or purlin roofs

When the span of a roof is more than 6·10 m and requires in a couple type roof rafters much greater than 100 mm in depth it is cheaper to introduce some support to the rafters along their length, thus reducing their effective span, rather than to use large rafters. This support could be in the form of a strut to the centre of every rafter resting on a suitable bearing below, such as a partition or wall but, as in the case of ceiling joist hangers referred to above, it is more economical in timber to introduce a longitudinal beam on which all the rafters bear and to support this member at intervals greater than the rafter spacing (figure 7.13 *A*). The introduction of this beam, or *purlin* as it is called, as a second stage of support brings the structure into the double roof classification (see page 150). Although this introduces extra members into the construction the total cube of timber in the roof (and the weight of the roof) rises less with increase in span than if the rafters were increased in size.

The purlins may be supported directly by cross walls or partitions at sufficiently close spacing along the length of the purlins (*B, C i*) or by struts off any suitably placed walls, partitions or chimneys (*A, C ii*). The size of the purlins will be governed by the weight of the roofing system, the spacing of the purlins (ie the length of rafter supported) and their span. As with rafters an increase in span results in increased size and cost of purlins and the span should, therefore, be kept within economic limits. Depending on the combination of weight and rafter length a 225 mm × 75 mm purlin will span from about 2·50 m to 3·70 m. If the spacing of available supports is such that purlins much larger than this are required it may be better to select an alternative method of construction.

Purlins may be placed vertically or normal to the rafters. The former is preferable when the purlin bearing is directly on walls (*B*) or on vertical struts, the latter is sometimes more convenient when inclined struts are used, which is the case when supports do not occur immediately under the purlins (*A*). Where possible inclined struts should be paired so that those to opposite purlins meet at the same point and bear against each other over the support. If this should result in struts at an excessively low angle a spreader piece nailed to the top of a ceiling joist may be used to increase the angle of the struts as shown in (*A*). Collars at the inclined strut positions enable the purlin to be positioned, provide it with initial support until the struts are fixed and overcome any tendency of the loaded purlin to force down the heads of the struts when positioned as in this example.

When suitable cross walls exist, such as separating walls to terrace houses, and the purlins bear on these, the size of the purlin may be minimised by providing strut support in line with the purlins, as shown in *C iii*, provided the roof pitch is great enough to allow sufficient depth above the ceiling plane for the struts.

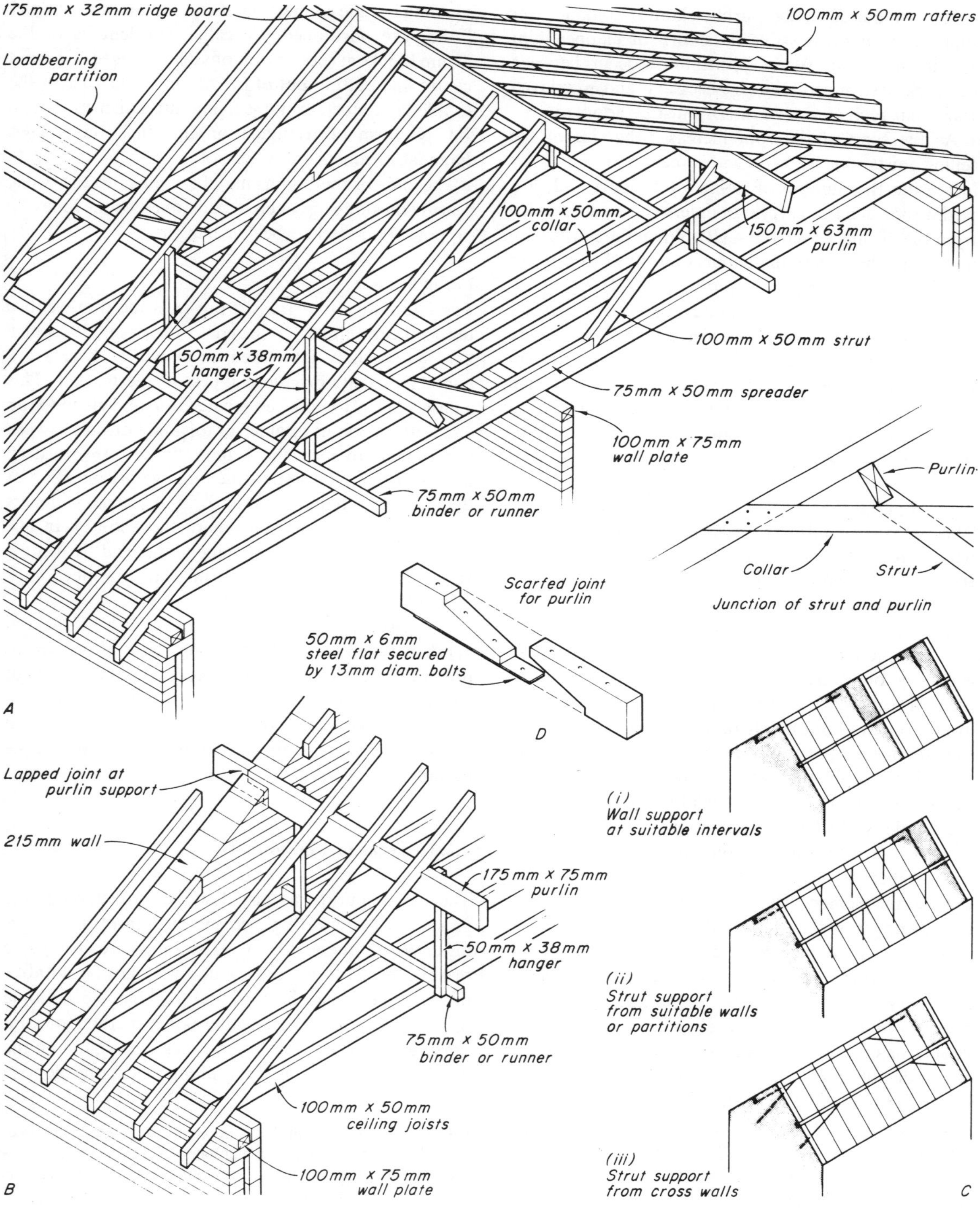

7.13 *Purlin roof*

Joints required in the purlins should be made over supports wherever possible in the form of a lapped joint as shown in (*B*). Where joints must occur between bearings a stronger joint is necessary and a splayed *scarf joint* such as shown in (*D*) must be adopted.

As the span of the roof increases the size of the ceiling joists can be kept within economic limits by increasing the number of points of support and in a purlin roof hangers carrying binders can be suspended from the purlins as shown in (*B*). When the purlins are normal to the rafters the hangers are fixed to a rafter face immediately above the purlin as in (*A*).

Where no support exists at intervals over which solid timber purlins of an economic size can span, but where suitable widely spaced cross walls exist, then deep beam purlins may be used. The maximum span over which they may be used in these circumstances depends to a large extent on the depth available for the beam. Two types are discussed below:

Trussed purlin

This is a *trussed, lattice* or *framed beam* or girder all of which are synonymous terms for a beam built up of triangulated members (figure 3.35). As pointed out in chapter 3 for a given load and span as the depth of a beam increases the bending stresses at top and bottom decrease and less material is required in the beam. This economy of material can be developed further by concentrating the majority of the material in the beam at the top and bottom where bending stresses are at a maximum (see page 44). In the trussed beam structural depth is obtained with a minimum of material at the centre or *web* by means of relatively thin triangulating members which connect the top and bottom *flanges* or *booms*. For maximum economy bending stresses in the members should be avoided as far as possible. To this end the members should be arranged on the 'centre line' principle as far as is practicable, that is to say at each junction of members their centre lines[1] should intersect at one point (figure 7.14). For the same reason loads should be applied only at the node points (see page 51). With trussed purlins, however, the rafters are closely spaced along the top boom and do not all bear at a node point; some bending therefore occurs and the boom size must take account of this.

The trussed purlin can be constructed wholly in timber or in timber with steel rod tension members as shown in figures 7.14 and 7.15. In the example with steel tension members the compression members are diagonals and are relatively long. This, as explained in chapter 3, is a disadvantage from the point of view of buckling but the connection of rods to booms is simpler if they are positioned vertically. The lengths of the 50 mm blocks at the connections are arranged to permit 'centre line' setting-out of the beam. In the case of the double-lap timber[2] example fixings are satisfactory with the 'N' form of triangulation in which the longer, diagonal members are in tension.

The ends of the booms may be built-in to the walls, bearing on stone or concrete padstones, or they may be supported on steel corbels or hangers.

Purlin beam

The alternative to a trussed purlin is the thin-webbed timber beam already referred to on page 139, which may be specially fabricated or of which there are a number of mass-produced types on the market. This consists of a plywood web rebated into and glued to top and bottom booms or glued at top and bottom between two timbers to form the booms (figure 7.16). In deep beams of this type some stiffening against buckling of the thin web is required in the form of vertical stiffeners glued at intervals on each side of the web (*A*). In one proprietary beam this stiffening is obtained by using a vertically corrugated ply web instead of applied stiffeners (*B*).

A trussed purlin invariably makes use of the full depth between rafters and ceiling joists as shown, to provide direct support to the latter without hangers but when ply-webbed purlin beams are used they are unlikely to be as deep as this, except in very low-pitched roofs, and hangers for the ceiling joists would be required.

7.2.3 Triple or trussed roofs

The use of purlins as just described presupposes the presence of supporting elements at appropriate spacings. Where these do not exist or where, for some reason, this form of construction may not be suitable, for example, when the roof span is large and multiple purlins are necessary, an alternative method of supporting purlins is by structural members spanning the width of the roof at intervals along its length, the tops

[1] The term *centre line* in this context, although generally used, can apply strictly only to members of symmetrical cross-section. What is implied is the line passing through the centroid of the cross-section of the member.

[2] See section on roof trusses, page 169 *et seq*, for methods of jointing in this form of construction.

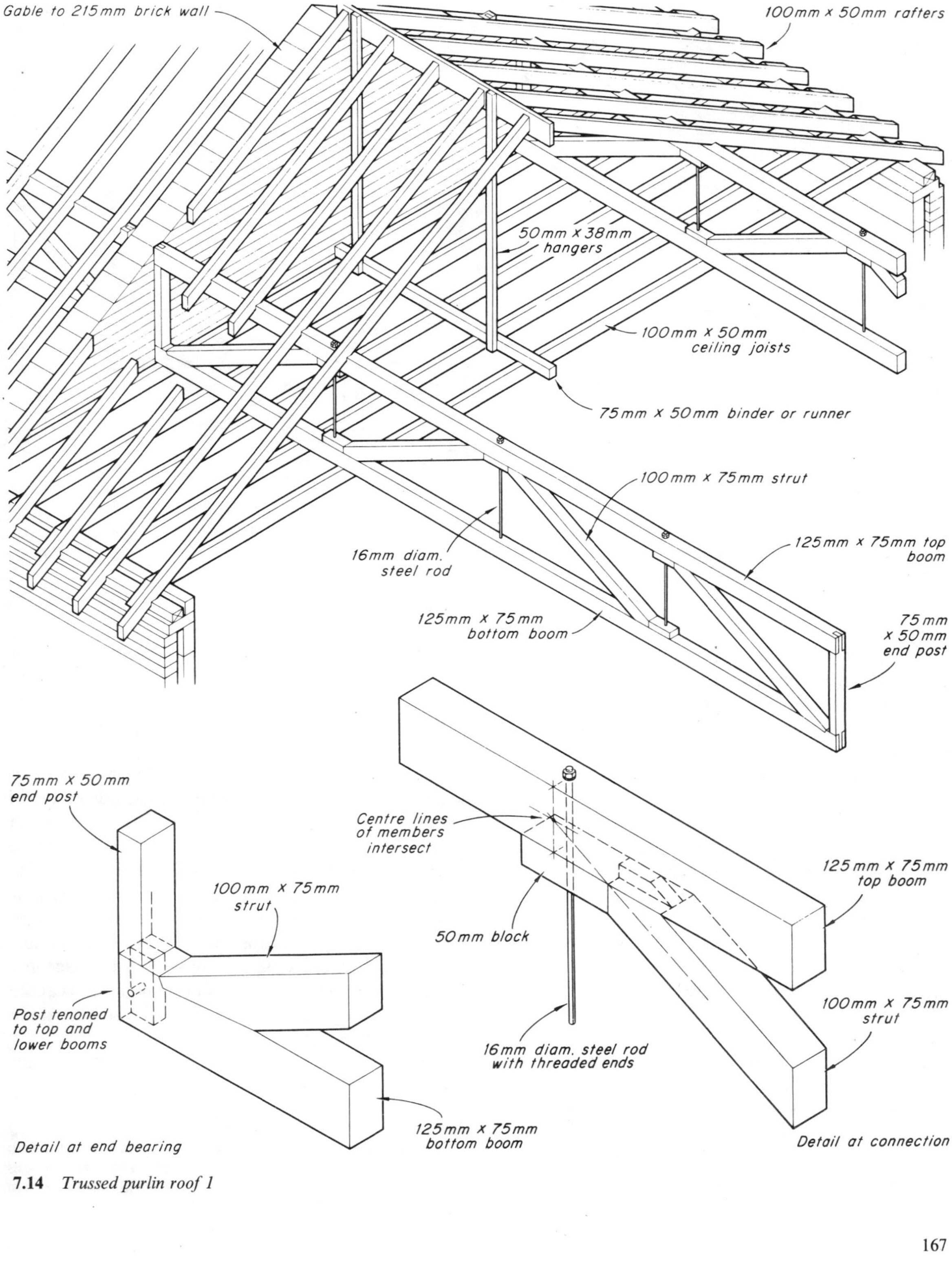

7.14 *Trussed purlin roof 1*

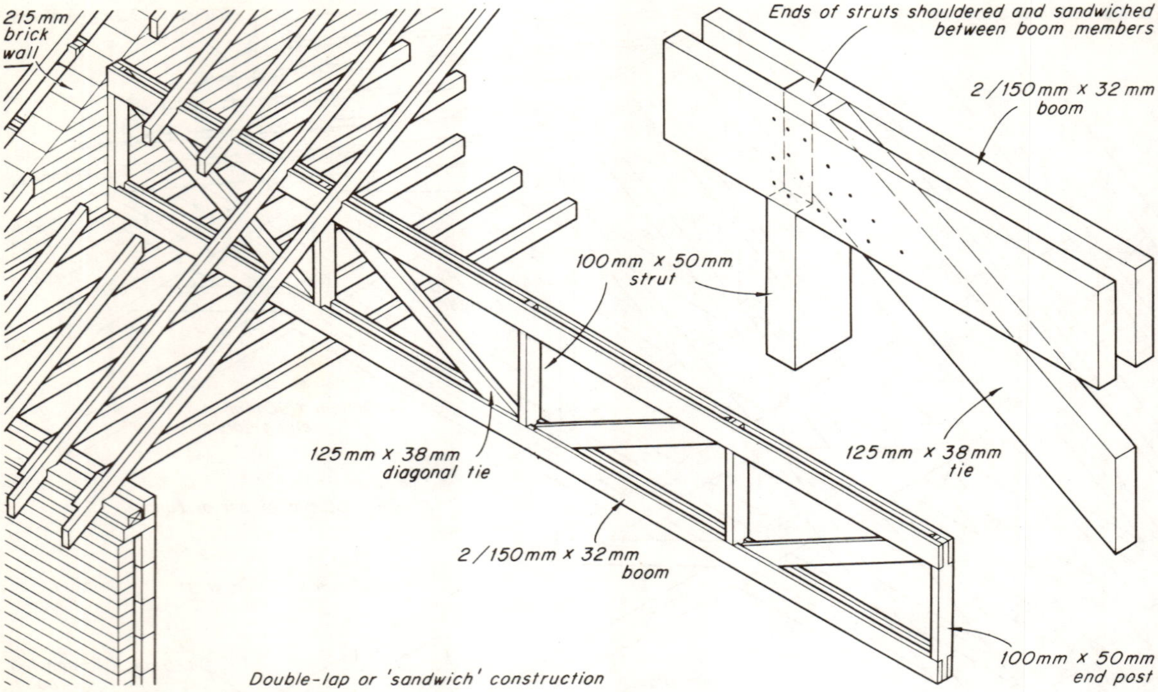

7.15 *Trussed purlin roof 2*

of which follow the pitch of the roof. These may be in the form of either a triangulated structure known as a *roof truss* (see figure 7.17) or of deep rafters fixed at their feet rigidly to a pair of supporting columns to form one structural component. The latter are called *rigid frames* and are discussed later (see page 184).

A roof truss consists essentially of a pair of rafters (or a single rafter in a monopitch roof) triangulated to provide support for the purlins, preferably at the node points (figure 7.17 *A*, *B*). For short span roofs two rafters lying in the same plane as their neighbours may be triangulated to carry purlins which are fixed immediately under them, so that the purlins are in the same relative position to the other rafters which they in fact support (see figure 7.20). These trusses are placed at relatively close centres. For wider spans resulting in large loads on the truss members the size of a normal rafter is usually too small to be used in the truss and separate rafters are triangulated and carry the purlins on their backs. These rafters, therefore, lie below the level of the normal rafters and do not directly support the roof covering. The rafters of the truss are called the *principal* rafters and the normal rafters the *common* rafters.[1]

Considerations affecting the form of triangulation or trussing are referred to on page 51. This depends to some extent on the span, which determines the number of purlins required and thus the number of nodes in the triangulation through which the purlin loads should be applied (figure 7.17 *B*, *C*). The bracing members vary in length and there is sometimes economic advantage in choosing arrangements which reduce the length of the struts (compare *D* and *E*, figure 7.17). This will reduce their tendency to buckle and thus reduce their size although problems of jointing may override this consideration as indicated under *Trussed Purlins*. Where support to ceiling joists is provided through runners carried by the trusses there is obviously advantage in a form of triangulation which results in node joints at the points where the runners bear at equal distances along the tie member (*E*).

Some of the forms of triangulation adopted are illustrated in this volume and in Part 2.

[1] In certain circumstances over short spans it is economic to fabricate *every* pair of common rafters into light trusses without the use of purlins in the roof. These are called *trussed rafters* and are discussed on page 175.

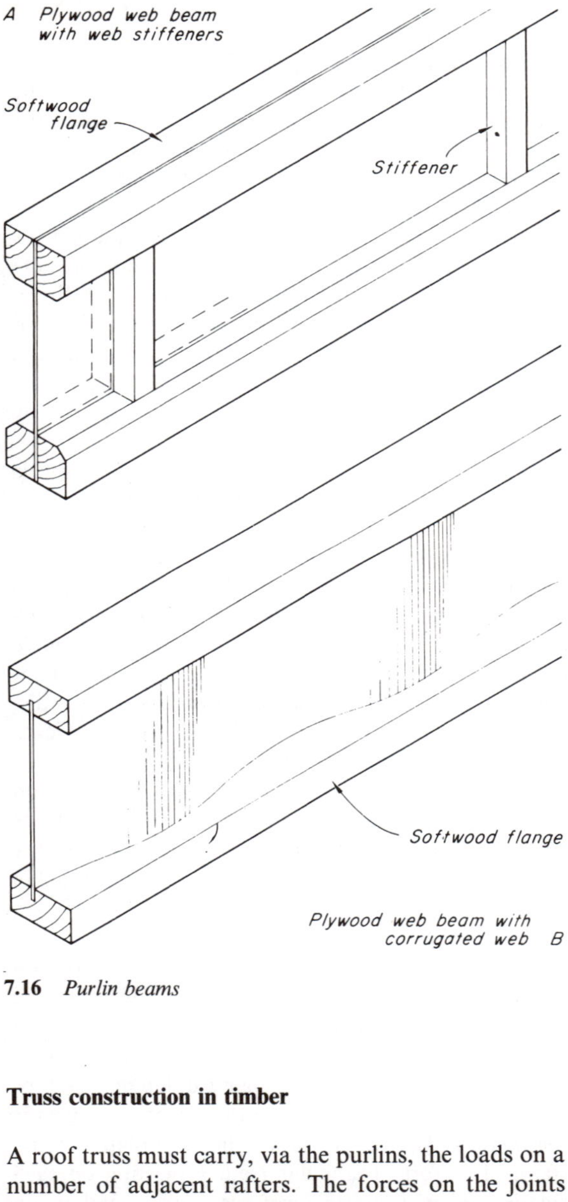

7.16 *Purlin beams*

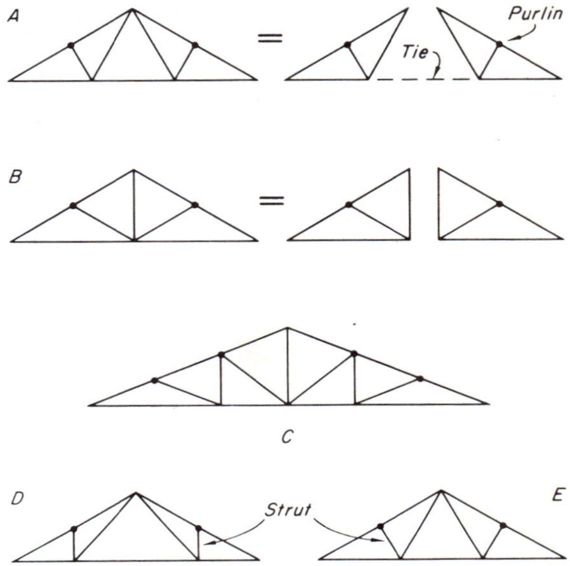

7.17 *Roof trusses*

Truss construction in timber

A roof truss must carry, via the purlins, the loads on a number of adjacent rafters. The forces on the joints between its members are, therefore, greater than those on the joints in a single or double roof structure and the use of one or two nails commonly used to secure members in the latter is insufficient in a truss.

The detailed construction of a truss depends largely on the method adopted for joining the parts. Earlier methods, involving mortice and tenon joints, necessitated relatively large amounts of timber at the junctions and, therefore, large heavy members (often larger than justified by the stresses in them) and the incorporation of large metal straps particularly at the tension points since the mortice and tenon joint is efficient only in compression. This type of truss is exemplified by the traditional king-post and queen-post trusses which, for these reasons, are now obsolete.

Modern methods permit efficient joints to be made with greater economy of material, the amount of timber at the joints often being no more than that required for the actual members being joined. With all modern methods the design of the joint is invariably based on calculations. The methods include bolt and connector joints, nailed joints and glued joints and, sometimes a combination of two.

These methods require the members to be laid one against the other, or *lapped* as it is termed, in order to make the joint as in figures 7.15 and 7.18 (*A*) or, alternatively, require the use of cover plates or *gussets* when the members butt one against the other (figure 7.18 *B*).

If two members lap the joint is called a single lap joint. If one is lapped by two other members, that is sandwiched between them, it is called a double lap joint (figure 7.15). When the latter technique is used it is, therefore, often termed 'sandwich' construction.[1] In a single lap joint the joint is under eccentric loading. For

[1] This term is preferable to 'laminated construction' which now implies timbers built up of thin boards glued together (see page 139).

small span trusses carrying light loads this is not significant but when the joints carry large loads eccentricity should be avoided by the use of double lap joints. Double members are also used in order to obtain a satisfactory arrangement of members in the truss as a whole for jointing purposes.

Sandwich construction enables the necessary sectional area of a member to be obtained by the use of relatively thin members, any double members in compression being blocked apart and fixed together to provide the necessary stiffness as illustrated in figures 7.19 *B* and 7.21 *B*.

The use of gussets permits members to butt against each other in the same plane, avoids eccentric loading on the joints and provides, where necessary, greater jointing area than is possible with lapped members, often an important factor in nailed and glued joints (7.18 *B*). Arrangement of members on the 'centre line' principle is usually possible with gussets; with lapped members eccentricity at some joints cannot always be avoided.

Nailed trusses Jointing by nails is the least efficient of the three methods referred to above, but it is a traditional and simple method. By preboring nail holes and using wide, thin members to provide ample fixing area efficient structures may be obtained, particularly where lightweight roof coverings are used.

An example of the application of nailing in this manner is shown in figure 7.18 *A*, where sandwich construction is used to carry corrugated asbestos cement sheeting over spans up to 6·10 m. The principal rafters and horizontal tie are each formed by two boards, 32 mm thick, and the struts and secondary ties are 100 × 50 mm scantlings sandwiched between, the joints at these points being made by direct nailing between the members. As the rafter and tie members lie in the same plane and butt against each other at the feet of the truss it is necessary to use gussets to effect a joint at these points. The gussets here are formed by 25 mm boards on each side set normal to the rafters and securely nailed to each member. The extension of the gusset by two vertical boards increases the rigidity of the whole truss. The double members at the feet are blocked apart by 50 mm packing pieces and at the ridge the rafters are secured to each other by a 25 mm board on each side.

Struts and secondary ties project beyond the rafters and 50 mm cleats are fixed at the intermediate purlin position to form seatings for the purlins. By joining them together the struts and cleats also serve to stiffen the thin rafter members which, being in compression are liable to buckle.

These trusses would be spaced 3·00 m to 3·60 m apart depending on the weight of the roof covering and the size of the purlins used. The point loads from the truss at its bearings are spread on to the walls by steel bearing plates as shown or by concrete templates built into the brickwork.

The purlin spacings shown in this example are for small section corrugated sheeting. The intermediate purlins impose a point load on the rafters and, therefore, induce bending stresses. Since, however, the roof covering is light these stresses will be small and it is more economic to allow for them in the size of the rafters rather than to form nodes at these points by extra bracing members.

When self-supporting coverings such as these sheets are used they are laid directly on the purlins as in this example, but when the roofing requires a base such as battens, boarding or other roof decking needing support at closer intervals it is then cheaper to support the base on common rafters at the required spacings carried in the traditional way on purlins at the node positions only. This usually results in less timber content than if the purlins are placed at very close intervals.

When loading and span conditions require thicker members and where lapped joints do not provide sufficient nailing area,[1] single thickness construction with gussets throughout may be used. By this means larger areas are available for nailing and all joints may be laid out on the 'centre line' principle as illustrated in the details shown in figure 7.18 *B*.

Bolted and connectored trusses Timber connectors are metal rings or toothed plates used to increase the efficiency of bolted joints. They are embedded half in each of the adjacent members and transmit load from one to the other. There are many different types, of which the most commonly used for light structures is the *toothed plate* connector, a mild steel plate cut and stamped to form triangular teeth projecting on each side which embed in the surfaces of the members on tightening the bolt which passes through the joint. For greater loads *split ring* connectors are used, but these require accurately cut grooves to be formed in each piece of timber (figure 7.19 *A*).

[1] This is governed by the number of nails required and the edge and spacing distances which must be allowed — see Part 2 chapter 9.

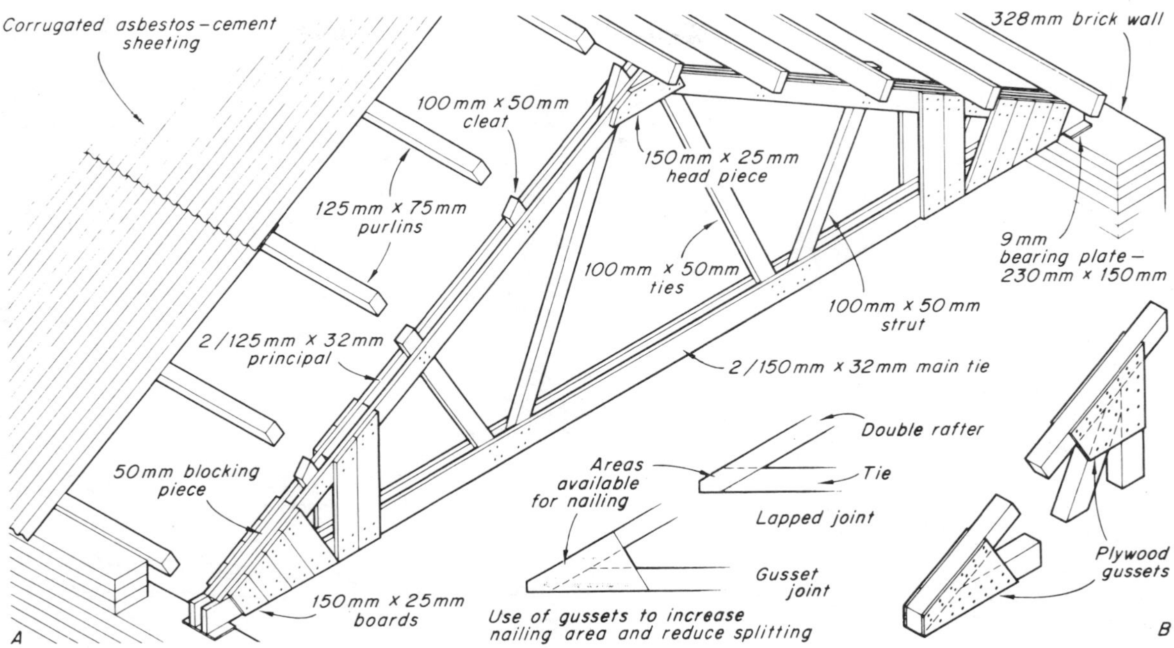

7.18 *Nailed timber truss*

Jointing by connectors and bolts permits thicker timbers to be used and its application is illustrated in figure 7.19 *B*. This truss is for a span of 7·60 m and is designed to be spaced at 3·90 m centres and to carry large section corrugated asbestos cement sheeting, which is self-supporting over a span of 1·40 m, and a ceiling.

Rafters and horizontal tie are of double members with single member secondary ties sandwiched between. Struts are of double members placed on the outside of rafters and tie. This arrangement permits 'centre line' setting out at all joints where three members meet. It also permits a single bolt to effect the joint.

Gussets are required at feet and ridge, firstly, because the main members do not overlap and, secondly, in order to obtain a greater fixing area for the number of bolts required at these joints. Since only one bolt is required at the foot of the rafter the gusset here need be no deeper than the tie, with a packing piece of the same size in the central space. To avoid the use of very long timbers the members of the main tie are joined or *spliced* at the centre using a central splice plate and four sets of bolts and connectors.

The joints in this truss are made with split ring connectors at each interface on 13 mm diameter bolts, with 50 mm square washers under bolt head and nut to prevent them sinking into the wood when the nuts are tightened. The projecting ends of struts and ties are necessary in order to obtain the minimum end distances beyond the connectors (see Part 2). It will be noted that the double members in the rafters and the long struts, which are compression members, are stiffened between the node points by 50 mm packing blocks securely spiked in position.

A variation of this type of truss is shown in (*C*). This is designed to be supported by columns the connection with which is stiffened against lateral movement by the triangulated and, therefore, stiff junction created by a *knee-brace* joining truss and column head. This is formed by extending the lower secondary tie to connect with the column some distance below the truss bearing thus rigidly uniting the two. In order to obtain a satisfactory junction with the column and to provide the necessary cross-sectional area for the knee-brace the secondary ties in this example are made of double members placed on the outside faces of the truss, and the struts are single members. As this truss is not designed to take a ceiling load the struts and ties are smaller, except those forming the knee-braces which must resist wind stresses. To provide for the greater number of bolts required at the feet, due to wind loads transferred to the truss, larger gussets are necessary at

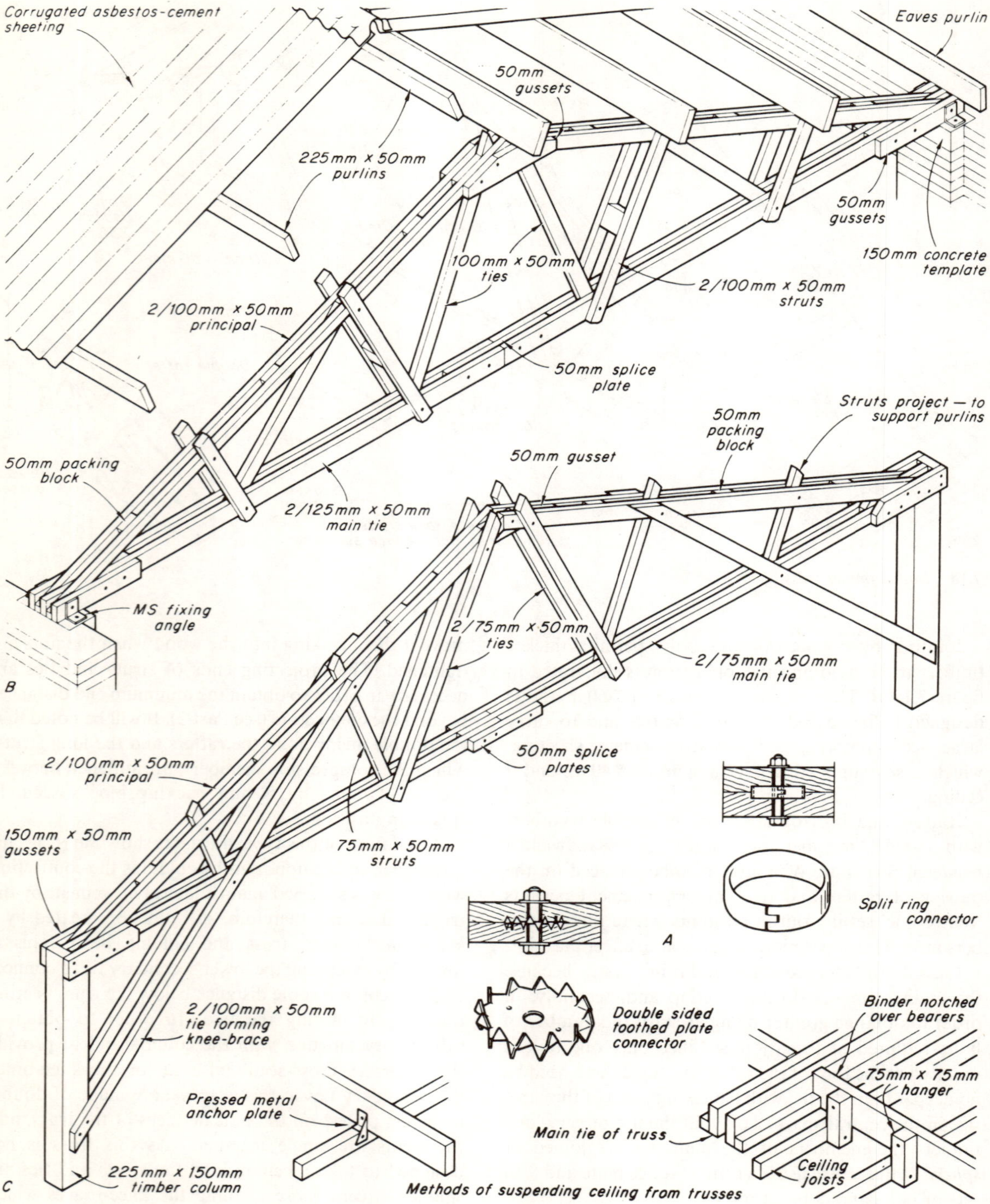

7.19 *Bolt and connector trusses 1*

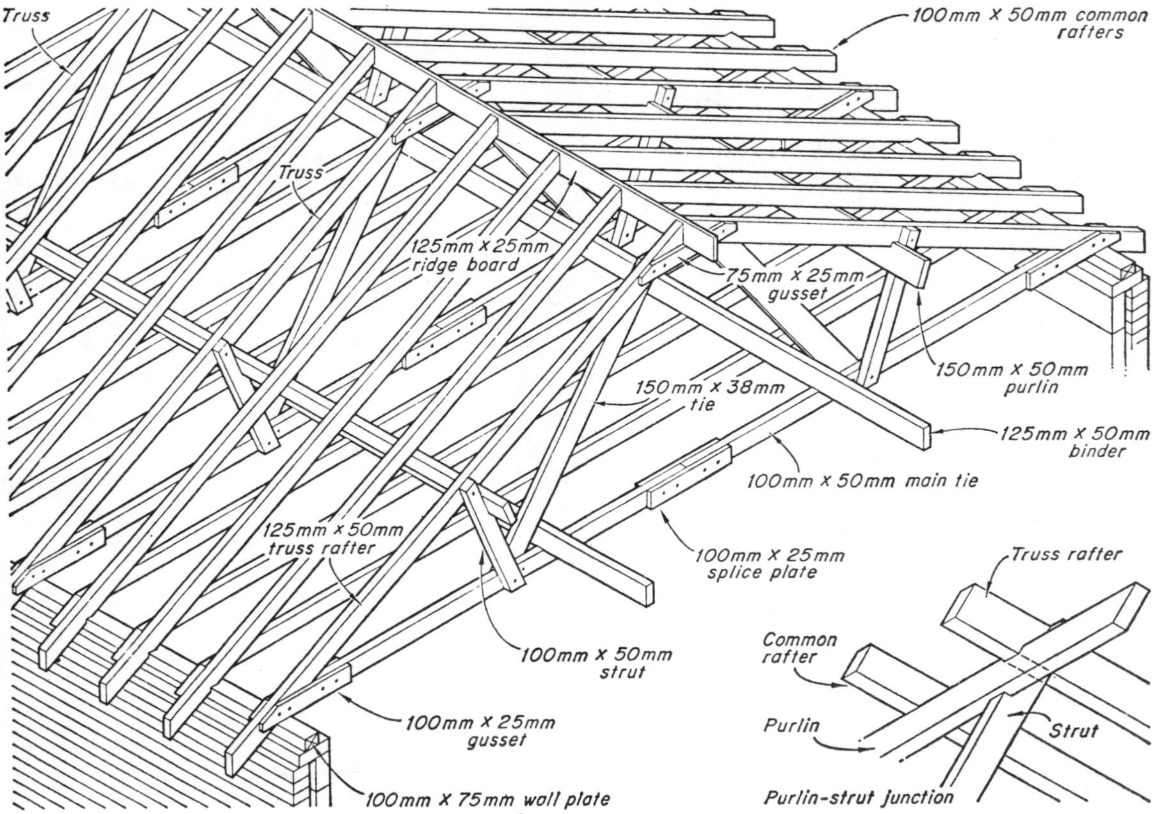

7.20 *Bolt and connector trusses 2*

these points. A single central gusset is provided at the ridge which also acts as a packing between the rafter members.

The two previous examples of bolted and connectored trusses are designed for self-supporting sheet coverings. Tiles, slates and similar coverings commonly used in domestic work require a sub-structure of battens supported by common rafters at 400 mm to 450 mm centres. A form of connectored truss for this type of work developed by the Timber Research and Development Assocaitation is illustrated in figure 7.20 and is essentially a pair of framed common rafters thus eliminating the need for separate principal rafters. The rafters of the truss therefore lie in the same plane as the adjacent rafters and the purlins, as a result of this, lie below the truss rafters and not on their backs as in a normal truss.

The truss is fabricated from single members, the joint between the rafters and main tie, which lie in the same plane, being made with gussets and the other joints by lapping the members. Binders to support the ceiling joists bear on the main tie near the lower nodes.

The trusses are designed to be placed not more than 1·80 m apart, that is at every fourth rafter where these are at 450 mm centres. The reactions at the feet are, therefore, not excessive and can be transferred adequately to the wall by the normal wall plate without a template or thickening of the wall. The example shown is for a span of 6·00 m.

Glued trusses Glues made from synthetic resins produce the most efficient form of joint, as strong as or even stronger than the timber joined, and many are immune to attack by dampness and decay. With this type of joint it is necessary to plane smooth all contact surfaces, and the necessary pressure during setting of the glue is provided by cramps or by bolts or nails which act as cramps. These are usually left in position.

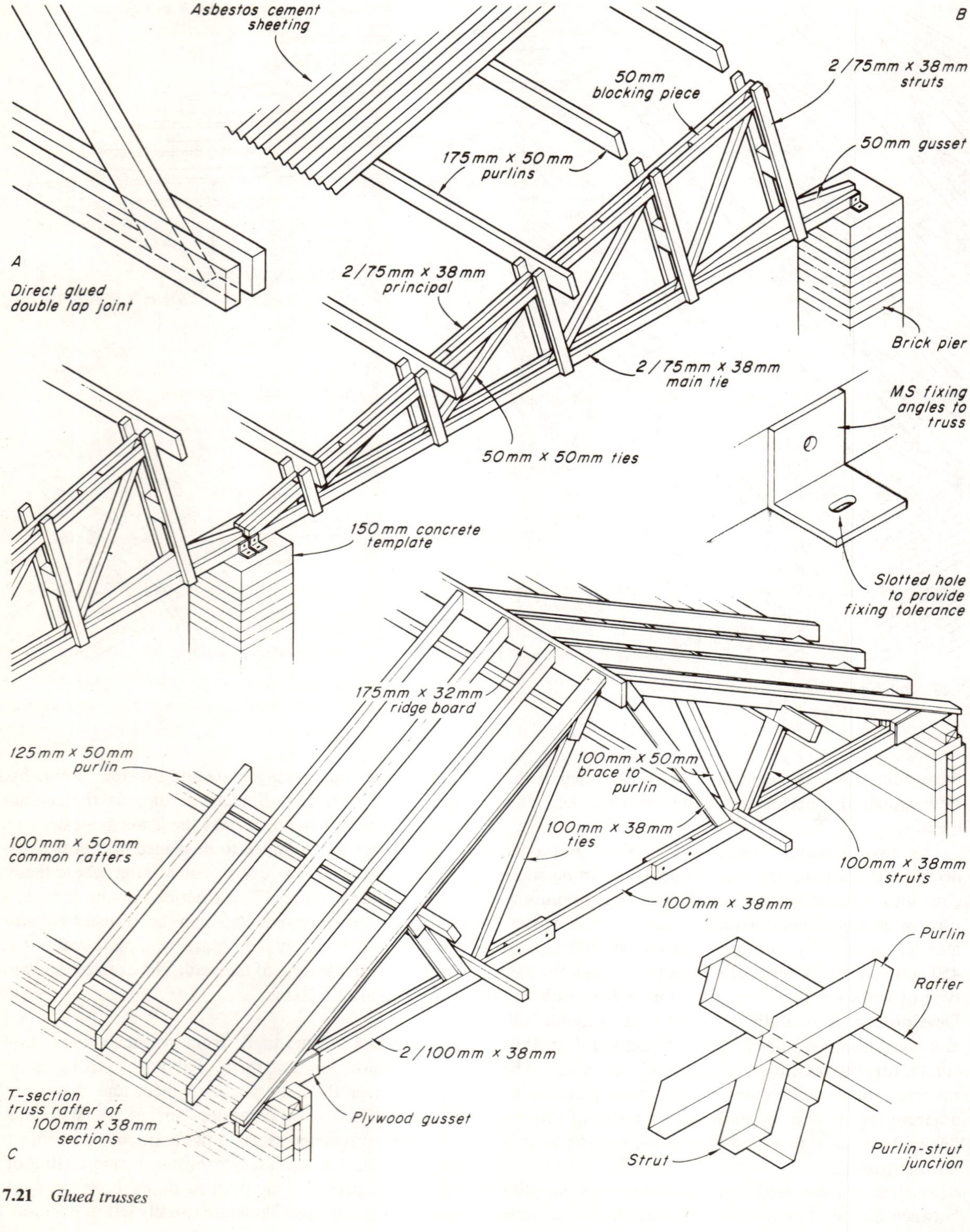

7.21 *Glued trusses*

The members may be glued directly to each other using lapped joints (figure 7.21 *A*) or single thickness construction may be used by the adoption of gussets. As with nailed joints, in certain cases lapped members may not provide sufficient gluing area even with double lapped joints and gussets must then be used to provide this.

An example of direct gluing is shown in the small northlight[1] truss of 5·20 m span in figure 7.21 *B*, in which single diagonal ties are sandwiched between double rafter and main tie members and the struts are formed by two thin members glued on the outside faces of the truss. This enables 'centre line' set-out of the members to be adopted. It should be noted that the two longest struts are packed out at the middle point to give increased stiffness to these compression members. Three nails driven in prebored holes act as cramps to each joint during setting of the glue.

Gluing not only produces very strong joints which result in quite small members, but also a very rigid structure which makes the truss easy to handle in transporting and fixing.

An example of a glued and gusseted truss is shown in (*C*). This is a factory made, standardised truss framed from 38 mm thick members, fabricated in two halves and requiring only site bolting of the main tie and site nailing to the ridge board. Rafters, struts and diagonal ties are single members joined by gussets, the compression members being formed into T-sections to stiffen them against buckling by the addition of 38 mm 'tables' glued and nailed on. Those to the struts form seatings for the purlins which lie below the rafters, so that the latter act also as common rafters as in the trusses developed by TRADA described above. The main tie is partially of double members between which struts and diagonal ties are sandwiched and secured by direct gluing. To provide greater gluing area the lapped joints between rafter feet and tie are packed out to allow the application of plywood gussets on each side. These trusses bear on the normal wall plate and are designed to be spaced up to 3·90 m apart for spans from 4·5 m to 9·0 m.

7.2.4 Trussed rafters

In domestic work the technique of triangulating or trussing *every* pair of rafters in roofs over spans which would normally require purlin construction produces what are known as *trussed rafters* (figure 7.22). These dispense with the necessity for purlins. The economic value of trussing every pair of rafters rests on the following factors: firstly, they are factory produced as prefabricated components resulting in the simplicity and speed with which this form of roof can be erected; secondly, the timber content of the whole roof structure is reduced by the elimination of purlins, ceiling hangers and binders and ridge board and by the use of the low pitches permitted by many forms of roof covering which results in short bracing members. It should be noted that since there are no purlins such a roof is a *single* roof construction.

Trussed rafters are fabricated from members of uniform thickness jointed by gluing or nailing, using plywood or, in the case of nailing, punched metal plate gussets. Punched metal plate connectors as they are usually called, fall into two groups (figure 7.22). Firstly, a thin-gauge plate with holes punched regularly over its surface to receive nails, called a *pierced plate connector* (*B*). Secondly, a similar plate with teeth punched from the plate and bent over 90 degrees, called a *toothed plate connector* (*C*).[2] The latter, in which the teeth are an integral part of the plate, must be driven in by a hydraulic press or roller and are used in factory production since they are not suitable for site fabrication.

An illustration of a typical trussed rafter roof is shown in figure 7.22 *A*. Trussed rafters, with members which may be as thin as 35 mm,[3] are not stiff components and in practice the rafter members may often be slightly curved along their length and the truss be out of plane. Unless adequately braced laterally the roof may deform by the trusses buckling over their entire length. It will be seen that no ridge board is incorporated since the heads of the rafter members are secured by a metal plate connector.

For these reasons the longitudinal bracing shown in figure 7.22 *A*, at ridge and tie levels and the diagonal bracing in the plane of the rafter members in each roof slope is the minimum necessary and essential to hold the trusses in position relative to each other, to avoid buckling of individual trusses and to provide the lateral stiffness to prevent 'racking' or sideways collapse of the whole assembly of trusses. The longitudinal bracing

[1] For definition see Part 2.

[2] This type of toothed plate connector should be distinguished from the double-sided toothed plate connector used in bolted joints between lapped members — see page 170.

[3] Such thin members require careful nailing of tiling or slating battens in order to avoid splitting the rafters, especially where the battens butt-joint over them.

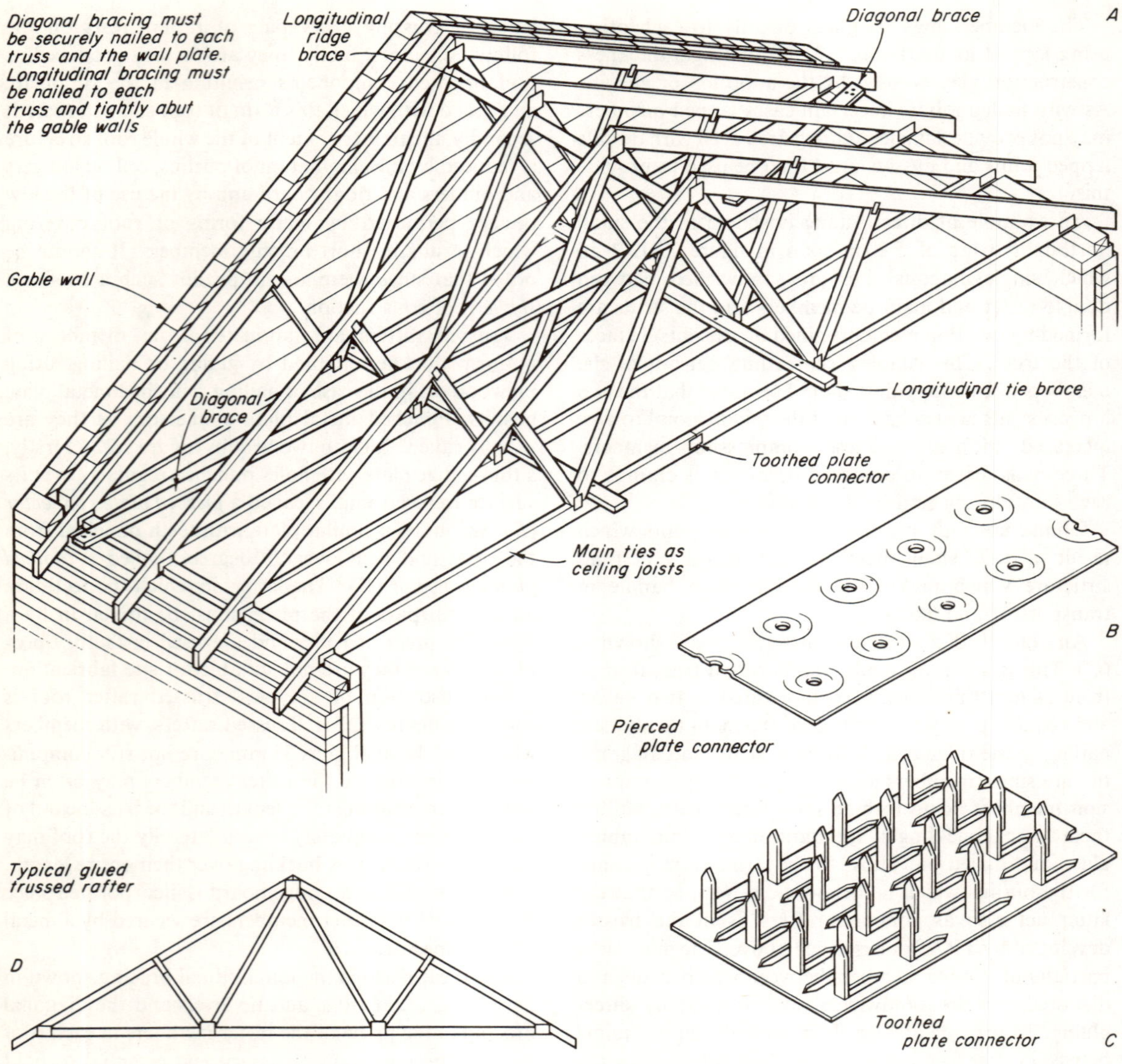

7.22 *Trussed rafters*

members should fit tightly at their ends against gables or walls and all bracing must be lapped at any joints in their length to give continuity. In some circumstances, for example in spans over 8 m, longitudinal and diagonal web bracing in the plane of the strut members may be required.

The tiling or slating battens (or sheet decking) are an important part of the bracing system, giving lateral restraint against buckling of the thin relatively deep rafter members which are in compression (see page 50). Their size must, therefore, be adequate for this purpose according to the spacing of the trusses.

The bracing described above is primarily to ensure the stability of the whole roof assembly and, since trussed rafters are invariably factory produced, is specified solely for this purpose by the manufacturer. It may or may not be sufficient to act as wind bracing which is necessary to provide the rigidity to enable a roof to act as a stiff frame or diaphragm when used to give lateral support to walls (see page 72 and Part 2 chapter 4). The need for and nature of wind bracing in any particular circumstances must, therefore, be considered apart from the stability bracing specified by the manufacturer.[1]

The essential difference between a trussed rafter and roof truss is that the former carries its own proportion of roof load directly on itself and only that load, whereas a truss carries the loads from a number of adjacent rafters via the purlins.

The use of low pitches, lightweight roof coverings and lightweight roof structures such as trussed rafters, by reducing the weight of the roof increases the danger of wind uplift and in these types of roof the provision of adequate anchorages on the lines discussed on page 159 must be considered.

7.2.5 Influence of plan shape and roof termination

The consideration of pitched roofs so far has been in terms of a straight run of roof without regard to the termination of the roof at the ends of the building or to the effect of the plan shape of the building upon the construction of the roof.

The roof may terminate in one of two ways (see figure 7.23):

(i) Against the end walls of the building which are carried up into the triangular section of the roof, forming what is known as a gable end roof, or simply a *gable roof*, derived from the triangular shaped area of wall at each end which is called a *gable*. This produces a simple roof shape of two parallel slopes (*A*).

(ii) By returning the roof slope on itself at the ends. This produces a roof shape of two parallel slopes with slopes normal to these at each end, which mitre or intersect at the junctions to form what is known as a *hipped roof*. These junctions are termed *hips* and the sloping ends *hipped ends* (*B*).

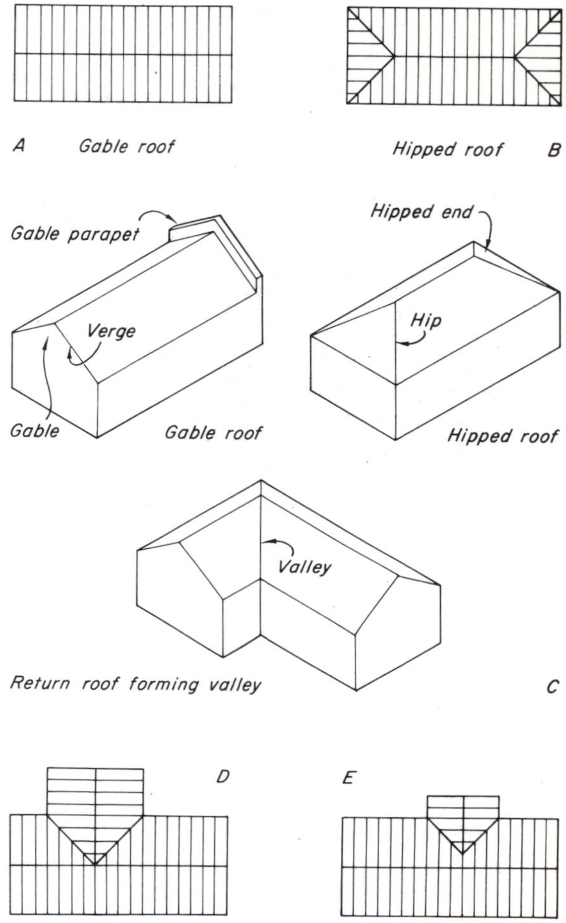

7.23 *Roof shape and termination*

The gable roof is simple to construct and cheaper than the hipped roof although a greater amount of walling is required at the ends and splay cutting in masonry gable walls is often necessary. The gables may project above the roof surface as parapets or may terminate at the roof surface in which case the roof covering is carried over the head of the wall to form a *verge*, as the sloping edge of the roof is called (see figure 7.23). Detailing at a verge will vary according to whether the roof structure terminates against the inside face of the gable wall with only the roof covering and its base extending over the wall head as shown in figure 5.27 (*C*) or whether the structure extends to carry the roof beyond the outside wall face as shown in figure 7.24. The projection beyond the gable face is formed by short rafters at 400 mm to 450 mm centres fixed back to the last common rafter. These cantilever over the wall head to carry an outer rafter to which is fixed an inclined fascia which is called a *barge-board*. In purlin roofs the purlins, ridge-piece and wall plates may extend beyond the gable face to carry the end rafter.

The hipped roof is more complicated in its construc-

[1] The design and construction of trussed rafter roofs for domestic buildings is covered by BS 5268: Part 3: 1985. See also articles in *Architects Journal* 9 and 16 September 1981.

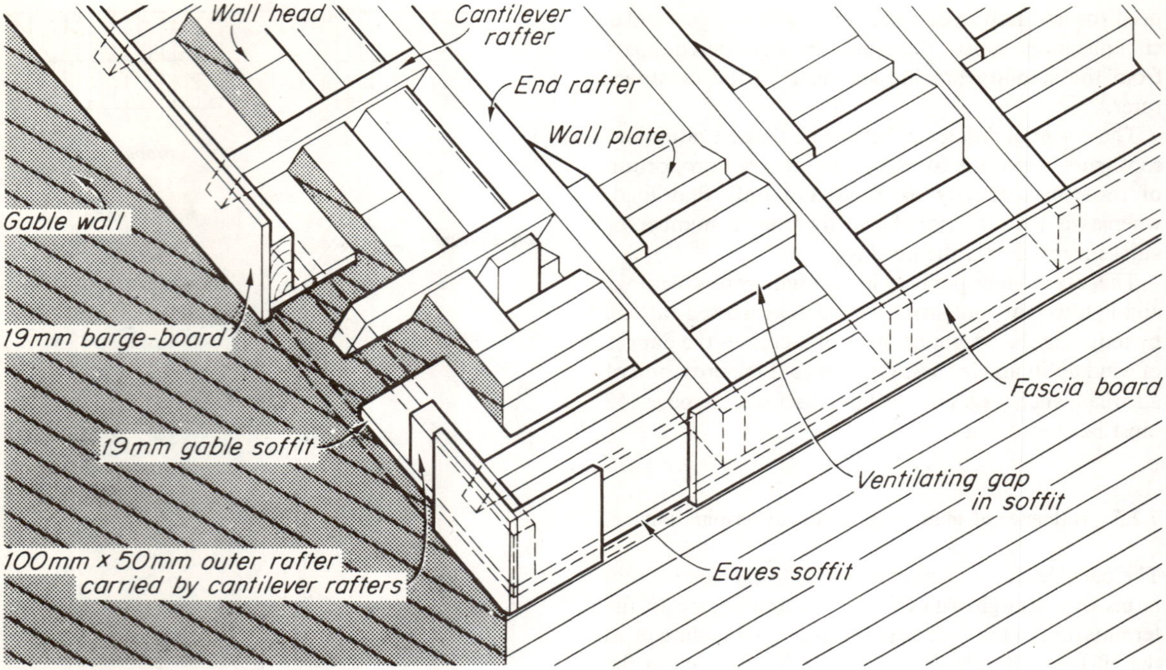

7.24 *Projecting verge*

tion, necessitating splay and skew cutting of all the shortened rafters (called *jack rafters*) at the intersections and the provision of a deep hip rafter running from ridge to wall plate to carry their top ends (figure 7.25 *A*). The hip rafter transfers its load to the wall plate and will, therefore, be 225 mm to 280 mm deep, depending upon its span and the depth of the rafters, and 38 mm to 50 mm thick. If the roof has purlins their ends will also be carried by the hip rafters which may then need to be 75 mm thick.

The tendency of the inclined thrust of the hip rafter to push out the walls at the quoin is overcome by tying together the two wall plates on which it bears by an *angle tie* dovetail notched or bolted to the plates (*B, C*). The foot of the hip rafter is notched over the wall plates which are half-lapped to each other. If the rafter carries purlins causing a greater thrust more resistance to this is provided by the introduction of a *dragon-beam* as shown at (*B*), (*D*), linking the ends of the wall plates to the angle tie, which would be larger in size. The dragon-beam is cogged over the plates and tusk-tenoned to the tie. A dragon-beam will in any case be necessary to provide a bearing for the hip rafter when the eaves are sprocketed and the feet of the rafters terminate on the wall plate (see figure 7.29 *A*).

When the plan shape of the building breaks out or returns the intersection of the roof surfaces results in a junction having an external angle less than 180 degrees which is called a *valley* (the hip has an external angle greater than 180 degrees). As at a hip jack rafters occur. These run from ridge to valley and their feet are nailed to deep valley rafters the function and sizes of which are the same as those of the hip rafters (figure 7.23 *C, D*).

If returns and projections produce roof spans equal to that of the main roof the valley rafters will extend to the ridge where they will gain support as in (*C* and *D*). If, however, a projection is less in span as in (*E*) the valleys will not meet the main ridge, and a support to the tops of the valley rafters and the lower ridge board must be provided in the roof space.

If the width of the projection is small valley rafters may be omitted and all the rafters of the main roof be carried down full length on to a suitable bearing with boards laid on them to take the end of the ridge board and the feet of the jack rafters to the projection.

A valley is finished with a triangular timber fillet or a valley board, as shown in figure 7.26 *A*, depending on the width required by the nature of the junction between the roof covering on the two slopes.

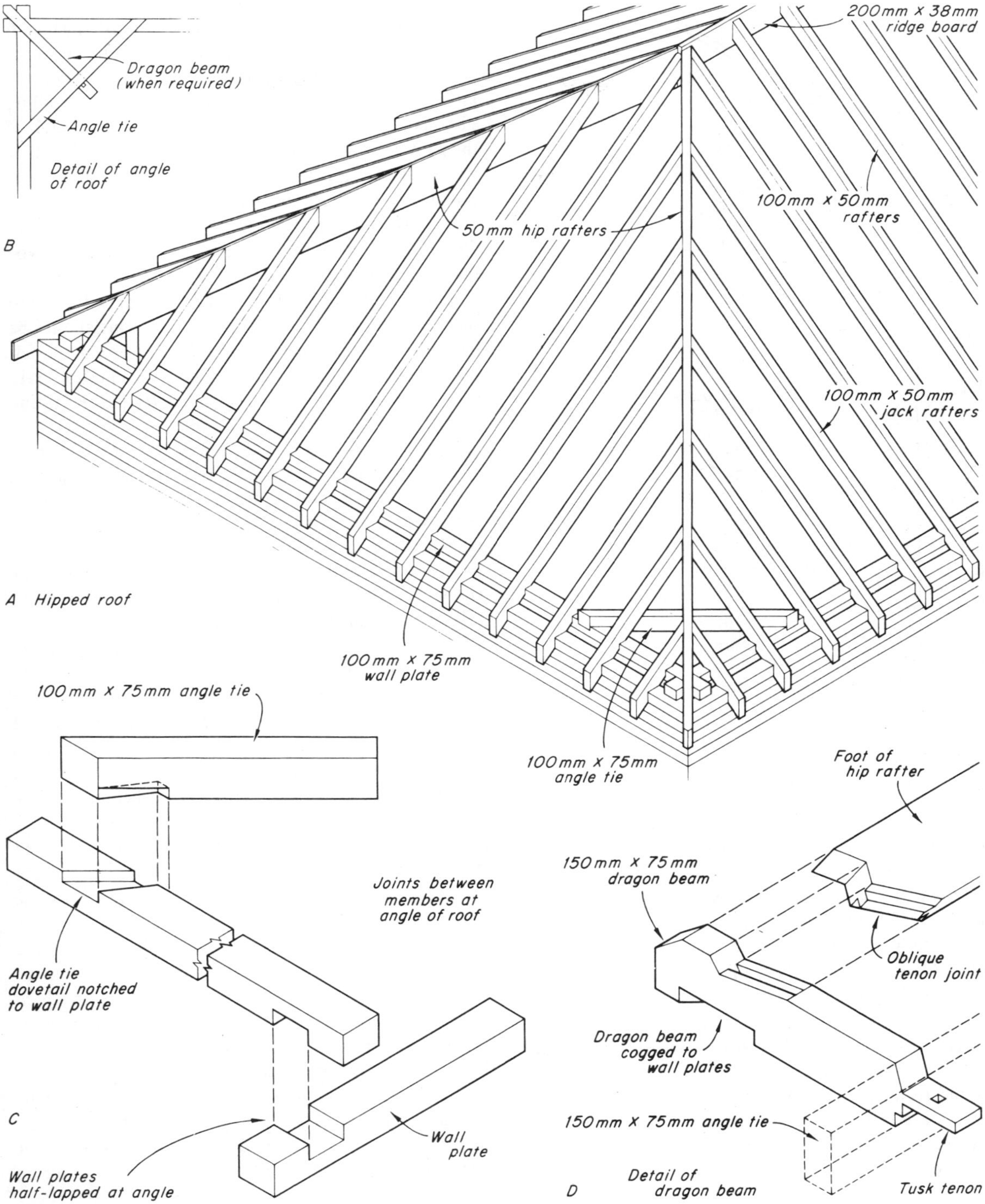

7.25 *Hipped roof*

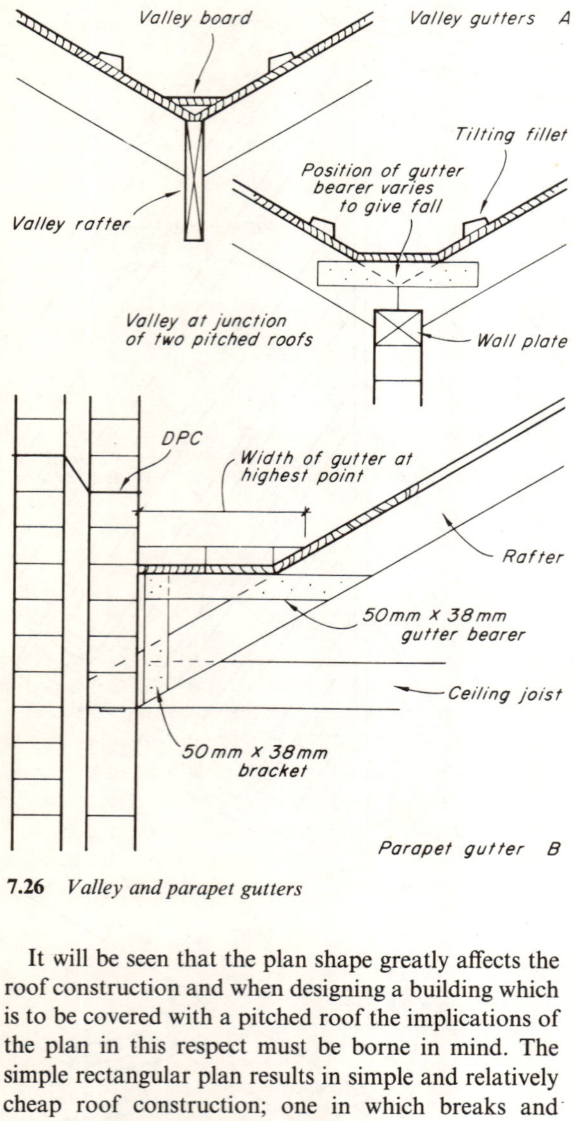

7.26 *Valley and parapet gutters*

It will be seen that the plan shape greatly affects the roof construction and when designing a building which is to be covered with a pitched roof the implications of the plan in this respect must be borne in mind. The simple rectangular plan results in simple and relatively cheap roof construction; one in which breaks and returns occur, especially if they are numerous, may result in most expensive construction. This applies not only to the structure itself but also to the roof covering. These considerations apply also to trussed rafter construction. The most economic application of this is to simple rectangular, gable ended roofs and if hips or valleys are introduced the complexity of construction involved can make it as expensive as traditional construction.

Eaves treatment

As with a monopitch roof, unless the roof is set behind a parapet, the eaves of a ridge roof may finish flush with or may project beyond the wall face, the former producing some economy in roof covering and timber, the latter providing some protection to the walls. Detailing of construction varies widely according to the pitch of the roof, the effect desired by the architect and whether an external or a hidden gutter is used. It is, therefore, possible to illustrate only some typical examples.

Examples of open projecting eaves are shown in figure 7.27 *A*, *B*. With tile or slate coverings of any type the fascia projects as shown 19 mm or so above the roofing battens (or boarding if slates are nailed direct to this) in order to tilt the eaves courses. This should be backed by a triangular tilting fillet to prevent the felt underlay forming into a trough which would hold rain water. Where no fascia is used as at (*B*) a larger tilting fillet is used to tilt the eaves courses.

Figure 7.28 shows closed projecting eaves. The variation in detailing necessitated by increased projection can be seen. The ends of the rafters are cut horizontally to provide some fixing for the soffit boards (*A*), but as a considerable portion of the boarding is not supported by the rafter, soffit bearers are fixed to the rafter ends as shown. The back of the fascia should be grooved to take the edge of the soffit. Greater projections necessitate longer soffit bearers and brackets are then required to support their inner ends as shown in (*B*). When plywood or asbestos or other fibre cement sheet is used for the soffit, as is quite common, the fascia must be grooved to take the front edge and the back edge should be given continuous support by a fillet secured to the wall (*C*). In this case the soffit bearers can be fixed to this rather than to brackets from the rafters. If the roof pitch is not too great the soffit can be fixed direct to the underside of rafters and, with a gable roof and projecting barge board, can continue up as the verge soffit (figure 7.24). In this particular case the barge-board will be slightly less in depth than the fascia, but with a horizontal eaves soffit it must be deeper in order to cover the end of the eaves, in which case the outer and cantilever rafters which support it must be deeper than the common rafters or a thicker barge-board must be used.

If a clear fascia, unobstructed by an external gutter, is desired an internal gutter may be formed as shown in figure 7.28 *D*. As emphasised under *Monopitch roofs* it is essential that the front edge of this type of gutter be at such a level that in the event of blockage of the outlet water will drain over the front rather then seep back into the roof structure and possibly into the building.

Roof ventilation must be ensured through the eaves

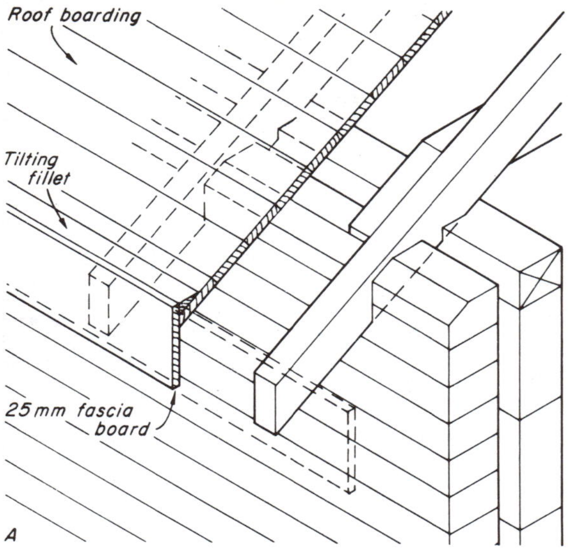

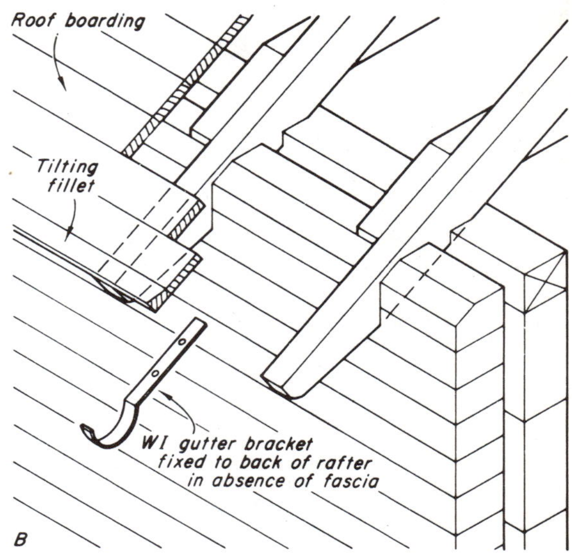

7.27 *Pitched roofs — open eaves*

as described on page 163. In the case of open eaves the walling must be kept below the tops of the rafters by splay cutting if necessary as in figure 7.27 or by arranging the brick coursing so that the required gap is left at the top of the wall. In closed eaves a continuous ventilation gap can be left in boarded soffits behind the fascia as in figure 7.12 *B*, or next to the wall as in figure 7.28, or individual vents of adequate area can be formed at intervals in thin soffit linings requiring edge fixing at all points as in figure 7.28 *C*. The ventilator tubes or panels referred to on page 164, which are used to prevent the roof insulation blocking the ventilation path, pass up the space between the rafters above the roof insulation from the open or enclosed eaves (see figure 7.12 *B*). In order to maintain the air path between the rafters in figure 7.27 the cavity barrier at the head of the wall could be asbestos free board equivalent to 9 mm asbestos building board laid across the cavity and fixed to the wall plate. 50 mm timbers fixed between the rafters could be used in the case of figure 7.28 (*A*).

Where a pitched roof with closed projecting eaves such as those shown in figure 7.28 is used over terraced or semi-detached housing the closed eaves usually passes across the line of the separating walls, constituting a passage for flame and smoke between the roof voids on either side of the separating walls. This passage must be blocked by the provision of cavity barriers in the eaves void in the plane of each wall. A cavity barrier should also be formed in the space between the head of each wall and the underside of the tiling or slating by means of fire stopping.[1]

When a gable roof finishes with a plain verge, that is with no barge-board, the end of any form of closed projecting eaves must be boxed-in or be closed by the gable wall supported either on corbelling or on a springer (figure 5.27 (*C*)). If the gable continues up as a parapet this is usually corbelled out for this purpose as shown in (*B*).

On wide, steeply pitched roofs the pitch may be reduced at the eaves in order to reduce the velocity of water during heavy rainfall and prevent over-shooting of the gutter. This is done by means of *sprockets* which are short lengths of timber the same size as the rafters, fixed to the sides of the rafter feet as shown in figure 7.29 (*A*) or to the backs of the rafters if the latter run over the wall plate as in (*B*). The reduced pitch must, of course, not be less than the minimum angle necessary for the particular roof covering.

As an alternative to framing up a projecting eaves in the ways described above proprietary precast concrete eaves or gutter units may be used as for flat roofs, bedded on the head of the external walls. The shape of the unit spreads the roof load over both leaves of a cavity wall and over openings of limited span a back

[1] See AD B, H3 for suitable materials.

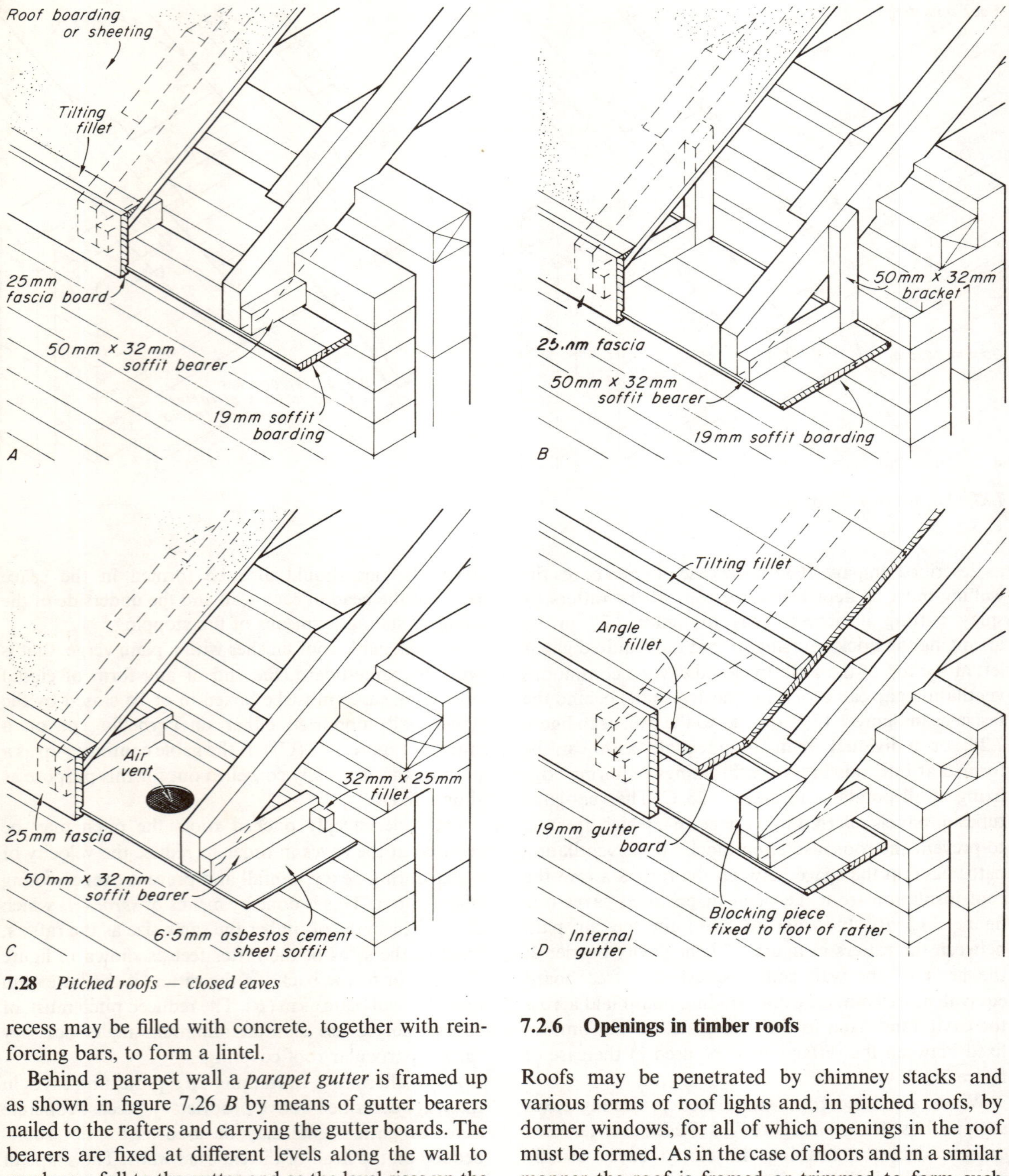

7.28 *Pitched roofs — closed eaves*

recess may be filled with concrete, together with reinforcing bars, to form a lintel.

Behind a parapet wall a *parapet gutter* is framed up as shown in figure 7.26 *B* by means of gutter bearers nailed to the rafters and carrying the gutter boards. The bearers are fixed at different levels along the wall to produce a fall to the gutter and as the level rises up the roof slope this results in a gutter which tapers in width on plan from a maximum at the highest point and is, therefore, termed a *tapered gutter* in contrast to the *parallel* or *box gutter* described under *Flat roofs*.

7.2.6 Openings in timber roofs

Roofs may be penetrated by chimney stacks and various forms of roof lights and, in pitched roofs, by dormer windows, for all of which openings in the roof must be formed. As in the case of floors and in a similar manner the roof is framed or trimmed to form such openings. Details of trimming to flat roofs are normally identical with those for floors (see page 199).

In pitched roofs openings may be required at any point between eaves and ridge, or at the ridge, as shown

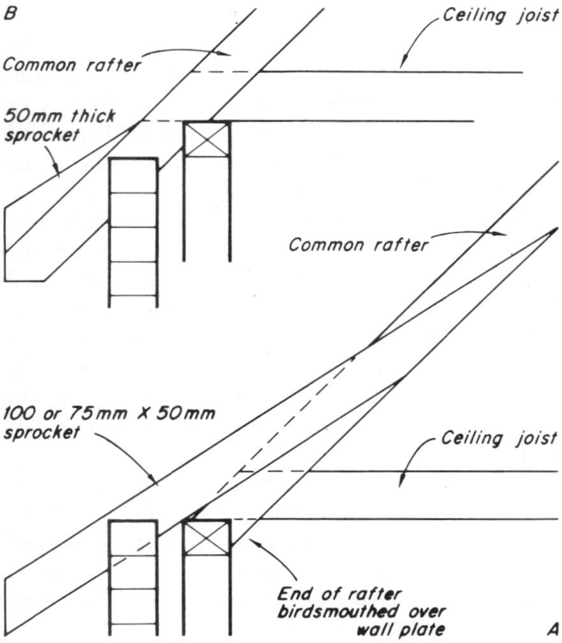

7.29 *Sprocketed eaves*

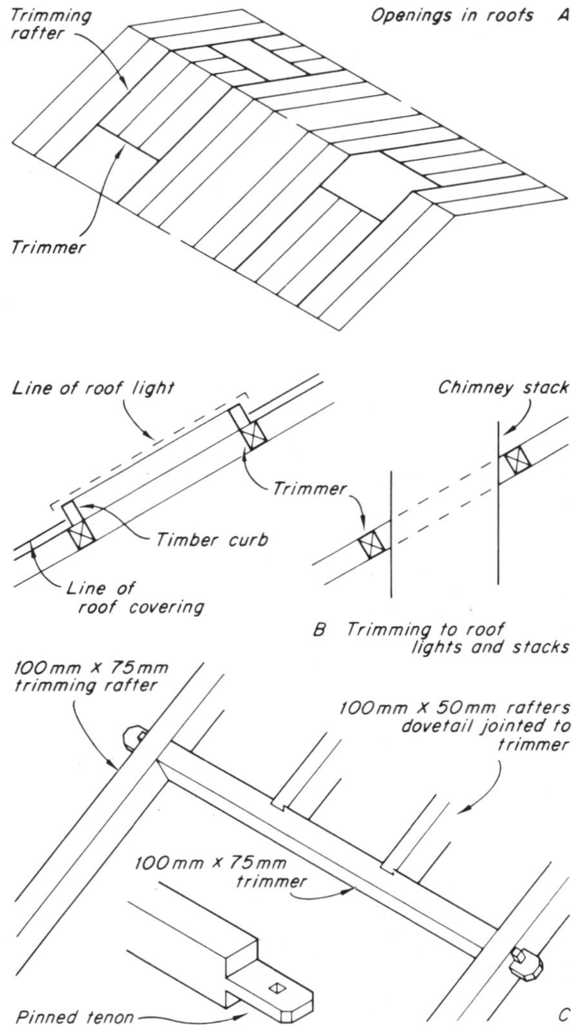

7.30 *Trimming to openings in roofs*

in figure 7.30 *A*. For stacks and skylights the trimmers are placed normal to the roof slope (*B*) and are fixed to the trimming rafters by *pinned tenons*. This joint has an extended tenon and is secured with a wedge (*C*). The trimmed rafters are fixed to the trimmers by any of the methods described for floors (figure 8.12).

Openings for roof lights are finished with a timber upstand or *curb* as indicated in figure 7.30 (*B*) which in a pitched roof, raises the light above the level of the roof covering and permits a watertight junction to be formed all round, and in a flat roof provides for a 150 mm upturn of the roof finish.

The positioning of trimmers for dormer windows varies according to framing requirements and is discussed below:

Dormer windows

The dormer window is a vertical window set in the slope of a roof as distinct from a skylight which is parallel to the slope. It may take various forms as shown in figure 7.31. The internal dormer which avoids a projection above the roof slope is less common and involves a small flat roofed area in front of the window.

For external dormer windows the lower or cill trimmer is fixed vertically to provide a seating for the dormer framework and window and to raise the window cill clear of the roof covering. It is 75 mm to 100 mm wide and its depth will vary with the roof pitch and type of roof covering (figure 7.32). The top or head trimmer may be fixed vertically or normal to the slope. If the dormer roof is flat a vertical trimmer provides a fixing surface for the boarding or other decking; if it is pitched a trimmer normal to the slope may be used and this simplifies jointing to the trimming rafters (compare figures 7.30 and 7.32 *B*). The cill trimmer is oblique notched over the trimming rafters and nailed in position (figure 7.32 *A*). The vertical head trimmer is

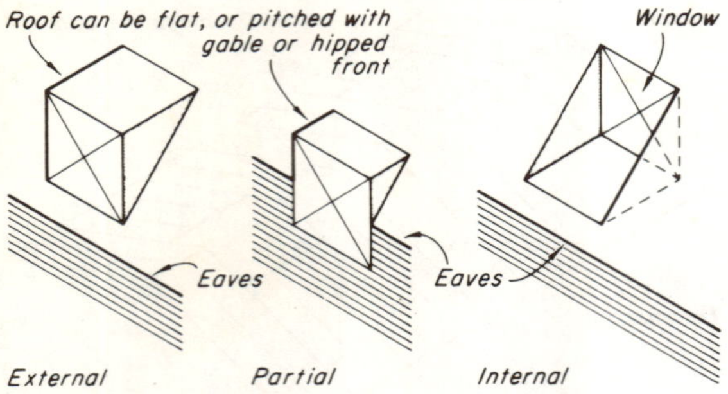

7.31 *Types of dormer windows*

oblique notched and tenoned to them (*B*), the tenon being necessary here in order to resist the thrust from the feet of the upper trimmed rafters. In the case of a partial dormer there is no cill trimmer since the window sits directly on the wall below.

The traditional method of forming the dormer front was to frame up 100 mm by 75 mm side posts and head on the cill trimmer, the posts being tenoned or dowelled to the trimmer, and within this to set the window. Nowadays, unless the dormer is large, it is usual to make the head and mullions of the window frame large enough to act structurally to support the dormer roof and cheeks as shown in figure 7.32. The cheeks are formed by a 75 mm × 75 mm side head running from the dormer front back to the trimming rafter against which it is splay cut and nailed, the spandrels thus formed being filled with 75 mm × 50 mm studs to which 19 mm T and G boarding is fixed externally. If the cheek is small studs can be omitted, the spandrel being covered with 25 mm boarding nailed to corner post and side head, running parallel with the roof slope.

The framing of an internal dormer varies slightly from this. The lower trimmer would be set vertically to form a front bearing for the flat roof below the window and the top trimmer set similarly to form a head over the window. Since neither may be notched over the trimming rafters, in order not to obstruct the roof covering, both must be tenoned into them. Two posts under the bearings of the top trimmer and running from floor to trimming rafters would support a cross bearer carrying the window and the members forming the flat roof. The spandrels would be formed in the manner described above.

In order to maintain the integrity of the thermal insulation of the roof as a whole the cheeks and roofs of dormer windows must be insulated to the same standard as the roof itself.

7.2.7 Steel roof trusses

Steel trusses are normally used with steel framed construction for industrial and similar building types although they may bear on masonry walls. Hot rolled mild steel sections are commonly used with welded connections between the members.

These trusses are normally built up from angles and tees. Flats, or bars, which are less stiff, were commonly used in the past for tension members but it is now usual to use angles throughout. This produces a stiffer component for handling in transport and erection and allows for the reversal of stress in members which may occur during these operations or through wind suction on the roof and which could cause buckling of the flat members.

Steel angle purlins are normally used for sheet coverings rather than timber, and are bolted to angle cleats welded to the rafters. When a truss is designed for, say, slate covering requiring common rafters, timber purlins would be used bolted to the angle cleats. Details of steel trusses are illustrated in Part 2 chapter 9.

7.2.8 Rigid frames

These have already been referred to as structural components in which the roof members are fixed rigidly to the supporting columns. The characteristic feature common, therefore, to all forms of rigid frame is the stiff or restrained junction, called the haunch, between the spanning and supporting members. The implications of this are discussed in Part 2, but it may be said here that one is the reduction in size of the spanning members whether flat or pitched, compared with that of members simply supported on the columns. As explained in chapter 3 a stiff joint can be utilised to obtain lateral stability in a framed structure which, together with the reduction in size of the spanning members is often an important requirement in large buildings. In the sphere of short span buildings, however, the value of the rigid frame lies primarily in the fact that the use of frames with inclined spanning members results in greater clear headroom within the building than with the use of pitched trusses springing from the same height. Where maximum headroom is a major design factor as, for example, in storage buildings, the pitched, or ridge type, frame is, therefore, relevant to short span construction even though its use may sometimes be rather more costly than the use of a truss.

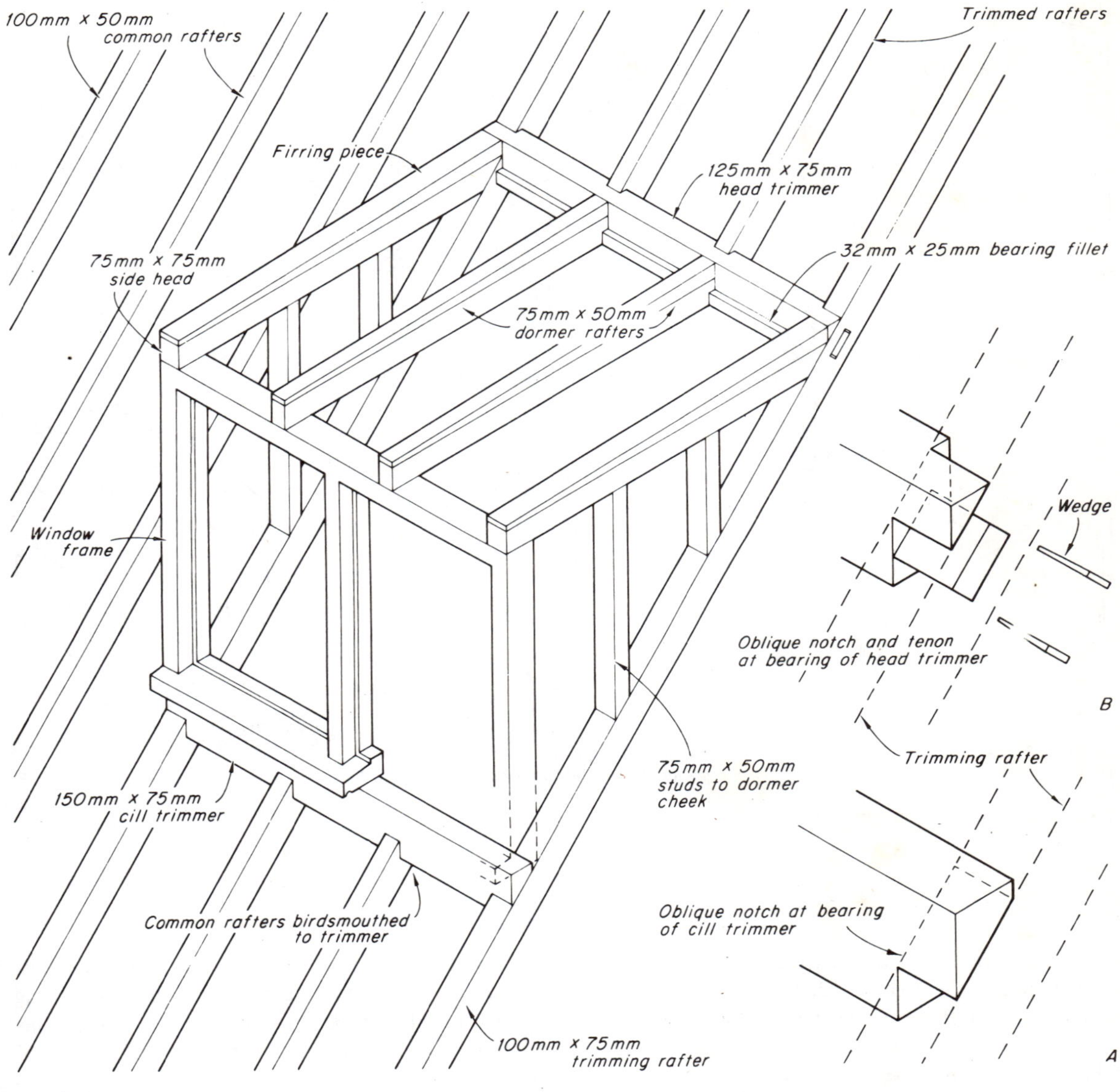

7.32 *Dormer window framing*

Precast concrete is probably most commonly used for short span work, especially where a fire-resisting and non-combustible material is required. These frames normally take the form of a fixed or hingeless portal, that is a frame with the feet rigidly fixed to the foundation slabs. They are cast in sections to facilitate transport and joined on site by bolting. The joints are usually made at the points of low bending stress in the spanning members or at the haunches where the required increase in section (due to the high stresses at these points) can be provided by the addition of one element to another. Figure 7.33 *A* illustrates this diagrammatically. (*B*) shows frames with the full depth of the haunch and the lower part of the roof member cast in with the column head. This terminates in a horizontal bearing at the point of lowest stress (*D*) on which the centre portion of the roof member sits and to which it is bolted. (*C*) show frames in which the joint is made

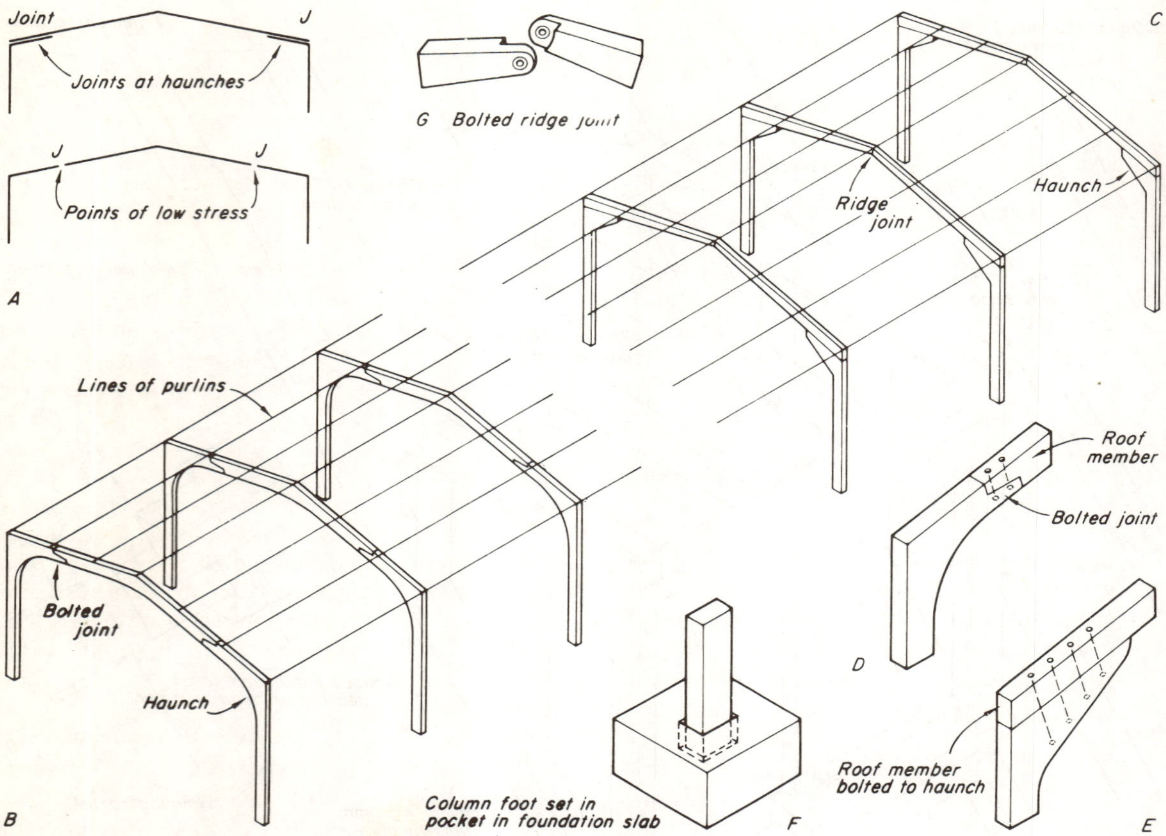

7.33 *Reinforced concrete rigid frames*

at the haunches, the lower part of the roof member contributing to the necessary depth at the haunch (*E*). In this example the roof member is cast as two components with a joint at the ridge which is non-rigid. This connection may be made as a true hinged joint as at (*G*) or, more simply, by steel connecting plates on each face bolted to each roof component. The feet of the frames are set in pockets formed in the foundation slabs, wedged in position and grouted in as at (*F*).

Spacing of the frames would be about 4·60 m and this is usually spanned by precast concrete purlins which are secured to the frames in one of two ways: either by bolting to precast blocks set in rebates on the back of the frame and which provide the bearings for the purlins, or by bearing the purlins directly on the frames and securing them by steel eyes and dowels cast in purlins and frames respectively, the joint being grouted in with mortar. Precast concrete angle purlins are used in conjunction with corrugated forms of roof covering fixed by hook bolts.

Steel frames for short spans are made from Universal sections (see page 128) cut and welded at the haunches and, in the case of pitched types, at the ridge. They are fabricated in three pieces and site jointed at the points of least stress with splice plates and bolts. By increasing the cross-section at the haunch, by welding into the angle a triangular shaped piece of steel tee, sections of smaller depth can be used since the stresses in other parts of the frames are often less than at the haunch as explained in chapter 9 Part 2.

8 Floor structures

The term *floor* in this chapter refers to the structural part of a horizontal[1] supporting element as distinct from the wearing surface.

At ground and basement levels full support from the ground is generally available at all points and a slab of concrete resting directly on the ground may be used. This is known as 'solid floor' construction. At upper levels the floor structure must span between relatively widely-spaced supports in order to leave unobstructed the floor area below. The forms of construction used in these circumstances are known as 'suspended floors'.

Functional requirements

The main function of a floor is to provide support for the occupants, furniture and equipment of a building. To perform this function and, in addition, others which will vary according to the situation of the floor in the building and the nature of the building itself, the floor must satisfy a number of requirements in its design and construction. These may be defined as the provision of adequate:

- strength and stability
- fire resistance
- sound insulation
- thermal insulation
- damp resistance.

Strength and stability

Problems of strength and stability are usually minor ones at ground and basement levels because of the full support available at all points. Where very heavy floor loads or the upward pressure of sub-soil water are involved however, the floor will need to be reinforced.

A suspended upper floor is required to be strong and stiff enough to bear its own self-weight and the dead weight of any floor and ceiling finishes, together with the superimposed live loads which it is required to carry, without deflecting to such an extent as to cause damage to ceiling finishes, particularly if these are of plaster. In framed buildings the floors are sometimes designed to act as horizontal 'struts' capable of transferring wind pressure to stiff vertical members in the structure and so provide lateral rigidity to the frame. They also serve to provide the necessary lateral restraint to loadbearing walls (see page 72 and Part 2, 4.1.3).

The dead load is usually based on the weights of materials specified in BS 648 and superimposed loads on average loadings per square metre for different types of buildings laid down in Building Regulations and Codes of Practice (see Part 2 chapter 6 in respect of superimposed loading on floors).

Fire resistance

Fire resistance is important in respect of upper floors which are often required to act as highly resistant fire barriers between the different levels of a building. The degree of resistance necessary in any particular case depends on a number of factors which are discussed in Part 2.

Sound insulation

Except when sources of excessive sound vibration, such as an underground railway, are in close proximity to a building sound insulation need not normally be considered in ground or basement floors. Contact with the mass of the earth damps out to a great extent sound vibrations originating at any one part of the floor. It is however, an important consideration in the design of upper floors.

The degree of insulation required will vary with the type of building and the noise sources likely to create a nuisance and the form of sound insulating construction adopted will vary with the type of floor used, particularly whether it is of timber or concrete construction. These considerations are discussed in *MBS: Environment and Services*, to which reference should be made.

Thermal insulation

Thermal insulation is normally not required in upper floors except in those the soffits of which are exposed to the external air or to an unheated space,[2] or in relation

[1] Theatres, concert halls, lecture theatres require a sloping or 'stepped' floor for sight requirements.

[2] Thus becoming *exposed* and *semi-exposed* elements respectively, for definitions of which see AD L2/3.

to certain forms of floor or ceiling heating. The Building Regulations require for ground floors the same standard of insulation as for external walls which in some cases may be achieved without the application of insulating material, depending on the area and dimensions of the floor (see page 192). In solid ground floors embodying heating pipes or cables the heat loss at the edges of the floor slab can be considerable and insulation against this may be necessary.

Damp resistance

The problem of damp penetration into the building generally arises only in connection with ground and basement floors. In the case of basements the problem becomes acute when the floor is below sub-soil water level and its solution involves the use of waterproofing methods resistant to water under pressure. These methods are discussed under *Waterproofing of basements* in Part 2.

Types of floor structure

Solid floors

These may be of plain or reinforced concrete. In most buildings without basements the ground floors are of solid construction, of concrete on hardcore resting directly on the ground.[1] They are invariably so in the case of basement floors and in floors taking heavy loads or traffic.

The thickness of the slab will vary according to the loading which the floor is to carry and the bearing capacity of the ground. When the latter is uneven or when the ground is weak or made-up, the slab is reinforced over the whole of its area with mesh reinforcement. Reinforcement is also required when a basement floor must resist the upward pressure of sub-soil water (see Part 2).

A concrete floor slab designed as a reinforced element to transmit the whole of the building load to the soil becomes a 'raft' foundation. This may take a number of forms which are described in the chapters on *Foundations* in this volume and in Part 2.

Suspended floors

These may be constructed in timber, reinforced concrete or steel and, as in the case of roof construction, may be in the form of single, double or triple construction according to the loads and spans involved.

The choice of floor type for small-scale buildings will usually be governed by considerations of loading and span, cost, sound insulation and speed of erection. For large-scale buildings and multi-storey buildings other factors such as the nature of the building structure, accommodation of services and fire protection will also need consideration. In these building types the floors are normally main structural elements closely related to the general structure of the building, and they must be considered at the design stage in relation to it.[2]

Various types of floor systems are most economic within certain span ranges for different superimposed loadings and some indication of this is given in table 8.1.

Although the span of a floor may sometimes be fixed by plan requirements, it may often be varied to fall within the economic range of a less complex and cheaper type of floor system.

Span range	*Loading*		
	Up to 2·00 kN/m²	2·00 to 4·00 kN/m²	Over 4·00 kN/m²
Up to 3·05 m	Timber	Timber	RC slab
3·05 m to 6·10 m	Timber (4·90 m max)	RC slab	Beams and RC slab
6·10 m to 9·10 m	Double timber floor (above 4·90 m)	Beams and RC slab	Special floor types

Table 8.1 *Economic range of spans for floor systems*

Timber floors

The timber floor has the advantage of light self-weight and of being a dry form of construction. It is simple to construct and this, together with the savings effected in the supporting structure because of its light weight, make it economical particularly where the imposed loads are small.

In itself it is a combustible form of construction and has a relatively low fire resistance which depends on the

[1] Unless circumstances, eg accommodation of services below floor, require the provision of a suspended floor to provide space under or an excessive depth of fill beneath the floor makes solid construction inadvisable.

[2] See Part 2 chapter 6 *Choice of Floor*.

thickness of the boarding or other flooring, size of joists and, especially, on the nature and thickness of the ceiling lining, but it is sufficient for many forms of two-storey small-scale buildings including houses. There is scope, however, for the use of the timber floor in many types of building higher than two storeys, where the means of escape is good and the building is sub-divided by compartment walls into sections of limited area or cubical content (see reference on page 109 to current Building Regulations and the use of timber).

The degree of sound insulation provided by a boarded timber floor is much less than that of a concrete floor and is generally acceptable only in the floors of a house. In most other buildings it is inadequate.

It is however, common practice to use single timber floors as the intermediate floors in multi-storey maisonette blocks, with reinforced concrete separating floors between the maisonettes.[1] This results in a considerable economy in the cost of each maisonette and is possible because the floors within the dwellings are not required to provide so high a degree of fire resistance as the separating floors and the required degree of sound insulation may be lower than that desirable in the separating floors.

Concrete floors

The concrete floor has the advantage of strength and good fire resistance. Its use is now normal in most forms of multi-storey building, particularly because of the requirements in respect of fire resistance which apply to such structures. In addition it provides better air-borne sound insulation than the timber floor and for this reason it is used where the sound insulation provided by a timber floor would be inadequate.

The choice of a concrete floor can be made from a wide variety of types including in situ solid concrete floors, in situ hollow blocks and precast floors of numerous forms.

The in situ cast concrete floor is a wet form of construction incapable of bearing loads until quite set and hardened and requiring shuttering which must be left in position with all the supporting props until the concrete has gained sufficient strength. This, with ordinary Portland cement and normal temperatures, is usually a matter of three days at least, followed by a further four or more days of strutting by a reduced number of props. As the floor must be kept free of traffic until it has attained strength and as the props underneath take up space and interfere with other work, the rate of progress of the job as a whole may, as a consequence, be reduced.

This type of concrete floor can be used to provide lateral rigidity to the structural frame of a building against wind forces (see Part 2) and is used in monolithic concrete construction to achieve the economies inherent in continuity of structure referred to on page 46.

Precast concrete floors have been developed in order to reduce or eliminate shuttering and to reduce site work and the use of wet concrete as far as possible, these being factors which lead to speedier erection.

The large variety of precast floor systems can be divided into two basic categories:

(i) precast beams placed close together

(ii) precast ribs or beams with filler blocks or slabs between.

Those in the first category can be erected rapidly to give protection to following trades below and almost immediately form a working platform as the non-structural top screed may be laid afterwards. Many of those in the second category require an in situ structural topping of concrete to be placed before they can take their working load. Because maximum economy in precast work results from a maximum repetition of standard units, the precast floor is not suited to irregular plan shapes requiring a large number of special-size units to cover the varying spans of floor. Nor is it so well suited to circumstances where monolithy is required in the structure because of the difficulty of obtaining an efficient rigid junction between the floor and the supporting beams or walls, although continuity over supports in adjacent floor panels themselves may be obtained easily.

Steel floors

Apart from open metal flooring used in industrial buildings steel floors always involve the use of concrete in their structure, often acting structurally with the steel elements. Different types of floor in this category are described in Part 2 chapter 6.

In this volume will be described those forms of floors suitable for the span range given in table 8.1 and for loadings up to 4·00 kN/m^2. As will be seen from the table these involve single and double floor construction in timber and reinforced concrete. For wide span floors and special types of construction including basements, reference should be made to Part 2.

[1] See reference to this under *Cross wall construction* Part 2.

8.1 GROUND FLOOR CONSTRUCTION

Level of floor relative to ground

In the design of ground floors consideration of the level at which the floor shall be placed relative to the surrounding ground is important. A number of factors govern the floor level, including the nature of the site and the form of floor construction.

For reasons given in the chapter on *Walls* the damp-proof course in an external wall is usually placed 150 mm above ground level. If the floor surface is lower than this and the wall is solid a vertical damp-proof course must be provided to prevent damp penetration through the wall into the building (figure 8.1 *A*). Whenever possible therefore, the level of the floor is fixed slightly higher than the DPC in the walls and, in the case of solid floors, at a level such as to permit the damp-proof membrane in the floor to line through with the damp-proof course in the walls, with which it should connect (*B*). This avoids the necessity for a vertical DPC to connect the two as in (*C*). In the case of cavity walls the floor membrane is related to the DPC in the inner leaf which need not be at the same level as that in the outer leaf, in which case the cavity acts as a vertical damp barrier (*D*). The structure of a hollow timber ground floor should be quite separate from the wall and the DPCs in each may be at different levels provided the floor timbers are above the level of the wall DPC (see figure 8.6).

On all sites it is necessary to remove turf, vegetable matter and topsoil, usually to a depth of 150 mm to 230 mm below ground level. This space is filled with hardcore (see page 191) as a bearing for the floor and in order to reduce the quantity needed the level of the floor is kept as low as possible consistent with other factors.

When suspended timber ground floors are used air vents through the walls to the underfloor space must be provided. These should be well clear of the ground in order to remain unobstructed and to avoid the entry of water through them and the floor level should be such that the vents may be about 150 mm and certainly not less than 75 mm above ground level (figure 8.6).

On slightly sloping sites it may be possible to position the floor about 150 mm above ground at the higher side and make up with a greater depth of fill at the lower (figure 8.2 *A*). With steeper gradients however, the amount of fill becomes excessive and uneconomic, in which case it is necessary to sink the floor into the slope, using the technique of *cut and fill* (*B*). The floor level should be fixed so that a greater volume of excavation ('cut') is required than hardcore fill at the other end. By reducing the volume of fill this not only reduces the quantity of hardcore required but also the height of foundation walling to retain it. A reasonable guide is to arrange the depth of excavation to depth of fill to be in the proportion of $1\frac{1}{2}$ or 2:1. In such circumstances a further economy will result if the excavation can be extended into the ground, clear of the building and sloped back to be self-supporting (*C*). This will usually be cheaper than making the external wall of the building a damp-proofed retaining wall.

In some circumstances the planning of a building on a sloping site may permit the stepping of the floor (*D*). By this means the floor may be kept above ground level and the necessity of vertical DPCs between those in external walls and floor may be avoided.

Considerations of drainage may also affect the level of the ground floor. It may be necessary, for example, to fix the floor at a level higher than that required by other considerations in order to obtain satisfactory gradients in the drains connecting to an adjacent sewer at a high level.

Solid and suspended ground floors

As already indicated, in most buildings the ground

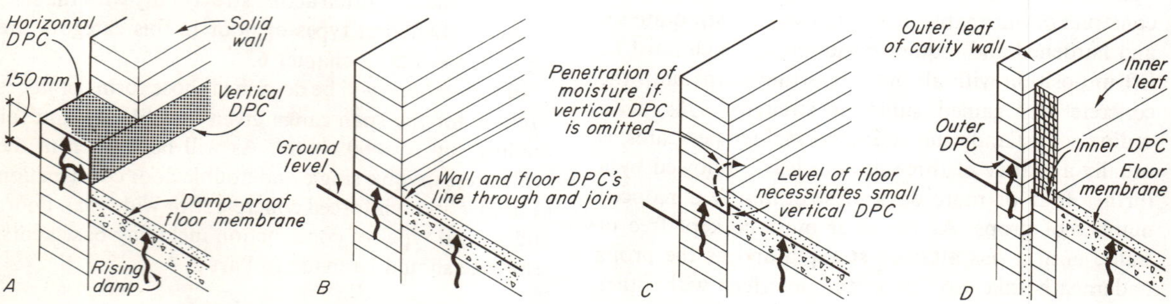

8.1 *Level of floor relative to ground — damp-proofing*

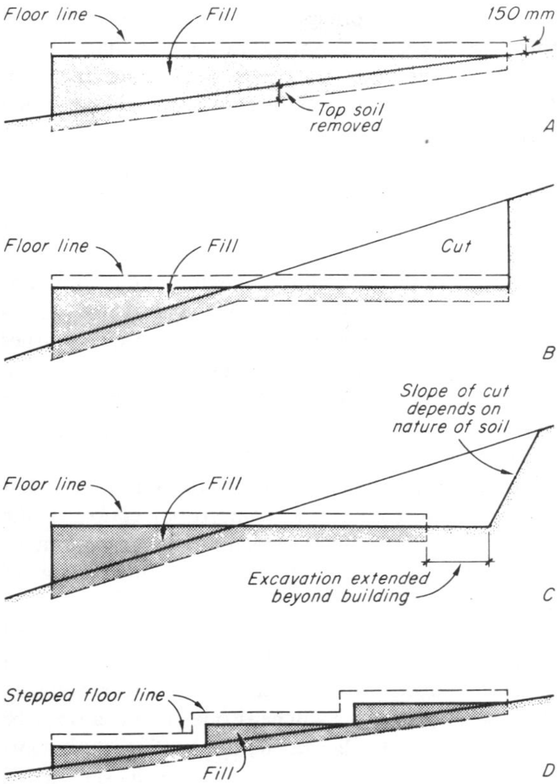

8.2 *Level of floor relative to ground — sloping sites*

floors are usually of solid construction. This is mainly due to economic reasons. In the past, suspended timber ground floors were in common use for domestic type buildings to overcome rising damp by the disassociation of floor structure and ground. This form of construction is generally dearer than solid construction but sometimes for small scale work on steeply sloping sites its use will prove more economic than a solid floor laid direct on cut and fill. In any circumstances where a fill beneath the floor is more than 600 mm thick the use of a solid floor may be considered inadvisable for the reason given below and a suspended floor would then be used. Apart from these considerations, solid floors are more suitable for heavy loadings and they also permit a wider choice in the selection of floor finishes.

8.1.2 Solid ground floor

The site within the walls of the building is prepared by removing all turf (which, if of satisfactory quality, is stacked for re-use or for sale) and excavating the vegetable or topsoil to a depth of at least 150 mm to 230 mm. Too great a depth of excavation may increase the amount and, therefore, the cost of hard filling needed to bring the floor to the required level.

A bed of well consolidated suitable hard material known as *hardcore* is then generally put down (figure 8.3). The purpose of this is to act as a filling to provide a horizontal surface at the appropriate level for the concrete slab and to form a firm, dry working surface, especially on soft or wet sites. On a firm, dry site there may be no need for any hardcore unless it is required for making up to levels.

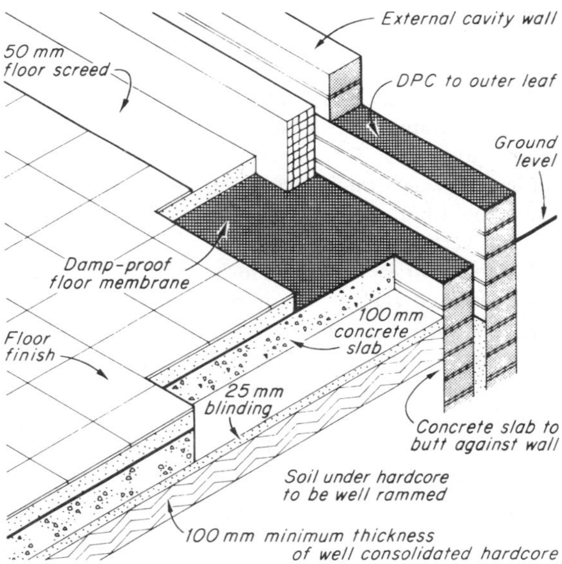

8.3 *Solid ground floors*

Hardcore consists of brick or concrete rubble, broken stone or other inert, coarsely graded material such as hard, well-burnt furnace clinker. On wet sites, that is those liable to high ground-water level or surface flooding, materials which swell on wetting, such as underburnt colliery shale, should not be used for hardcore, and those which may contain sulphates, such as clinker, colliery shale and gypsum plaster in brick rubble, if they must be used, should be isolated from the concrete which the sulphates would attack. This is done by laying waterproof paper or other impermeable sheeting over the hardcore before the concrete is laid. If of considerable depth the hardcore may consolidate further after building work is complete resulting in cracking of the floor slab by the removal of its support over parts of its area. Where the depth of hardcore would be greater than 600 mm a suspended floor is generally preferable.[1]

The hardcore is laid to a minimum thickness of 100 mm and is usually blinded on the top surface with ashes or other fine material before the concrete is laid (figure 8.3).

The concrete slab is not less than 100 mm thick and the top surface is finished with a power float[2] or is spade finished to take a screed according to the floor finish to be applied. In large areas of unreinforced floor slabs shrinkage cracks are liable to form. To minimise this and to ensure cracking along regular lines it is desirable to lay the concrete in relatively small areas at one time, say in alternate squares of about 83 m^2, laying the remaining squares after the initial shrinkage of these has occurred.

The edges of the floor slab should not be built-in to the surrounding walls nor should they rest directly on foundation slabs. To avoid cracking due to unequal settlement which might occur through doing this the slab should butt against wall or foundation to form a straight vertical joint.

Where a variety of floor finishes of different thicknesses is used each will require a different sub-floor level obtained either by varying the thickness of the screed or by varying the level of the slab. The former method, unless the variation is great, is usual and the most economic.

Some floor finishes are themselves damp-resistant, some let moisture through without deteriorating and some are adversely affected by moisture.[3] When finishes in the latter category are to be used it is essential to incorporate a damp-proof membrane in the floor. Materials which may be used for membranes are mastic asphalt, bituminous felt, hot-applied pitch or bitumen, cold-applied bitumen solution, pitch or bitumen/rubber emulsion and polythene film.[4]

Polythene film in roll form may be laid directly on 25 mm minimum of sand blinding on the hardcore immediately under the floor slab (figure 8.5). Since evaporation of moisture can then take place only through the top of the slab adequate time must be allowed for this to occur. For a 100 mm floor slab with, say, 25 mm screed at least five months[5] should be allowed for drying out before the application of moisture sensitive finishes. Other membrane materials, apart from asphalt and pitch mastic which may be applied to the slab surface and avoid delay in finishing, are sandwiched within the concrete. To minimise shrinkage cracking and curling as the concrete dries out a mix as dry as is practicable should be used and the thickness of the layers should not be less than 50 mm. In practice the lower layer is made 100 mm thick and the upper constitutes the finished screed, 50 mm thick (figure 8.3). Polythene film can, of course, also be sandwiched in this way but should be placed on a 13 mm blinding of sand on the lower layer of concrete. This is essential to avoid the possibility of puncturing the film.

Because of the difficulty of avoiding slight curling or unevenness at the junctions of bays, which shows through thin floor finishes, screeds under such finishes are now laid continuously. When screeds are to contain underfloor heating cables or are to be finished with an in situ flooring they should be laid in bays not exceeding 15 m^2 in area.

Whatever the membrane or its position in the slab it must be connected to the damp-proof course in the walls, if necessary by means of a short section of vertical membrane, in order to prevent the passage of damp at the edges as described on page 190.

Although in some circumstances a damp-proof membrane may be omitted when floor finishes not adversely affected by moisture are used (for example, clay tile flooring to a kitchen) one should always be provided under the slab when floor heating is incorporated. This is necessary in order to keep the slab dry to avoid excessive heat losses and to keep the necessary insulation dry.

The loss of heat through solid ground floors on dry ground is relatively small and most loss occurs near the edges (figure 8.4, inset). Provided the area of floor and its cross dimensions are great enough the standard of thermal insulation required by the Building Regulations may be achieved without the addition of insulation. AD L2/3 gives a graph showing the dimensions of floors which achieve this standard and for which insulation is not required. On damp sites, however, irrespective of size insulation should always be provided over the full extent of the floor. Materials suitable for floor insulation are dense resin-bonded mineral or glass fibre slabs and polystyrene and cork slabs. These should be

[1] For a more detailed consideration of hardcore see BRE Digest 276.

[2] See Part 2 *Contractors' Plant*.

[3] See table 4.3 *MBS: Finishes*.

[4] For relative efficiency of these see table 2 BRE Digest 54.

[5] ie 1 month per 25 mm thickness at least.

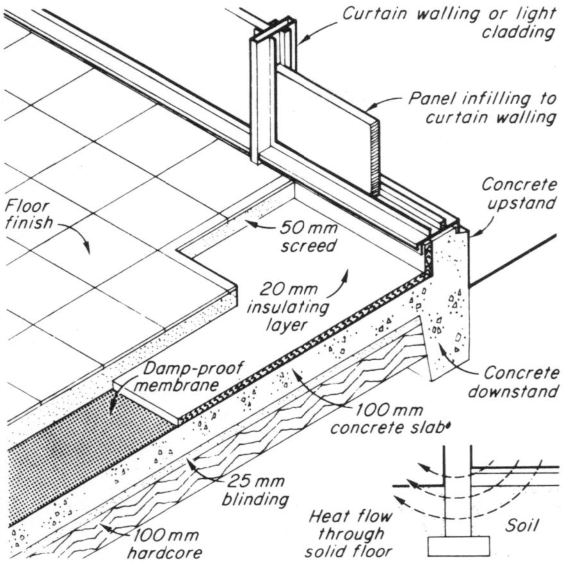

8.4 *Solid ground floors — edge insulation*

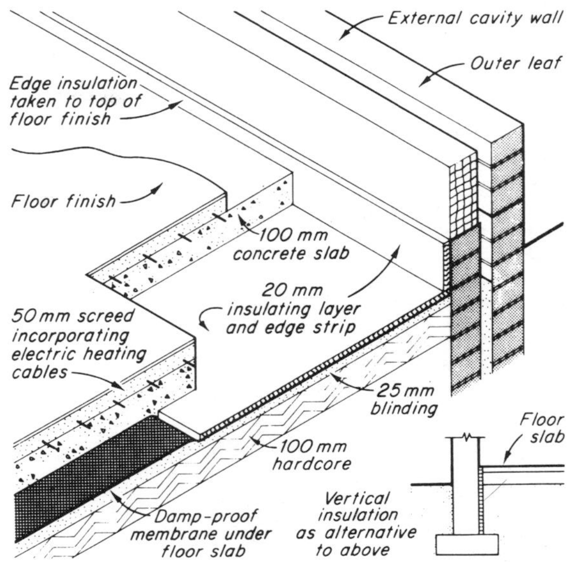

8.5 *Solid ground floors — edge insulation with floor heating*

placed above the damp-proof membrane and be turned up at the edges of the floor slab to prevent heat loss through the wall as in figure 8.5.

Where insulation is not provided throughout the floor but where the walls are thin, such as curtain walling or other light cladding, and bear on the edge of the floor slab (figure 8.4), or where the floor forms a raft projecting beyond the walls, some insulation at the edges is necessary to break the potential cold bridge and avoid condensation at the floor edges. It is always necessary in conjunction with floor heating. Edge insulation may be a horizontal layer about 900 mm wide together with a vertical edge strip, as shown in figures 8.4 and 8.5, or a vertical strip against the wall extending at least 300 mm below the underside of the floor slab (figure 8.5, inset). The latter is rather more difficult to place. It is also difficult to keep dry and, therefore, non-absorbent expanded polystyrene or rubber should be used. An alternative to this would, of course, be cavity insulation in the wall carried down to foundation level. When floor heating is incorporated in the screed the horizontal layer is usually placed below the floor slab as in figure 8.5.

8.1.2 Suspended ground floors

Timber floor

This type of floor is a suspended floor of limited span in which the floor structure is supported on short walls built off the ground, and is thus disassociated from the ground and out of direct contact with ground moisture (figure 8.6).

After topsoil removal the ground must be covered with a 100 mm layer of site, or surface concrete 1:3:6 mix on hardcore to exclude ground moisture and ground air and to prevent vegetable growth. On wet sites the mix should be 1:2:4 or an oversite covering of asphalt, pitch or bitumen or of 1200 gauge polyethylene sheeting overlaid by 50 mm of weak concrete should be used laid on a base of hardcore blinded with ashes or sand. The surface of this covering or of the surface concrete should not be lower than the surrounding ground unless it is certain that the site is such or is so drained that the underfloor area cannot be flooded at any time.

Dwarf half-brick walls called *sleeper* walls are built off the surface concrete to support the floor structure; if concrete is not used the walls are built on small concrete foundations.

The floor structure consists of timber bearers called *joists* (more properly *bridging* or *common joists*, especially in double floors) bearing on the dwarf walls and their size will depend upon their span and spacing and the loading on the floor.

Joists are 38 mm to 50 mm thick. If less than this nailing of the floor finish tends to split the joists, especially where boards or sheets butt-joint over a joist,

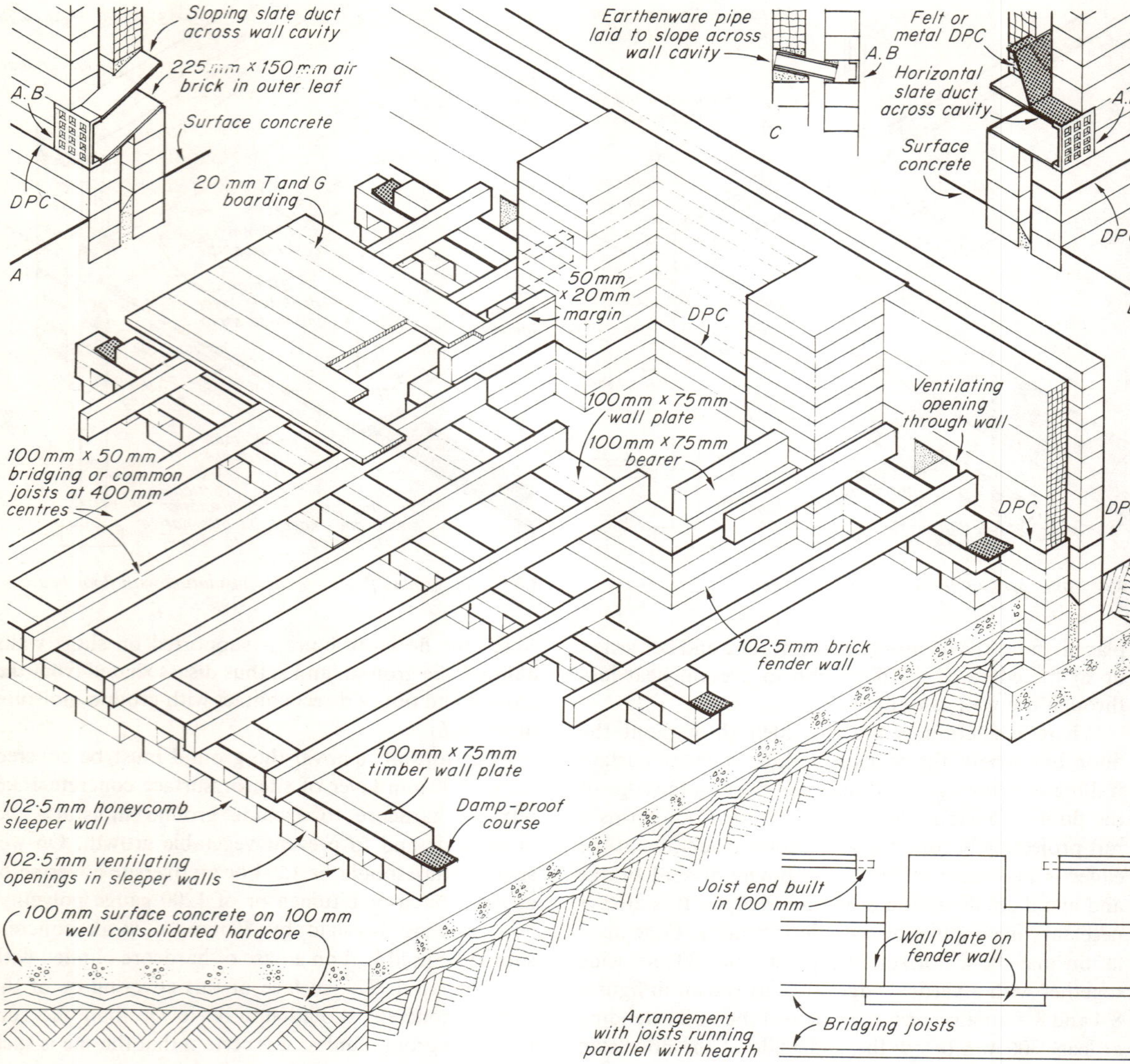

8.6 *Suspended timber ground floor*

requiring two lines of nails close to the faces of the joist. A spacing of 400 mm to 450 mm centre to centre of joists is suitable for 19 mm thick tongued and grooved boarding or 400 mm for 18 mm thick chipboard as a floor finish.

A joist depth of 100 mm is economic for domestic loading (1·50 kN/m^2) requiring the span, that is the spacing of the sleeper walls, to be kept to about 1·20 m to 2·0 m. Actual depths should be calculated or taken from tables in the Building Regulations.

The joists bear on timber members, called *wall plates*, bedded in mortar on top of the sleeper walls. These are 75 mm × 50 mm or 100 mm × 75 mm[1] and serve as a bearing for the joists and distribute their loads uniformly to the wall. They also provide a means of fixing the joists by skew-nailing to them through the joist

[1] When related to brickwork the approximate brick sizes of the latter dimensions are useful in detailing.

sides. As in wall plates for roofs longitudinal, or running joints are made by half-lapping. A damp-proof course in the sleeper walls immediately below the wall plate prevents rising damp reaching the floor timbers.

The height of the underfloor space above the surface concrete must be at least 75 mm to the underside of any timber wall plate and at least 125 mm to the underside of the suspended timbers.[1] This space must be adequately ventilated to prevent the air becoming humid and giving rise to conditions favourable to the growth of dry rot. Air bricks must be provided, in all external walls if possible or at least in two external walls on opposite sides of the building, and all sleeper walls must be in *honeycomb* construction to permit a free flow of air. For the same reason ventilating holes must be formed in all partition walls passing through the underfloor space. Pipes or ducts leading to air bricks should be laid in any adjacent areas of solid floor which might otherwise create stagnant areas in the underfloor space. Air bricks should be provided on the basis of 3000 mm^2 of *open* area for every metre run of external wall. The air bricks should be placed well clear of the ground and free of obstructions and when they are situated in cavity walls ducts must be formed to prevent the ventilation of the wall cavity or, when the wall insulation is by cavity fill, to prevent the airways being obstructed by the fill. Methods of accomplishing this are shown in figure 8.6 *A*, *B*, *C*.

In order to avoid the building-in of joist ends to external walls they should bear on independent sleeper walls 38 mm to 50 mm from the main walls as shown in figure 8.6. Where this is not possible the joist ends should be treated with preservative and a small air space left at sides and top of each joist.

T and G boarding or sheeting should be used for the floor finish to avoid discomfort due to draughts.

Suspended floors of small dimensions are unlikely to achieve the required standard of thermal insulation without adding insulating material. This may be provided as shown in figure 8.7, supported between the joists on wire or plastic mesh or on battens in the case of rigid slab materials. Care should be taken to ensure that insulation is not omitted from the space between walls and the adjacent joists in order both to avoid draughts at the junction of wall and floor finish and to maintain the required standard of insulation at that point.

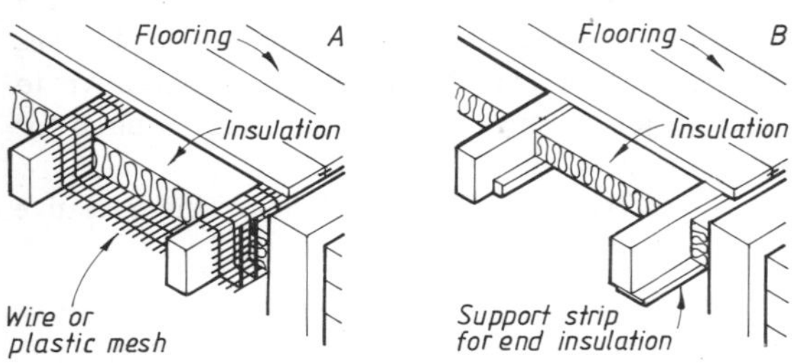

8.7 *Thermal insulation to ground floors*

If a fireplace occurs in a room having a timber ground floor the concrete hearth must be at the floor level, necessitating a *fender wall* (figure 8.6) built up off the surface concrete to provide support for the hearth and the floor round the hearth. This will be 102.5 mm or 215 mm brickwork depending on whether the hearth is supported on hardcore or on the edge of the fender wall (see page 209).

The floor joists may run parallel with or at right-angles to the front of the hearth as shown in figure 8.6. It will be seen that in the latter case a short length of joist is required on each side of the hearth to act as a bearer for the floor finish at these points.

Reinforced concrete floor

Precast concrete rib and filler block construction (see page 189 and Part 2 chapter 6) is increasingly being used for suspended ground floors in houses. These consist of precast concrete inverted T-beams or ribs spaced apart and carrying light-weight concrete blocks or proprietary insulating slabs between them to provide a degree of thermal insulation and a flat upper surface to take the floor finish. The greater span possible with the concrete ribs results in the elimination of or reduction in sleeper walls.

8.2 UPPER FLOOR CONSTRUCTION

Timber floors

An upper floor in timber construction differs in some respects from a timber ground floor. In the latter there is no restriction on the number of supports in the form of sleeper walls so that the span and size of the joists may be kept small. Relatively large unobstructed areas are, however, required under upper floors resulting in wider floor spans and larger joists. Furthermore, support for the joists round a hearth on a ground floor is

[1] The Building Regulations, 1985, require this space, if it is accessible for persons or is more than 1 m high, to be subdivided by cavity barriers to prevent the passage of flame or smoke — See AD B, G7.

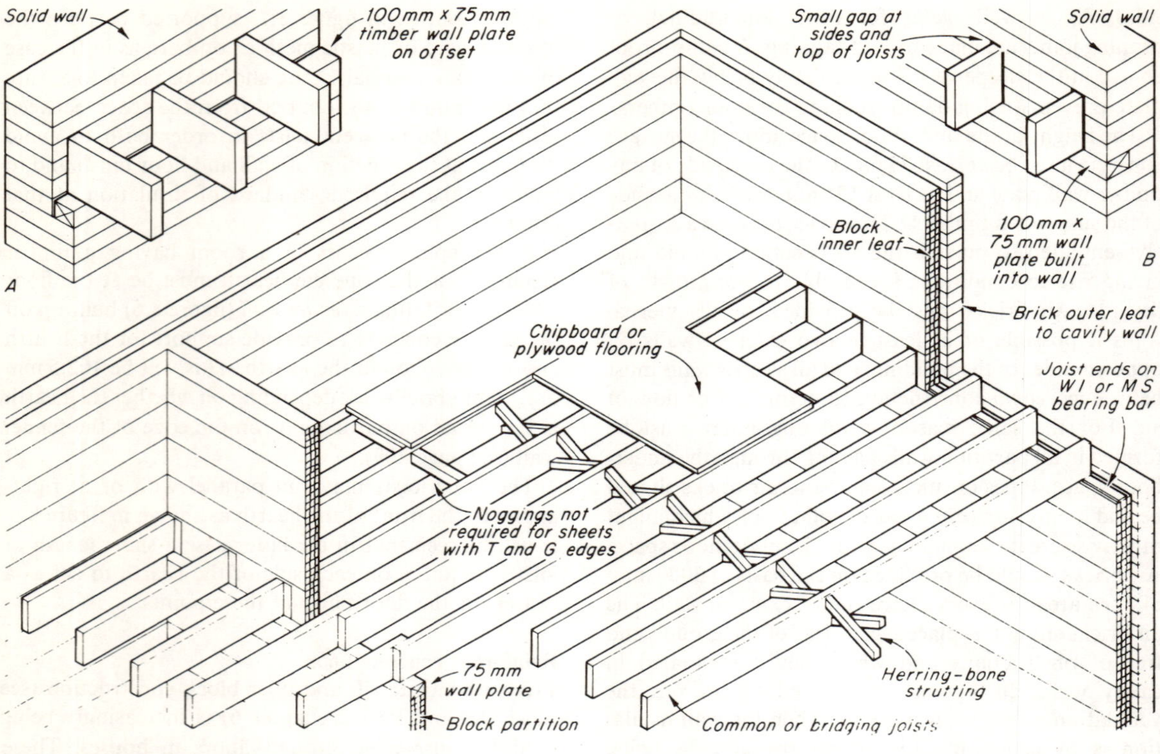

8.8 *Timber upper floor construction*

direct on to fender walls; in an upper floor the timbers must be framed to be self-supporting, as they must be around any other opening in the floor.

8.2.1 Single floor in timber

This consists of common or bridging joists spanning between walls or partitions and bearing usually on wall plates or other members to distribute the load (figure 8.8). Material and minimum thickness of joists are the same as for ground floor joists. The depth of joist will depend upon loading, span and spacing and it may be calculated, taking account of these factors. Regard must be paid to the need for adequate stiffness in the floor as this may require scantlings larger than those necessary to avoid collapse of the floor. When calculations are made deflection should, therefore, be considered.[1]

For economic reasons the clear span of softwood joists is usually limited to about 4·90 m which will require, for domestic loading with joists at 400 mm centres, 225 mm × 50 mm or 63 mm joists. For heavier loadings and greater spans the required size of timber soon falls outside the range of stock sizes and becomes uneconomical.

It is common practice to space timber floor joists at 400 mm centres with 19 mm tongued and grooved, or 22 mm plain-edge, boarding. In terms of labour and materials this usually gives an economic floor since it can be shown that as the spacing of joists is increased the timber content of the joists reduces at a slower rate than the rate of increase in the content of the boards. Chipboard sheet may be used as an alternative to boarding, joists centres at 400 mm being required for 18 mm sheets.

In situations where the cost of labour is high relative

[1] For domestic loading only (1·50 kN/m²) in houses up to three storeys in height sizes may be taken from the tables in AD A, Part B.

to the cost of timber, it can be cheaper to reduce the amount of labour involved by using larger, widely spaced joists spanned by thick boarding. This produces what in America is called plank and beam construction. The thickness of the boards would be calculated in the usual way as beams spanning between two joists. This gives a board of sufficient stiffness for most purposes.

An extra 3 mm over the calculated board thickness is usually added to allow for wear.

Apart from preventing the passage of dust and draughts through the joints, tongued and grooved boarding is preferable to plain-edge because it is able to transmit point loads on one board to adjacent boards. This reduces the intensity of load on individual boards. This does not occur with plain-edge boarding and for this reason BS 5268: Part 2 recommends that each individual plain-edge board shall be designed to carry the full superimposed load given in BS 6399 as the minimum load per 300 mm width of boarding. When chipboard or plywood is employed the use of tongued and grooved sheeting, in addition to preventing passage of dust and draughts through the joints, avoids the need for cross noggings to provide support at the ends of the sheets.

The ends of the joists may be supported in various ways, the adoption of any particular one depending on a number of considerations.

When there is little risk of damp penetration a 100 mm × 75 mm timber wall plate is commonly used, built-in to the wall or bearing partition. When the plate is built-in to an external wall it is, however, a wise precaution to treat it with preservative together with the ends of the joists and to leave a small air space at the sides and top of the joists (figure 8.8 *B*). Occasionally it may be convenient to use an *offset*, where a wall reduces in thickness, as a bearing for the plate (figure 8.8 *A*) although this will be less common than in the past now that walls for economic reasons are usually designed to be the same thickness throughout even the height of a multi-storey building. As an alternative to building-in, the plate may be carried on MS corbel brackets (figure 8.9 *A*). This, however, results in a projection under the ends of the joists, unless the depth of the joists permits full notching over the plate.

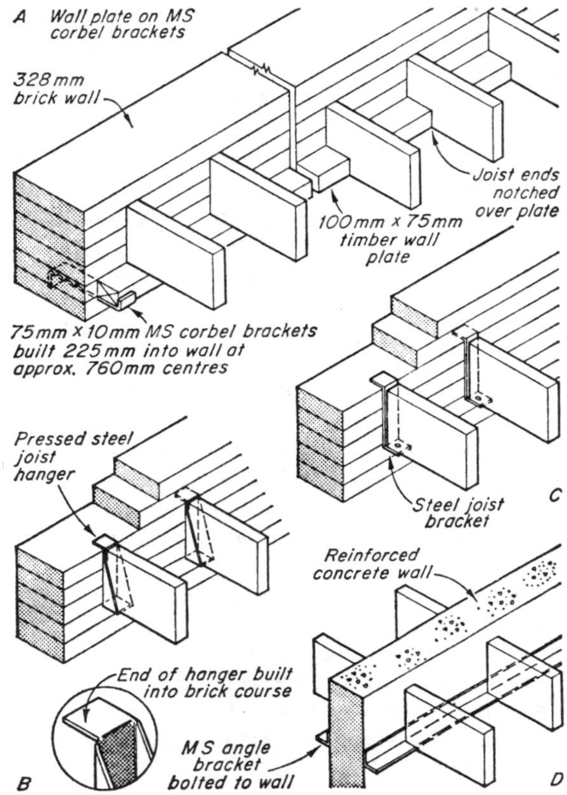

8.9 *Support of joists*

In terrace houses and in multi-storey maisonette blocks of cross wall construction in which, for reasons given on page 189, the intermediate floors may be of timber construction, the joists may span between the separating walls. In order to ensure a minimum of 90 mm of solid non-combustible material between the ends of joists on opposite sides of a 215 mm masonry wall, so that the necessary degree of fire protection is maintained, the bearing of the joist must be limited to about 62 mm. This will be ample for normal softwoods and domestic floor loadings. A MS bearing bar can be used to spread the load, 50 mm or 57 mm by 8 mm tarred and sanded for bedding into the brickwork. As an alternative to building-in the ends of joists may be carried on steel hangers or brackets as shown in figure 8.9 *B*, *C*, *D*. Hangers or brackets should be bedded tight against the wall face and only a very small clearance should be left between the joist end and hanger.

Joists supported on the inner leaf of an external cavity wall should bear on a MS bearing bar as in figure 8.8 rather than on a timber plate so that as little timber as possible is exposed permanently to the unventilated cavity of the wall, the air in which will often be relatively moist. The joist ends, which should not project into the cavity, should be treated with preservative and a small air space left at sides and top of each joist. The use of metal hangers to support the joist ends on an inner leaf should normally be avoided because of the excessive eccentricity of load which is caused.

Should the floors be required to provide positive lateral restraint to the walls for design purposes, the ends of the joists must in certain circumstances be secured to the wall by metal ties (see Part 2 chapter 4).

To take advantage of the crane now normal on most jobs of any size the timber floors may be prefabricated in sections in the workshop and be lifted into position by the crane on to angle or bracket supports. Experience has shown that when this is done it is better to prefabricate the joists only as panel units and to fix the boarding or sheeting in the normal way after the building is covered in or at least, in maisonettes, after the reinforced concrete floor above has been completed. Otherwise in wet weather the panels become wet and the boards swell and rise.

Joints in the length of joists must be made over walls and partitions. If the lining up of the joists is not important they may be laid side by side over the bearings and spiked together as in figure 8.8. This is particularly useful where the joints occur over thin partitions as it avoids the scarfed or halved joint necessary to line up the joists.

Lateral support and cutting of joists Stiffening is required when joists are deep in order to avoid winding or buckling at the top or compression zone (see page 50). BS 5268: Part 2 makes recommendations in respect of this for solid and laminated members up to a depth to breadth ratio of seven. Unless special calculations are made these recommendations, which are tabulated in table 8.2, should be followed. The 'bridging or blocking' referred to in the fifth item may be provided by herring-bone or solid strutting (see figures 8.8 and 8.10). Should the joists be slightly in-winding when laid, solid strutting is difficult to fit. It is used mainly for heavy floors with a bolt passing through the centres of the joists close to the strutting which, on completion of the strutting, is tightened to give a rigid result. It can be argued that in practice, to avoid winding which sometimes occurs in the timber for reasons other than buckling, it is desirable to use bridging for depth to breadth ratios less than six and it is still, for example, often used for 225 mm × 50 mm joists.

Timber is easily cut and drilled for pipes and conduits but this must be done with care. Notching near the centre of the span should be avoided, particularly if a number of adjacent joists are notched in line, since this cuts through the fibres at the point where, in uniformly loaded joists, they are most heavily stressed in bending. Notching should, therefore, be done near the bearings of the joists where, over simply supported single spans, bending stresses are at a minimum. This does reduce the section available for resisting the shear forces, but if the joists carry uniformly distributed loads only and the notches are limited in depth to one-eighth of the joist depth and are kept within the limits shown in figure 8.10 their effect need not be calculated. Holes for pipes and conduits should not be greater in diameter than one-quarter of the joist depth and should be drilled on the neutral axis within the limits shown in figure 8.10.

Degree of lateral support	*Maximum depth to breadth ratio*
No lateral support	2
Ends held in position	3
Ends held in position and member held in line, as by purlins or tie rods	4
Ends held in position and compression edge held in line, as by direct connection of sheathing, deck or joists	5
Ends held in position and compression edge held in line as by direct connection of sheathing, deck or joists, together with adequate bridging or blocking spaced at intervals not exceeding 6 times the depth	6
Ends held in position and both edges firmly held in line	7

Table 8.2 *Lateral support to joists*

Partition support When timber or other light-weight non-loadbearing partition bears on a timber floor the joists of which run parallel with the partition the joists are commonly doubled up under the partition, a pair being spiked together if the ceiling is of plasterboard or similar sheet lining. If the ceiling is plaster on lathing they are blocked apart about 50 mm, the distance blocks being kept about 19 mm up from the underside so that the plaster key is not broken for too great a distance. If the partition runs at right angles to the joists, a 75 mm deep timber sole piece, the same width as the partition, is used to span over the joists. A check

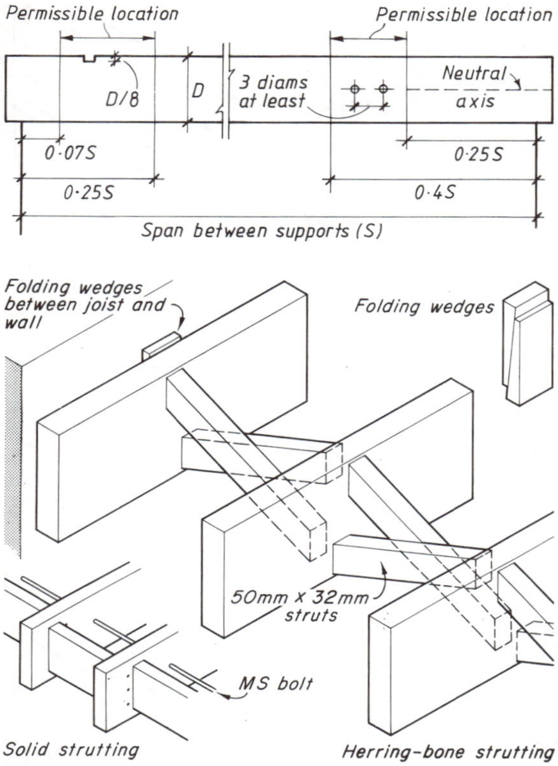

8.10 *Lateral support and cutting of joists*

should be made to ensure that excessive deflection will not occur under the extra weight of any partition carried by the floor.

Trimming round openings and hearths

As indicated at the beginning of this section the floor around any openings within it, such as for stairs or hearths, must be so constructed as to be self-supporting at these points. This is accomplished by cutting short some of the bridging joists to form the opening, these are then known as *trimmed* joists, introducing a thicker joist or a pair, called *trimmer* joists, to carry the ends of the trimmed joists and thickening one, or a pair, of the bridging joists to carry the trimmer joists, when they are known as *trimming* joists as shown in figure 8.11. The framing of a floor in this manner is called *trimming*.

Both trimmer and trimming joints are made thicker than the bridging joists as they carry greater loads and are often cut to accommodate the joists they carry. They are made 25 mm thicker than the bridging joists, with the trimmers supporting not more than six trimmed joists (figure 8.12). If the loading conditions are worse than this, the members should be calculated.

The following carpentry joints shown in figure 8.12 are used in the trimming of openings:

Tusk tenon joint This is a strong joint and is used to frame the trimmer to the trimming joist. It is cut to the proportions shown, the tenon being cut on the end of the trimmer and extending 100 mm to 125 mm beyond the outer face of the trimming joist. The *tusk* below the tenon transfers most of the weight to the trimming joist and the bevelled portion above strengthens the tenon. The function of the wedge is to draw the trimmer tight up to the trimming joist and to ensure this the hole in the tenon must be long enough to permit this to take place.

Dovetailed housed joint This is used to join trimmed joists to trimmer. The form of the joint holds the two members together but the two are usually also secured by nails at an angle through the trimmer into the end of the trimmed joist.

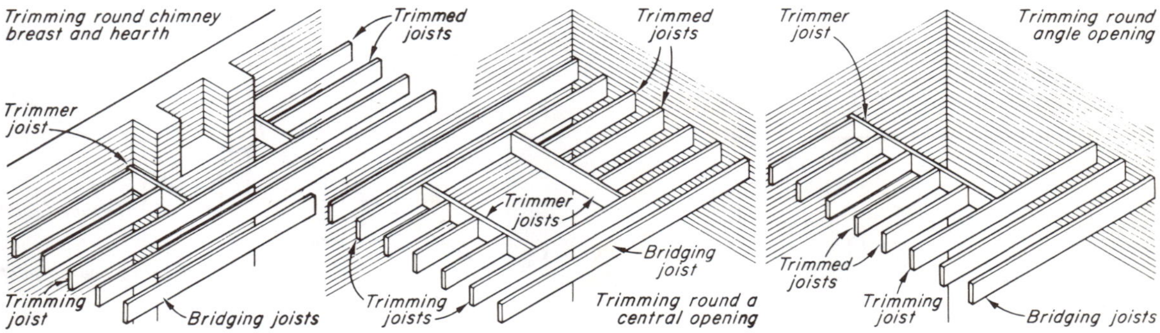

8.11 *Trimming round openings and projections*

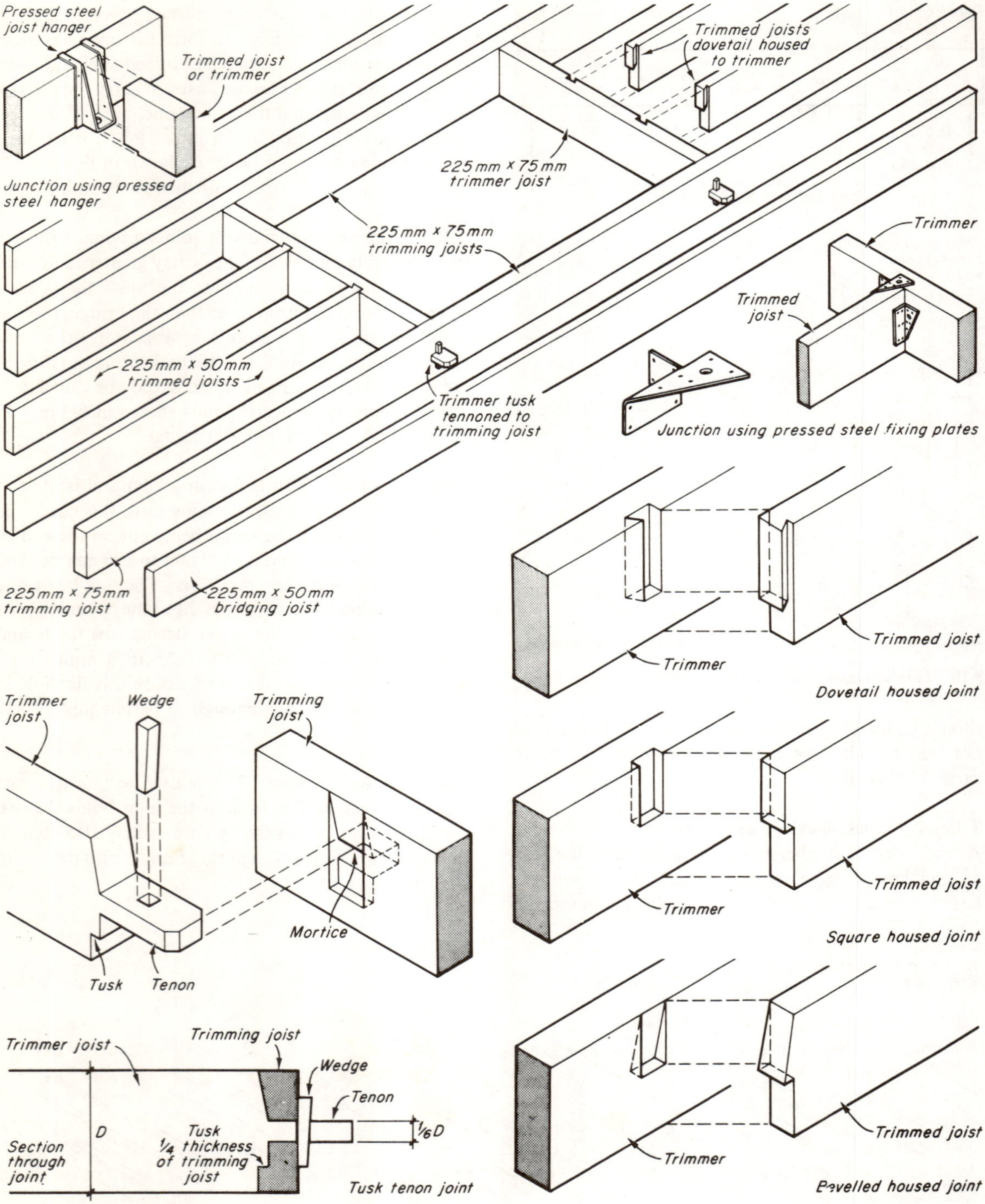

8.12 *Trimming to timber floors*

Squared housed joint This joint extends only half the depth of the members joined and is satisfactory for joining short trimmed joists to trimmers. It requires secure nailing.

Bevelled housed joint This is similar to the square housing but due to its shape less timber is cut from the compression zone of the trimmer.

As an alternative to carpentry joints the junction between trimmed and trimmer joists may be made with metal hangers or fixing plates as shown in figure 8.12, requiring less labour in forming the joints.

In trimming round an upper floor hearth the opening in the floor is usually wider than the hearth itself, because of limitations on the building-in of timber near fireplaces and flues (see figure 8.13). The actual arrangement of the trimmers and trimming joists depends on the direction of the bridging joists relative to the hearth and is shown in figure 8.13. In order to provide fixings for floor finishes and timber margin at each side of the hearth a short joist, called a *cradling piece*, is housed at one end into the trimmer or trimming joist as the case may be and bears at the other end on a brick or steel corbel since it must not be built into the chimney breast or flue (figure 8.13). When the bridging joists are at right-angles to the hearth a cradling piece only is required to provide both end fixing for the floor finish and fixing for the timber margin. When the joists are parallel with the chimney breast another piece of joist is housed between the cradling piece and the trimmer next to the chimney breast to provide the fixing for the floor finish. For details of hearth construction and the support given by the floor structure see chapter 9.

An upper floor in timber with its soffit exposed to the external air or to an unheated space must be insulated to comply with the requirements of the Building Regulations and the methods described for timber ground floors can be used. Since, however, the soffit will be ceiled with a lining probably impervious to water vapour, insulation such as shown in figure 8.7 should be positioned near the floor finish so as to leave a void immediately below the insulation which can be vented at each end to allow ventilation of the space. A vapour barrier must be placed under the floor finish, that is on the warm side of the insulation. This construction is, in fact, a cold deck roof in reverse.

8.2.2 Double floors in timber

The single joisted floor is rarely used for spans above about 4·90 m because the rapidly increasing depth of timber required makes it uneconomic. When timber floors are suitable but spans are large, cross-beams are introduced to carry the ends of the joists. By this means the span of the joists themselves can be kept within the limit of 4·90 m.

The beams are normally of steel or timber although reinforced concrete can be used. Methods of bearing the joists on beams are shown in figure 8.14. Timber beams are usually in the form of plywood box beams or of laminated timber. If the joists cannot bear directly on the tops of these beams the most suitable method of support is by metal hangers.

Triple, or framed, floors in timber are now obsolete. If the span requires this type of construction steel or reinforced concrete is normally used, often for functional reasons other than, or in addition to, that of strength and stability.

8.2.3 Reinforced concrete floors

In small-scale work a reinforced concrete suspended floor would be used mainly because of its greater fire resistance and better sound insulation than a timber floor.

In its simplest form it consists of a solid in situ cast, one-way spanning slab with the reinforcement acting in one direction only between two supports. The reinforcement may be either mild steel main rods and distribution bars wired together at right-angles, fabric reinforcement, consisting of main bars and distribution bars electrically welded at the crossings and supplied in sheets and rolls, or expanded steel or ribbed metal lathing. The last form can act as permanent shuttering requiring temporary support only by timber posts and beams. These alternatives are illustrated in figure 8.15.

Increase in span and load lead to an increase in thickness and a consequent rapid increase in the dead weight of this type of floor. It is economic only over small spans up to 4·60 m. When spans much above this are required it is usually cheaper to introduce secondary beams to keep the slab span within these limits.

The slab is cast on formwork or shuttering which, for a small job will normally be in timber (see chapter 11),

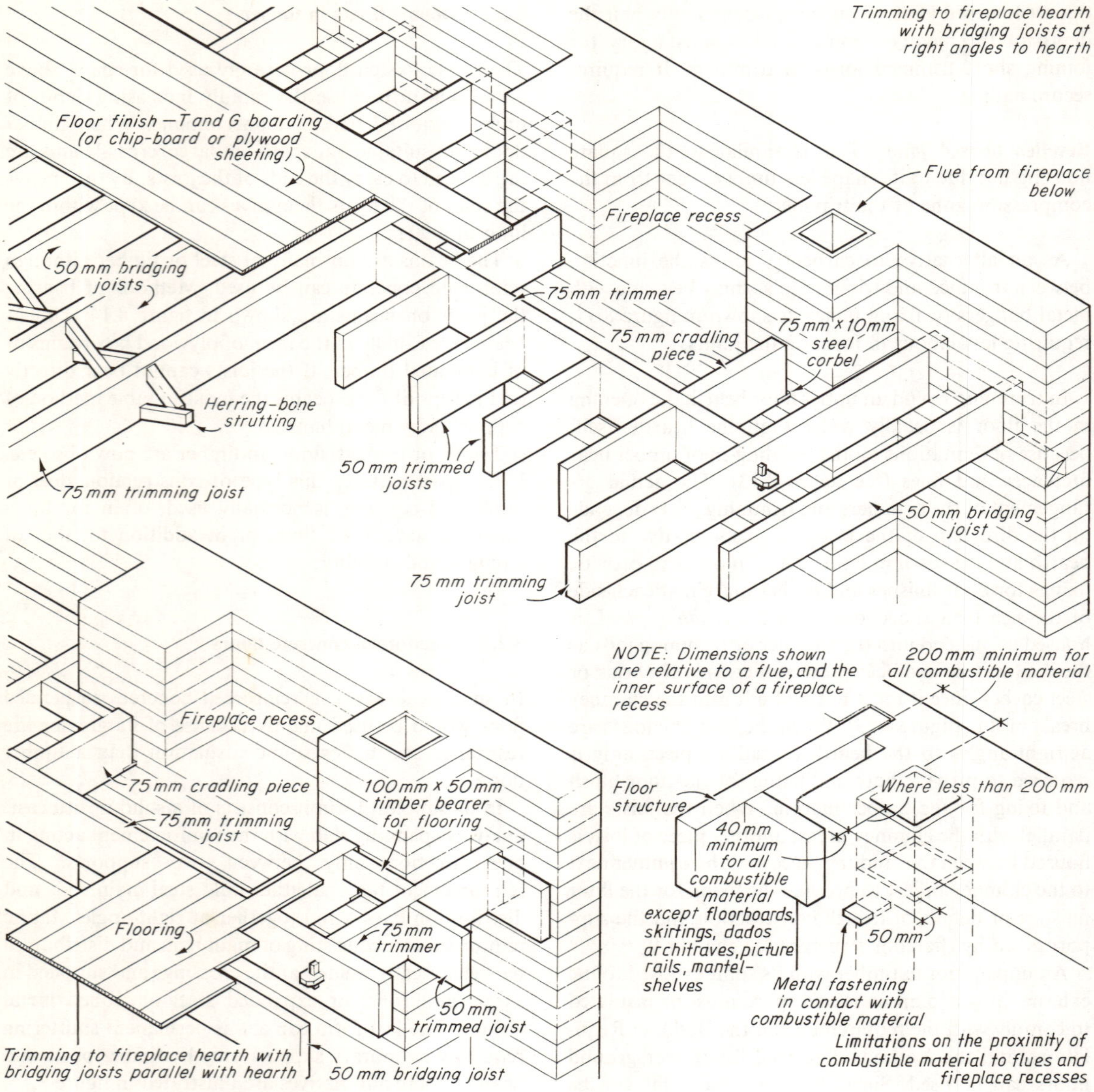

8.13 *Trimming round upper floor hearths*

the top surface of the shuttering being painted with mould oil to prevent the set concrete adhering to the decking, making removal of the latter difficult. Steel reinforcement is placed in position and supported at intervals on small blocks of high grade concrete or special plastic supports to keep it 19 mm to 25 mm above the shuttering according to the thickness of cover required.

The concrete is then poured, tamped or vibrated and screeded level and to the correct thickness. As indicated on page 189 the shuttering and props must not be removed until the concrete has gained sufficient strength and in reasonably warm weather this will generally be about seven to ten days.

In order to reduce construction time and avoid the use of shuttering precast concrete can be used for floor

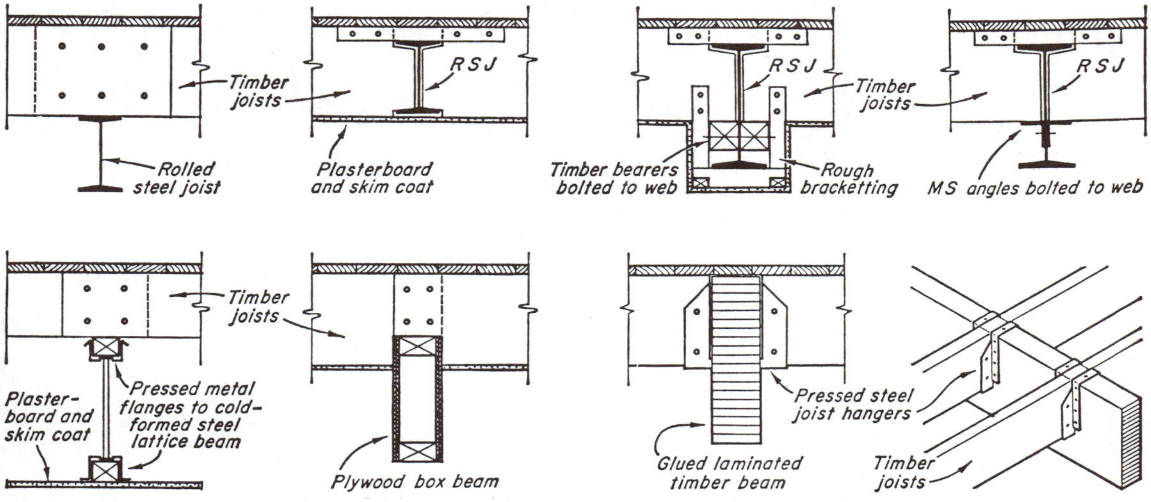

8.14 *Double floor construction*

construction as indicated on page 189, although for small jobs it is not usually as cheap as in situ cast concrete. Precast concrete floors, of which there is a great variety of types, are, however, very widely used where the size of job justifies it and these are described in detail in Part 2. Two types are illustrated in figure 8.15 (overleaf). In one the units are in the form of hollow beams which are placed individually and, if not too long, can be manhandled into position. In the other a number of 'trough' units are cast to form wide slabs by means of which a given area of floor can be covered more quickly. The size and weight of these requires the use of a crane to place them in position.

Where the soffit is exposed to the external air or to an unheated space a concrete upper floor must be insulated to comply with the requirements of the Building Regulations. The insulation, the thickness of which will depend upon the material used, may be applied above or below the floor slab. When the insulation is above the slab a vapour barrier must be incorporated below the floor finish on the warm, upper side of the insulation.

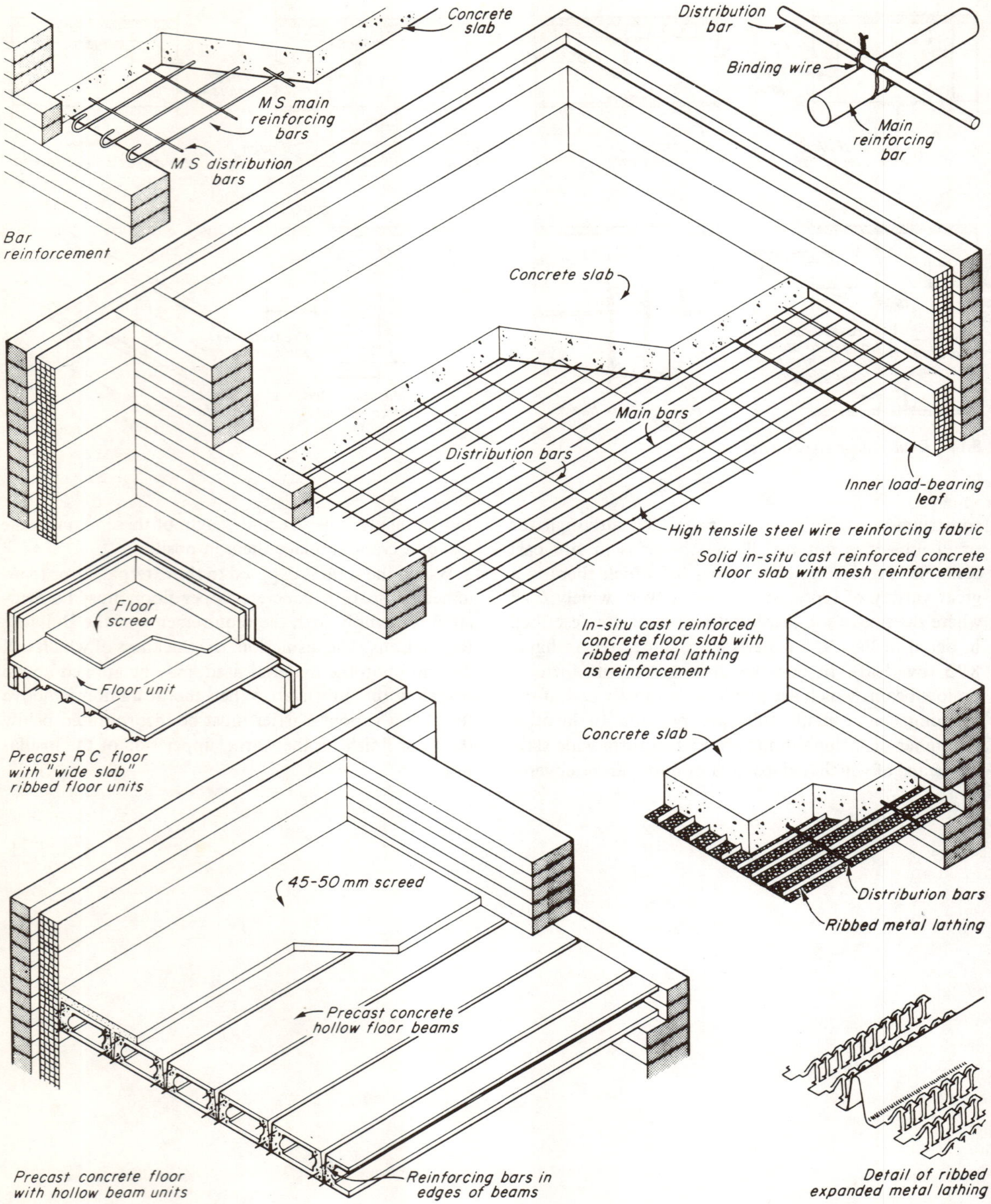

8.15 *Suspended concrete floors*

9 Fireplaces, flues and chimneys

The open fire burning solid fuel is still widely used in houses as a means of space heating and often, in addition, for heating water for domestic purposes. A *fireplace* is a space in a wall, or formed in a free-standing position, to accommodate an open fire from which the smoke and gases pass to the open air through a duct or *flue*. The structure enclosing a flue or flues is called a *chimney* and where this rises above the roof, a *chimney stack*. A projecting part of a wall in which a fireplace and flues are constructed is called a *chimney breast*. A tall, free-standing chimney, usually required for large heating plants, is called a *chimney shaft*.

This chapter is concerned with the construction of fireplaces and flues serving solid-fuel and oil-burning appliances of a domestic scale and with flues for domestic type gas heaters. Larger flues and chimney shafts are discussed in Part 2.

Function of fireplace and flue

The function of a fireplace is to burn fuel efficiently and safely, and to transfer effectively the heat generated into the room.

An adequate supply of air is necessary for the efficient combustion of any fuel. The domestic fire, burning coal, relies for its air supply on an upward air movement which is caused by cooler air flowing through and over the fire bed to replace a volume of heated air rising in a flue.

This cooler air is made up of two components — primary and secondary (see figure 9.2 *A, B*). The primary air supply is that air which feeds the fire bed and contains the oxygen necessary for combustion. The secondary air supply is that required to cause the column of air heated by the fire to rise up the flue carrying away with it the products of combustion. An efficient flue promotes this upward air movement, or 'draught', and a suitably designed fireplace establishes a proper balance between the primary and secondary supplies so that efficient combustion may occur. Since the secondary air must be supplied to the fire via the room, which it enters through cracks, windows, doors or controlled vents, a measure of air change or ventilation results.

The primary function of the flue, therefore, is to contain the rising warm air and gases above a fire in a manner which will promote a natural upward flow of air, the power of which will depend on the difference in weight between the column of light, warmed air in the flue and a similar column of cool, heavier external air. Its secondary function is to ventilate the room in which the fire is situated.

Functional requirements

In order that fireplaces and flues shall satisfactorily fulfil these functions a chimney and chimney breast, which are also structural parts of the building, must satisfy certain requirements. These are the provision of adequate:

- strength and stability
- weather resistance
- thermal insulation
- fire resistance.

Strength and stability

The factors affecting the strength and stability of a chimney are the same as those for a wall and it can be designed in the same way. A chimney rising through a building usually forms part of the wall structure and will receive a measure of support from the floors through which it passes. It will usually be thicker and, therefore, heavier than the wall and allowance must be made for this in the foundations. Above the roof the stack is self-supporting, subject to wind pressures at a considerable height. It must be stable enough safely to resist these pressures and this must sometimes be ensured by calculation. Building Regulations do, however, lay down limitations on the height of a stack above a roof relative to its width, and in most cases it is sufficient to comply with these without the use of calculations.

Weather resistance

The requirements of weather resistance are the same as for external walls of which the chimney often forms part. The prevention of wind and rain penetration is particularly important because of the adverse affect on the functioning of the flue caused by the cooling of the flue gases. Special care must be taken to prevent damp penetration at the point where a stack passes through a roof and flashings and damp-proof courses are required at the junction of the two. The top of the stack must also be protected to prevent saturation of the chimney.

Thermal insulation
Adequate thermal insulation must be provided to the flue by the chimney in order (i) to avoid the cooling of the flue gases and the consequent slowing down of the upward air flow or draught, (ii) to prevent condensation of flue gases on the walls of the flue which, particularly with slow-burning appliances, can cause considerable damage to the chimney. Where possible flues should be located internally rather than on external walls in order to reduce heat loss to a minimum.

Fire resistance
The construction of a fireplace and its chimney must be such that combustible materials within and outside the building cannot be ignited by the fire or hot flue gases. This is ensured by the provision of adequate thickness of non-combustible material around flues and fireplace and by keeping all combustible materials a sufficient distance away from a flue or fireplace.

Fireplaces must have a bottom or hearth of non-combustible material of adequate thickness and extent on or above which the fire bed will rest.

The outside surface of a chimney should not become hot enough to ignite timber or other combustible material which may be near it. A temperature of 70°C is considered to be a safe maximum which should not be exceeded and is achieved by the use of suitable materials of adequate thickness for the walls of the flue, such as 100 mm of brick, stone or concrete. Where this is not possible, as in the case of metal flue pipes from heating appliances, other precautions must be taken (see page 219).

The outlet of a flue should be well above the roof, especially if the roof covering is combustible, in order to avoid danger from sparks. Sufficient height for this is normally achieved when the top of the stack is arranged to be outside the zones of wind pressure referred to on page 215). Building regulations lay down requirements concerning heights of stacks, thicknesses of materials and proximity of combustible materials to flues and fireplaces and these are referred to in the following pages.

9.1 PRINCIPLES OF FIREPLACE DESIGN

The function of a fireplace as already defined is to provide conditions for fuel to burn efficiently and to transfer heat to the room. To promote the efficient combustion of fuel the shape of the fireplace must be designed to allow an adequate but not excessive supply of primary air to the fire bed and secondary air to the flue. To contain the fire safely and dissipate the heat the fireplace must be constructed of suitable materials, having high fire resistance but capable of storing and radiating heat. As will be seen later, the elementary open fireplace has been refined by the introduction of controls for both air supplies and by providing means of transferring into the room by convection much of the heat normally lost to the surrounding structure. In addition, the modern fireplace may have a small water-heating boiler incorporated in its design.

The fireplace consists basically of a rectangular builder's opening, or *fireplace recess*, of suitable height with means of supporting the chimney breast above and some means of reducing the width of the recess to that of its flue (figure 9.3). The back and sides of the recess are formed of material capable of radiating heat and the base of the recess must be of fire-resistant material extending beyond the opening at front and sides. A *surround* around the opening is often incorporated for aesthetic reasons or to increase the effective depth of the fireplace.

Originally fuel was burnt in a simple rectangular recess as described but during the course of time the efficiency of fireplaces has been improved in various ways and the means of doing so are briefly described before the actual constructional details of fireplaces are considered.

9.1.1 Traditional open fireplace design

At the end of the eighteenth century scientific principles were applied to the design of fireplaces and grates for burning solid fuel. These principles, formulated to improve efficiency and reduce smokiness remain basically sound and involve (i) the correct design of the junction of fireplace and flue, called the *throat*, situated perpendicularly over the fire; the entrance to the throat should be rounded: (ii) splayed sides to the fireplace on plan to obviate eddies of smoke entering the room. This occurs with fireplaces having the back and the front of

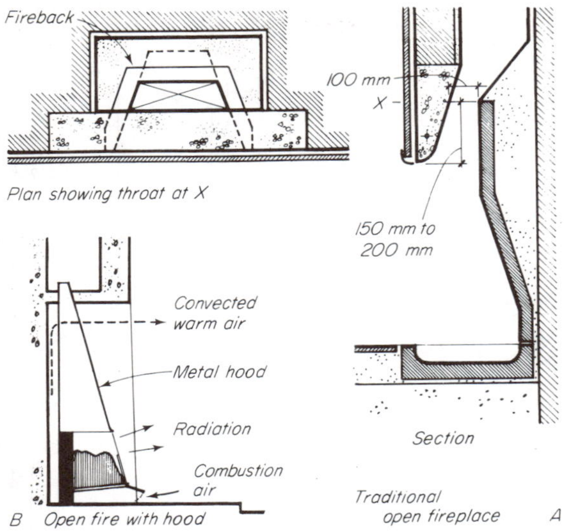

9.1 *Traditional open fireplace*

the opening equal in width: (iii) sufficient depth from the face of the chimney breast to the back of the fireplace to prevent smoking when a draught crosses the opening: (iv) the fireback sloping forward to direct radiant heat into the room and raise the temperature of the fire, thus assisting combustion.

A traditional open fireplace with these features incorporated is shown in figure 9.1 *A*. Such fireplaces, however, remain uncontrolled and tend to consume large amounts of fuel whilst promoting too large an air change. Control of the secondary air supply can be effected by a hood placed above the fire bed, in which case some heat transfer occurs by way of air circulating round the hot metal forming the hood (see *B*), or preferably by an adjustable metal throat restrictor.

When a stool grate to hold the fuel is used some control of the primary air supply to the fire can be effected by selecting a design with a solid front incorporating a variable inlet opening (*B*).

Open fires may be fitted with back boilers, in which case combustion air is controlled by a damper at the throat which, by closing the front aperture to the throat, directs flue gases around the back of the boiler when rapid water heating is required (figure 9.2 *A*).

9.1.2 Solid fuel appliances

The Clean Air Act 1956 empowered local authorities to require the installation of fireplaces capable of burning smokeless fuels. This, together with a general quest for improved fuel utilisation, led to the design of many improved solid fuel appliances.

Open fires

Normal open fires will burn a wide range of fuels including coal, wood and peat but they are unsuitable for burning smokeless fuels such as coke and anthracite and they will not burn throughout the night. The improved appliances incorporate suitably spaced fire bars and provide increased vertical depth in the fire bed which permits both smokeless fuels and bituminous coals to be burnt. A version of the open fire with a back boiler incorporated is shown in figure 9.2 *A*. The boiler provides hot water for domestic use or may heat a limited number of radiators situated near the fire. A removable front enables an extra deep fire bed to be laid for overnight burning.

Some open fires incorporate a heat exchanger which provides heat by convection in addition to the radiant heat of the fire. They operate by passing air through a convection chamber round a metal fire container and returning the warmed air to the room in which the appliance is situated. These are called *convector fires* and may be fitted with back boilers. The features incorporated in this type of appliance are indicated in figure 9.2 *B*.

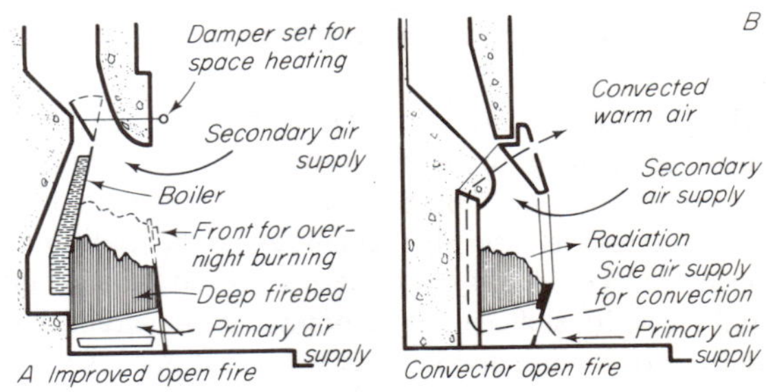

9.2 *Improved solid fuel appliances*

Room heaters

Despite the improvements in efficiency associated with open fires a certain amount of heat is inevitably dissipated in the surrounding structure of the chimney breast and flue. This heat is admittedly not lost entirely if the fire is situated on an inside wall, but there is a reduction in the heat available to the room which the fire serves and it is possible for the heat from an open fire appliance to be so widely dispersed through the structure or through radiators, that the room in which the appliance is situated is inadequately heated. Modern domestic solid fuel room heaters, however, are highly efficient in heating individual rooms and are of two basic types. The first, the *closed* room heater, is similar in construction to the Continental closed stove and has a very high efficiency, being restricted to smokeless fuels since its internal passages would soon become filled with soot if ordinary house coal were used. The second type is the *openable* room heater, with doors which hinge or slide open to reveal the fire. This can also burn coal but is not quite so efficient as the closed heater.

Both types warm by radiation of the heat from the body of the metal container and by convection. Further fuel utilisation can be obtained by fitting these appliances with back boilers (see figure 9.8).

The ordinary domestic boiler is simply a form of totally enclosed stove with a back boiler used primarily for domestic water heating. The burning efficiency of these boilers is usually high, particularly if the access door is well fitting.

A fireplace recess is not essential for room heaters and boilers. They may be placed against a chimney and be connected to the flue by a metal flue pipe from the back as shown in figure 9.8 *A*.

9.2 CONSTRUCTION OF FIREPLACES FOR SOLID FUEL APPLIANCES

All types of modern open fire appliances require a fireplace opening, or recess, in the chimney breast into which they may be built (figure 9.3) and the methods of forming this are basically the same as for the traditional open fire. The depth of the recess should be 350 mm and the width sufficient to accommodate the fireback if the size of the fire appliance is known. If not it should be 840 mm. The height to the lintel from finished hearth level should be 560 mm to accommodate a standard sectional fireback (figure 9.6 *A*). If a projecting surround is to be incorporated this height should be increased to permit the proper formation of a throat.

Minimum thicknesses of material at sides and back of the recess are laid down in building regulations and are indicated in figure 9.4. The jambs are required to be 200 mm thick. The back of the recess may be 100 mm thick when (i) it is set in an external wall and no combustible external cladding is attached behind it (*A*) or (ii) it is common to two fireplaces set back-to-back in a wall other than a party or separating wall (*B*). In all

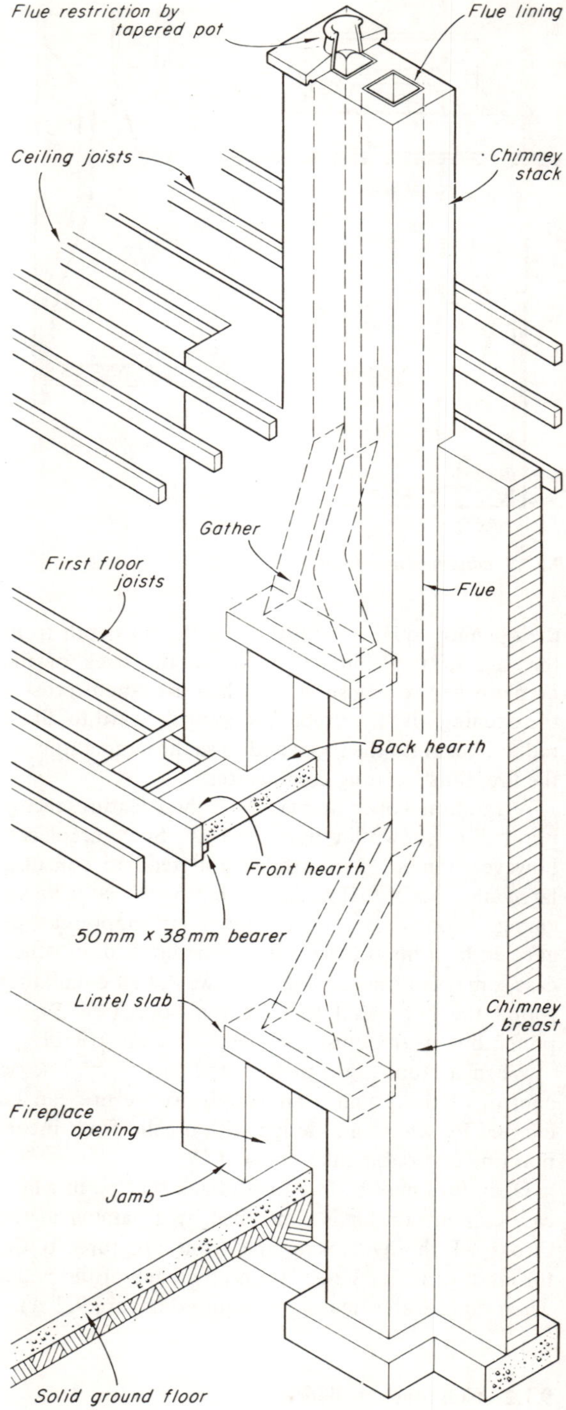

9.3 *Chimney breast and stack construction*

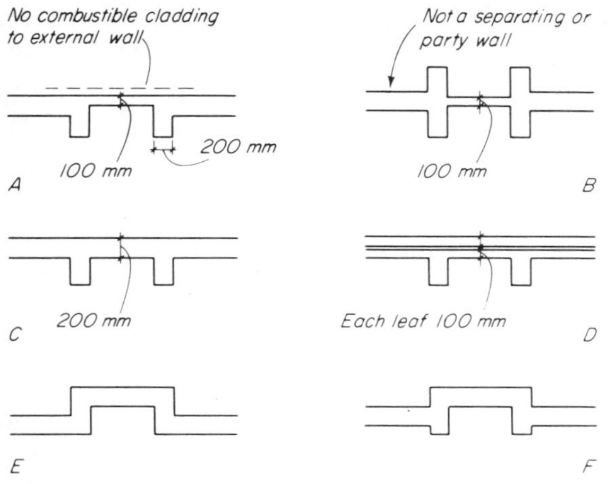

9.4 *Thickness of material round fireplace openings*

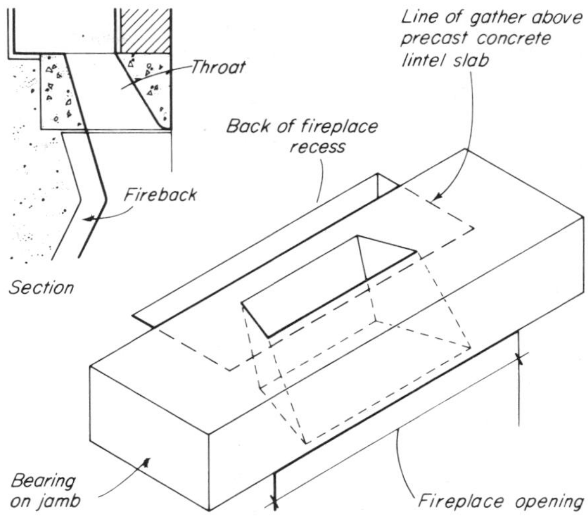

9.5 *Precast concrete lintel slab*

other cases the back must be 200 mm of solid walling (*C*) or cavity walling with each leaf not less than 100 mm thick (*D*). (*E*) and (*F*) show alternative ways of setting the chimney breast in the wall of which it forms part.

Where a wide chimney breast is required for sake of appearance and where a jamb carries a flue as on an upper floor (figure 9.3), the jambs are made wider than 200 mm.

The traditional method of forming the head of the opening was by a segmental rough brick arch but the arch form presents some difficulty in forming a smooth narrow throat and a reinforced concrete throat-forming lintel is preferable and is now normally used. This forms the front and entry to the throat and, as indicated earlier, should be smooth and rounded. The back and sides are formed by the top of the fireback and the filling above (figures 9.1 and 9.6 *A*). Alternatively, a precast concrete lintel block or slab may be used in which the throat aperture is formed (figure 9.5). At its smallest point the throat should be between 100 and 120 mm from back to front with a width approximately three quarters of the nominal width of the grate (figure 9.1 *A*). This size of throat is necessary in order to provide easy access for sweeping, but it tends to induce excessive air change in the room. This may be minimised by the provision of an adjustable throat restrictor.

The junction between the relatively wide fireplace opening and the narrow flue is made by corbelling or 'gathering over' the brickwork or stonework of the chimney breast. The funnel-shape produced is called *the gather* and provides a smooth flow from throat to flue (figure 9.3). This is discussed further in the section on flue construction.

The base of the fireplace recess is called the *hearth*. It is constructed of concrete and building regulations require a minimum thickness of 125 mm. The *back hearth*, within the recess, bears on the chimney breast. The *front hearth* must project at least 500 mm in front of the breast and 150 mm beyond each side of the opening. The full 125 mm thickness of the front hearth must be taken into the recess (figure 9.3).

In solid ground floors the floor slab itself forms the hearth of the fireplace. Timber ground floor construction requires the provision of a fender wall as explained on page 195. This wall may be 102·5 mm thick, providing support to the floor joists, the space within being filled with hardcore which carries the concrete hearth (figure 8.6) or it may be 215 mm thick to provide also a bearing for the front and side edges of a reinforced concrete hearth, the back edge of which is supported on the breast. The concrete in this case is laid on permanent shuttering of asbestos-cement or other fire-resistant sheet.

In upper floors of timber construction the hearth slab is of reinforced concrete supported on a 50 mm × 38 mm bearer nailed to the inner face of the front trimmer or trimming joist (figure 9.3), and sometimes to the cradling pieces. Temporary timber shuttering is used in forming the hearth or some form of permanent shuttering as for ground floor hearths.

Traditionally, as for the head of the opening, brick arch construction was used for upper floor hearths in the form of a 102·5 mm trimmer arch, a thin brick 'vault' spanning from chimney breast to front trimmer or trimming joist and topped with concrete to the floor level to form the hearth. This method is now obsolete.

Apart from the timber hearth bearers referred to above no combustible material may be placed under the hearth within 250 mm from the upper face unless it is separated from the underside by an air space of 50 mm. Limits on the proximity of combustible materials to the fireplace recess and to flues are shown in figure 8.13.

The constructional or builder's opening thus formed is fitted with a *fire interior*, consisting of a fireback and an inset *grate*, in the case of a non-convector open fire or with a complete fitting or 'appliance' if it is a convector type fire or a room heater.

The front hearth may be finished with tiles or stone laid flush with the floor or raised above it a few inches to form a *raised hearth*. If brick is used this will invariably result in a raised hearth. The level of the back hearth must be brought up to that of the front. If the raised hearth is a separate tiled precast concrete slab, or is of brick or stone, bedded on to the constructional hearth it is advisable to provide a fire-resistant string or tape expansion joint between it and the back hearth to prevent movement due to successive expansions and contractions of the latter. The fireplace surround, used to trim or finish off the front of the opening, may be of tiles, stone, brickwork or other material suitable for placing close up to the fire itself.

A number of constructional matters arise in the fixing of fireplace interiors and appliances and these will be discussed in the following pages, before proceeding to a consideration of flues.

9.2.1 Non-convector open fires

Modern inset open fires or 'all-night burners' comprise a grate with a front which is sealed into the fireplace recess and incorporates in its design some device for controlling the primary air supply such as a spin wheel or controllable flap. These grates are designed to fit British Standard firebacks which are made of firebrick or refractory concrete (aluminous cement and broken firebrick). The bend or knee at the back should be fairly high to permit the formation of a satisfactory throat (figure 9.6).

When set in position the space around the fireback should be filled in solid with vermiculite or other lightweight concrete. This provides insulation to prevent excessive heat loss to the structure of the fireplace

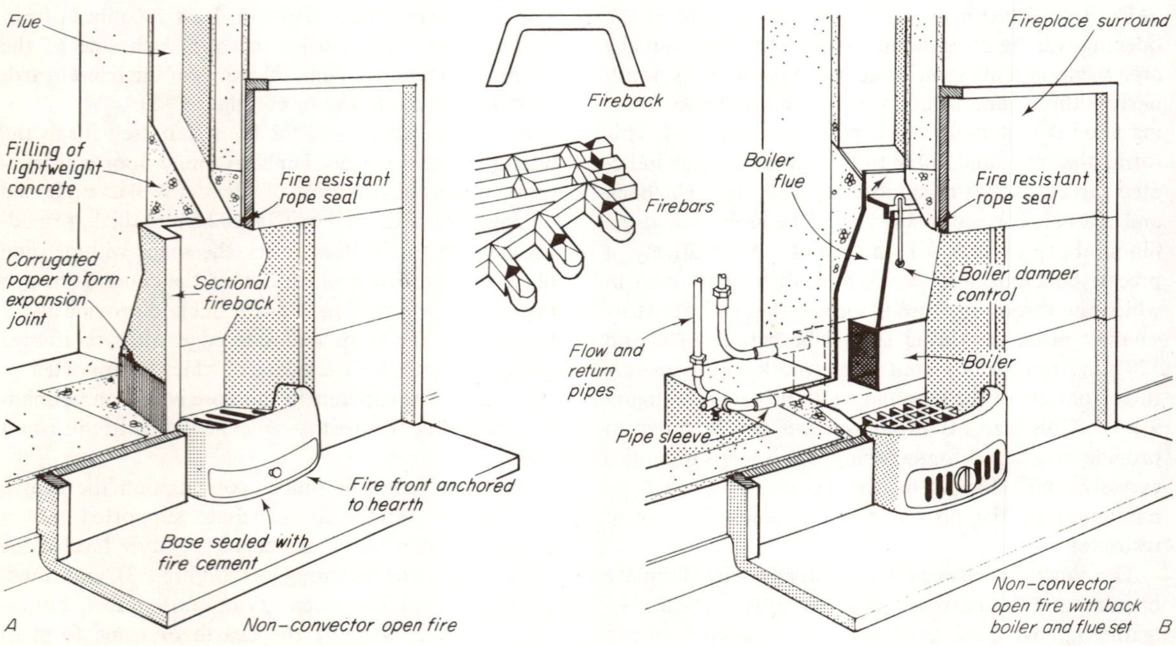

9.6 *Non-convector open fires*

recess.[1] The filling at the top of the fireback should be finished smoothly to an angle of about 45 degrees. The insertion of corrugated paper round the back of the fireback before filling in provides a small expansion gap. This need be placed only round the lower half of a two-piece fireback (figure 9.6 *A*).

The grate must be properly fixed if adequate control of the burning rate is to be achieved. It must be screwed or bolted to the back hearth and sealed at each side to the fireback and surround with 6 mm diameter fire-resistant string to exclude air. It should be bedded in fire cement on the hearth for the same reason. A string seal is made at the sides rather than a fire cement filling to provide for expansion movement.

Deep ash pits The ash container for the normal inset grate is placed above the hearth level. Consequently the fire itself is well off the floor, thus blanketing the heat which would be radiated to the floor. By lowering the ash container the fire can be at hearth level and by forming a suitably sized pit, a container can be provided large enough to hold between three and seven days' accumulation of ash. The builder's work necessary will require the building-in to the back hearth and sometimes to the front hearth, of a preformed ash pit of cast iron, firebrick or precast concrete, normally supplied by the makers. The ash pit will contain the ash pan which is withdrawn either by removing the fire bars when the fire is out (figure 9.7 *A*) or through an extension of the pit below a removable plate in the front hearth as shown in (*B*).

The primary air supply to the fire is below the floor level and is controlled by some form of valve. With a timber ground floor the air supply may be drawn from the ventilated under-floor space (*B*). With a timber upper floor or when the floor is solid it will be necessary to construct ducts to two outside walls at right-angles to each other, since single inlets are subject to suction effects in strong cross winds. These ducts should be not less than 7500 mm^2 cross-sectional area and should be gathered in a balancing chamber from which runs a pipe into the pit as shown in figure 9.7 *A*. With this type of fire having no firefront the fire is liable to smoke into the room and to reduce this risk the opening in the surround should be not more than 500 mm high.

Back boilers Non-convector open fires with back boilers are cast iron units incorporating a water container, flue and damper which are installed in place of the normal fireback as shown in figure 9.6 *B*. The same general methods of constructing the fireplace already described are used, but the height and depth of fireplace recess may need to be greater than for a normal open fire. Flow and return pipes should be sleeved with larger diameter pipes where they pass through the chimney breast, the gap between being caulked with fire-resistant string.

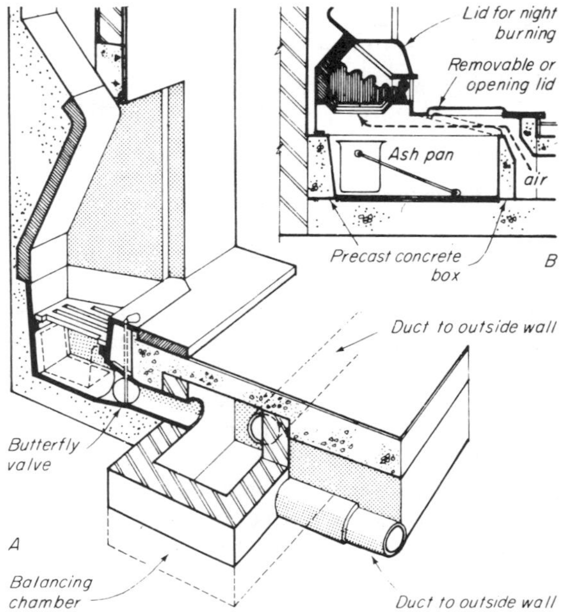

9.7 *Deep ash pits*

9.2.2 Convector open fires

These are open fires in which the fire is contained in a metal enclosure surrounded by a second metal jacket to form an integral convection chamber. The flue penetrates the outer jacket as in figure 9.2 (*B*). The junction of the front of the fire with the fireplace surround must be sealed with soft fire-resistant rope or string and the appliance must be screwed to the back hearth so that no movement takes place which might break the seal.

Convector fires are also available with back boilers, the boiler being built into the appliance by the manufacturers.

Chimney and flue cleaning Most open fires are swept through the front. Where adjustable throat restrictors

[1] Suitable mixes are given in BS 8303.

are installed, these are normally removable to allow cleaning brushes to be passed through the remaining opening.

9.2.3 Room heaters

Closed heaters generally have back flue outlets but openable heaters, which may be had as inset or freestanding types, may have either a back or top outlet. The fixing of a closed heater is similar to that of a freestanding openable heater with a back outlet (see figure 9.8 *A*). Both types can be installed either in front of the chimney breast or wall or within a recess.

When the heater is placed against the wall it is preferable to provide a cast iron flue box which contains the flue socket in a removable plate as shown in (*B*). The 225 mm × 225 mm brick flue is brought straight down to terminate at the flue box and the heater and plate can be removed for flue sweeping. The heater should be placed about 25 mm away from the plate to avoid heat loss by conduction and the spigot should be sealed where it enters the box with fire-resistant string and a clamping ring similar to that shown for the top flue outlet at (*C*). When the heater is placed against a wall in this manner care must be taken to ensure a hearth projection of a least 225 mm in front of a closed heater or 300 mm in front of an openable type, with 150 mm at the sides.

If a back boiler is incorporated the heater will not be removable for flue cleaning and a means of access must be provided either in the back or side of the chimney. This should be fitted with a double metal plate soot door as this avoids the chilling effect associated with single plate types.

Freestanding heaters fitted in a recess should, preferably, have top flue outlets, since this facilitates installation and keeps the heater as far back as possible (figure 9.9 *A*). The recess will need to be higher than that for an open fire and 1100 mm is usually suitable with a width of 690 to 840 mm. The top should be closed by a reinforced concrete 'raft' lintel with an opening into which the heater flue pipe should extend 50 to 75 mm. The connection should be sealed with fire-resistant rope and a clamping ring as shown in figure 9.8 *C*. Unless a soot door can be conveniently fitted at some point round the flue it will be necessary to fit one in the raft lintel. The dimensional requirements for the hearth are as given above. The Building Regulations require a minimum distance of 150 mm between an appliance and any combustible materials in any adjacent wall

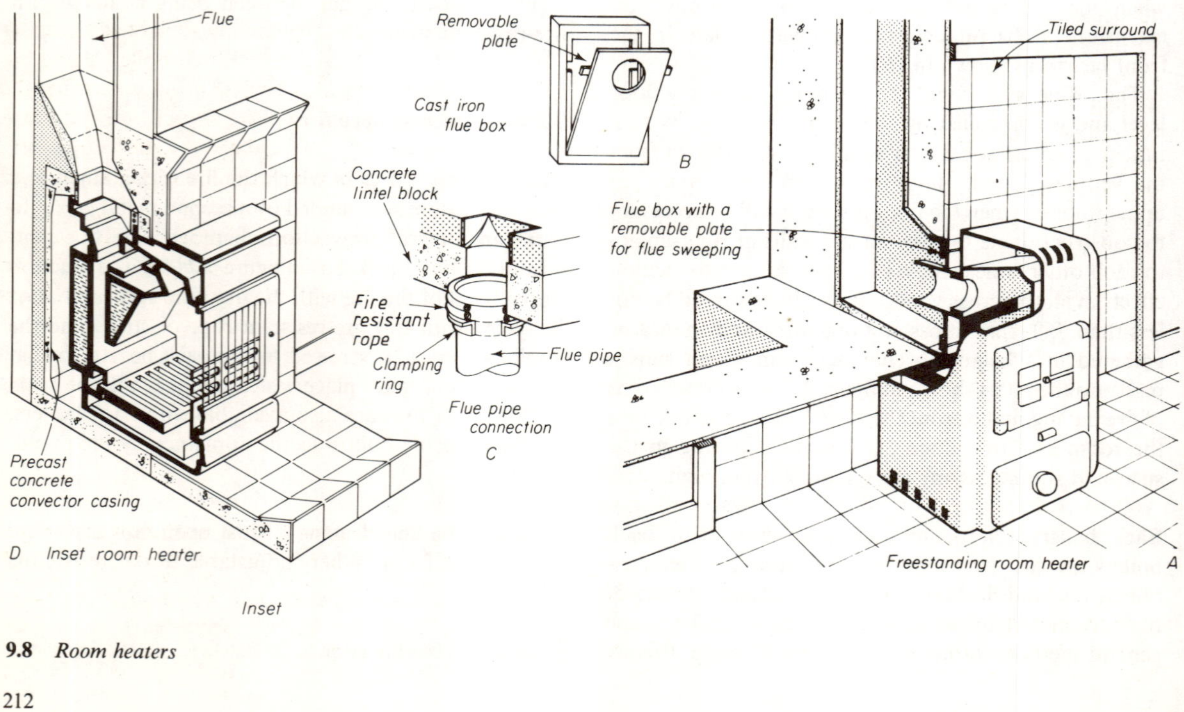

9.8 *Room heaters*

unless they are protected by stipulated thicknesses of non-combustible material.[1]

If the appliance has a back flue outlet the connection to the raft lintel is made by a flue pipe adaptor as shown in figure 9.9 *B*.

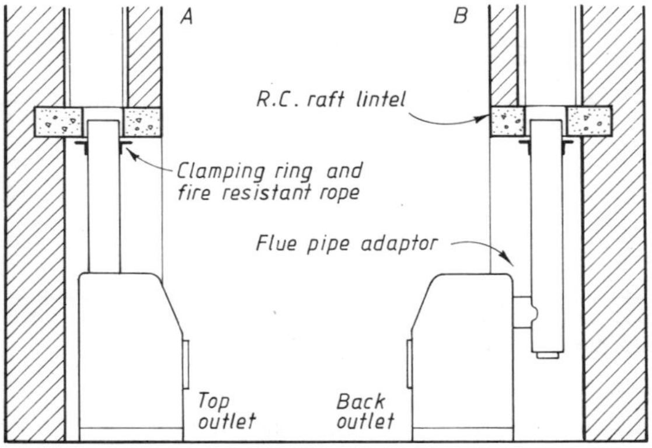

9.9 *Room heaters in recess*

Inset room heaters are openable heaters incorporating a convection chamber, the majority being fitted with a boiler. There are two broad types:

Single case The fire is contained in a metal enclosure and the convection chamber is formed by the space between this and a prefabricated convector casing of metal or concrete slabs.

Double case As in convector open fires the fire is contained in a metal enclosure surrounded by a second metal jacket to form an integral convection chamber. The flue penetrates the outer jacket, being sealed against loss of convected warm air through the flue.

Both types are built in. The space behind a double case appliance should be filled with mineral wool or other non-combustible insulation and the junction with the surround must be sealed with soft fire-resistant rope to prevent convected air being drawn back from the room to the flue.

Figure 9.8 *D* shows a single case heater with a precast concrete slab convector casing which must be carefully jointed with mortar. Careful sealing is essential where the flue outlet passes through the convector casing to avoid loss of warm air. Similar sealing is necessary where the pipes from the boiler pass through the casing, using fire-resistant string and clamping rings.

The constructional problems in fitting solid fuel cookers, combustion grates and domestic boilers are, in general, the same as those already discussed in connection with fireplaces and heaters.[2]

9.3 PRINCIPLES OF FLUE DESIGN

To ensure the proper functioning of a flue the following factors must be considered in its design.

9.3.1 Size and shape

Flues to domestic fires should not be less than about 4·50 m high measured vertically from the outlet of the appliance or fireplace to the top of the flue terminal in order to ensure an adequate difference in weight between the internal flue gases and the external air referred to on page 205. The entry to the flue should be restricted by a correctly designed throat to increase the initial velocity of the gases (figure 9.5) and a further slight restriction at the flue terminal is desirable to increase the velocity at the outlet. This reduces the danger of down draughts. The required size of the flue is affected by the type and size of the appliance and its setting. Table 9.1, based on Table 1 in BS 6461: Part 2, shows the required minimum sizes for various appliances. Flues of circular cross-section are the most efficient.

The flue dimensions and chimney heights that satisfy the requirements of a solid-fuel fired boiler are normally more than sufficient when oil firing is used for the same size boiler. Most engineers would use the same size as for solid fuel, particularly for domestic work, where there is always the possibility of installing solid fuel appliances at some future date.

Flues should be as straight as possible. Unavoidable bends should be limited to two and should be at an angle of no more than 45 degrees and preferably less than 30 degrees with the vertical.

[1] See AD J1/2/3, 1.33.

[2] For full details of all aspects of the installation of solid fuel appliances see BS 8303: 1986 *Installation of domestic heating and cooking appliances* and *Correct Installation of Domestic Solid Fuel Appliances* by W C Moss, Bsc, AMIHVE, FInstF, issued by the Solid Fuel Advisory Service from which much information has been drawn.

Appliance type	Minimum flue size — diameter
Flues serving a fireplace recess	
Open fire with opening up to 500 mm × 550 mm or any closed appliance	200 mm*
Flues not serving a fireplace recess	
Freestanding open fire	200 mm
Room heaters, stoves and cookers up to 20 kW	125 mm (150 mm if burning bituminous coal)
Independent boilers Up to 20 kW 20 kW to 30 kW 30 kW to 45 kW	 125 mm 150 mm 175 mm

* For larger sizes of open fire, or closed appliances that can be used as an open fire, the cross-sectional area of the flue should be between 14 to 16 per cent of the free, unobstructed area of the fire opening (BS 6461: Part 1).

All flues may be square cross section of equivalent area to those given in this Table.

Flue sizes should be increased by 25 mm on each dimension (i.e. diameter or side) for each offset in the flue.

Table 9.1 *Minimum flue sizes for solid fuel burning appliances*

9.3.2 Airtightness

A flue must be airtight in order to maintain the strength of the draught at the appliance and to prevent the escape of smoke. Air can enter through faulty jointing or faulty withes.

9.3.3 Insulation

Care must be taken to prevent the flue gases cooling, which might result in downdraught and condensation. This precaution is particularly important where slow burning appliances are used (see page 219). From the point of view of general heat conservation fireplaces should not be situated on outside walls.

Flues, as previously indicated, should be constructed with walls not less than 100 mm thick and should be lined (see page 216). The use of 215 mm brickwork in place of 102·5 mm does not afford much increase in insulation value and has the disadvantage of offering more surface area to the atmosphere, with consequent cooling of the flue. It also has a high thermal capacity which requires a longer preheating period before the flue is warm enough to encourage 'draught' action. The greater thickness may, however, be necessary for any external walls of flues to minimise damp penetration.

Flues situated internally only need special consideration where they penetrate the roof and become exposed to the weather. Thickening of the flue walls can be effected by corbelling out within the roof space, and particular attention should be paid to the arrangement of the damp-proof course and flashings to the stack. A suitable capping should be provided to prevent saturation of the chimney (figure 9.10). A projecting capping, in addition to throwing water clear of the chimney walls, helps to create a zone of low pressure at the flue outlet. (See *Position of outlet* below.)

9.3.4 Position of outlet

Where the chimney passes through the roof more than 600 mm from the ridge the Building Regulations, for safety in terms of fire, require the outlet to be at least 2·3 m horizontally from the surface of any roof with a pitch of 10 degrees or more. Where the chimney passes through the ridge, or within 600 mm of it, the outlet may be not less than 600 mm above the ridge. With pitches less than 10 degrees the outlet must be at least 1 m above the highest point of intersection of chimney and roof. It must also be this distance above any opening light or ventilating opening less than 2·3 m horizontally from the outlet, irrespective of the roof pitch.

If the ridge of any adjacent roof or an adjacent structure lies within 2·3 m of the outlet the latter should be at least 600 mm above the ridge or structure in order to minimise the risk of inadequate draught. These dimensions are exclusive of any chimney pot or other terminal. If the roof covering is of combustible material the outlet should be at least 1 m above the level of the ridge whatever the position of the stack.

These precautions do not, however, necessarily ensure the efficient functioning of a flue, the outlet of which must be positioned outside any potential zones of high wind pressure. The positioning of a flue outlet in a potential suction zone will assist in the removal of the smoke and gases, but should it occur in a high-pressure zone there is every likelihood of the gases being taken down the flue by air moving from this zone to an area of lower pressure within the room. It will be seen from figure 7.1, which shows the distribution of these zones, that in the case of flat and low-pitched roofs up to 30 degrees suitable positions for chimneys are almost anywhere above eaves level. In the case of roofs pitched greater than 30 degrees the ridge position is best unless the chimney is extended above the ridge level from a lower position or is fitted with a cowl. The latter may not be effective unless it takes the outlet out of the high-pressure zone, although there are on the market cowls which are claimed to prevent down-draught even in pressure zones.

Tall buildings, hills or trees close to a building can influence the distribution of wind pressures on the building and may cause downdraughts in chimneys which otherwise might be satisfactory. In these cases, some forms of cowl will divert a strongly directional wind current. The type of slab capping shown in figure 9.10 can also give similar protection.

9.4 CONSTRUCTION OF CHIMNEYS FOR SOLID AND OIL FUELS

9.4.1 Brick chimneys

Domestic flues are mostly constructed in brickwork, with walls not less than 102·5 mm thick. Bends and slopes in the flue are formed by corbelling. The back of the flue in a party or separating wall must be at least 200 mm thick, or be of cavity construction with each leaf not less than 100 mm thick, up to its intersection with the roof.

The chimney breast, immediately above the top ceiling is reduced in width to that required for the stack, allowing for at least 102·5 mm walls and *withes*, that is the walls between adjacent flues.

For safety in terms of fire stacks, as already indicated, must have certain minimum heights above the roof surface. For safety in terms of stability the height of a stack, including any chimney pot or other terminal, above the highest point of intersection with the roof must not exceed four and a half times the least horizontal dimension unless the stack is braced in some way or its stability under wind pressure is checked by calculation.

When a chimney breast or stack projects beyond the face of the wall below the total projection of the oversailing brickwork must not exceed the thickness of the wall below with a maximum projection of 60 mm in each course.

The gathering over of the flue above the fireplace recess, referred to under *Fireplace construction*, should be steep, not flat, with the entry to the flue itself, that is the top of the 'funnel' more or less central with the fireplace unless the flue has to pass to one side in order to clear an upper fireplace (figure 9.3). A 'dog-leg' bend once always formed in the gather is no longer considered essential although it may help to reduce rain splashes and soot falling on the hearth. The corbelling forming the gather must be rendered from the top of the fireplace recess to the entry to the flue to provide a smooth surface. Forming of the gather in this way may be avoided by the use of a precast concrete throat unit similar to that shown in figure 9.5 but designed also to make the transition from the throat to the flue entry. No gather is required when the flue to a recess rises directly from a raft lintel into which the appliance flue pipe extends (figure 9.9).

The top of a flue is usually terminated by a cylindrical fireclay pot. Tapering pots provide the slight restriction at the flue outlet to increase the velocity of the rising flue gases. The pot is bedded in one or two courses of brickwork or, if the pot is tall, one quarter of the pot length if this is greater, and the top of the stack round the pot is *flaunched*, that is weathered with mortar, to throw off water (figure 9.10 *A*). Instead of a pot the liner may be extended at least 20 mm above the chimney head (figure 9.15). The use of a perforated and weathered stone or precast concrete one-piece capping (9.10 *B*) as a terminal has the advantage of eliminating joints and dispensing with the need for flaunching which after a time, even with a cement-lime mortar, may crack and permit the penetration of rain. The advantage, in certain circumstances, of a slab cap as shown in (*C*) has been referred to under *Position of*

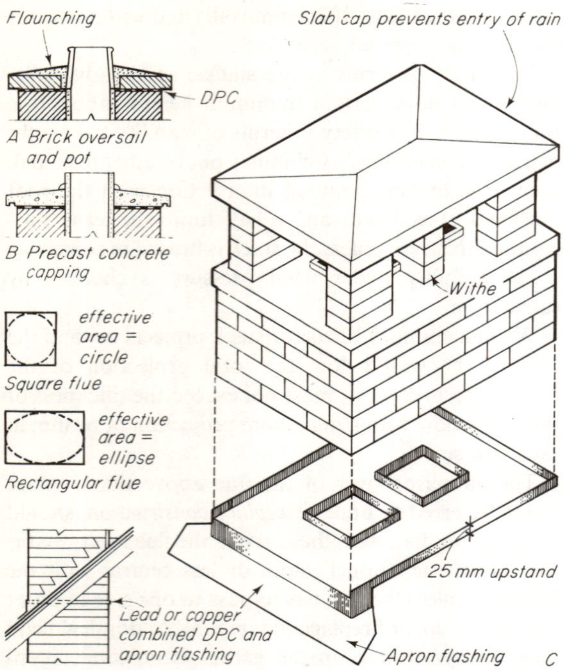

9.10 *Flues and chimney stacks*

outlet. Any withes should be carried up to the underside of the top slab and flue liners should project 20 to 25 mm. The total area of opening between the piers should be not less than twice the area of the flues in the stack.

The top twelve courses of a stack should be laid in cement or cement-lime mortar of a strength not less than 1:1:6.

In order to ensure a smooth surface to the flue and to seal possible cracks in the brick joints the flue is lined with a suitable flue liner. Flue liners, as well as ensuring a smooth, airtight flue of uniform section, permit added insulation to be provided around them. The Building Regulations require flues for solid fuel and oil-burning appliances to be lined with rebated or socketed liners. Linings may be in the form of fireclay or terra-cotta liners with rebated joints or of refractory concrete or vitrified clay pipes with socketed joints. Rebates and sockets should be placed uppermost to prevent possible condensate running out (figure 9.11 *A,B*). Socketed joints should be made with fire-resistant rope and high alumina cement: the rope allows expansion and the cement is acid resistant. The space between the lining and the flue wall is filled with loose rubble flushed up with mortar or with an insulating material such as lightweight concrete (*C*).

The need to prevent damp-penetration in order to avoid cooling of the flue has been emphasised. Means of protecting the top of the stack have been described above and the possibility of making the external walls to flues and those to stacks 215 mm thick for the same reason has been mentioned under *Insulation* on page 214. Penetration of damp down the stack to the interior is prevented by the incorporation of a damp-proof course at roof level. In a flat roof this will be placed in the stack level with the top of the apron flashing over the turn-up or skirting of the roof covering. In pitched roofs a similar horizontal damp-proof course may be built-in level with the top of the lowest apron flashing so that all brickwork exposed to the weather is above it (figure 9.10 *C*). Provided the brickwork is reasonably non-absorbent and the roof timbers are kept clear of the brick face this can be satisfactory. The small amount of damp which might pass below the roof line will evaporate into the roof space. Alternatively, a more expensive stepped damp-proof course may be used following the line of stepped apron flashing down the slope of the roof. This has the advantage that there is no damp brickwork at all below roof level and is a method suitable for exposed situations. Any of the materials described under *Walls* are suitable provided they can be worked to the necessary shapes.

Penetration of moisture through the joint between stack and roof covering is prevented in various ways by means of gutters, soakers and flashings which are described in *MBS: Components*.

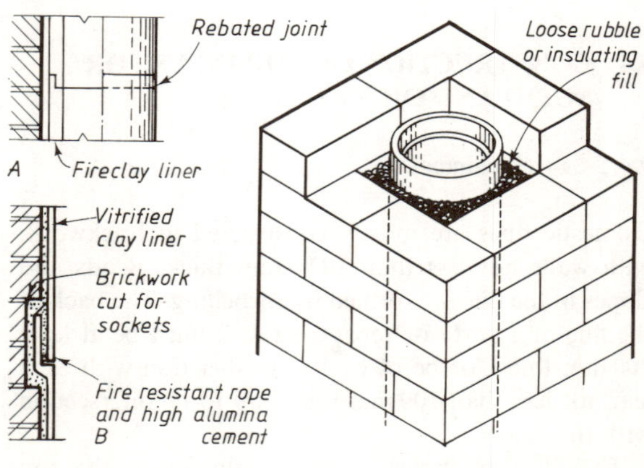

9.11 *Flue linings*

9.4.2 Stone chimneys

The temperatures encountered in a domestic flue are not likely to damage a good building stone except in the immediate vicinity of the fire and in this position sandstone should be used if protection is not to be provided by firebricks, as in the case of an open fireplace with stone-faced walls to the recess.

When the walling generally is of random rubble, in which bond is poor, squared rubble either uncoursed or coursed should be used above roof line. With flint walling a change to coursed stone or brickwork should be made at a short distance below the roof line.

The flue walls throughout the chimney and its stack should be at least 200 mm thick inclusive of any brick or concrete backing, except that where stone alone is used the wall thickness in that part of the stack above roof line may be not less than 150 mm. As in brickwork the flues in stone chimneys should be lined. The maximum height of stack relative to its least horizontal dimension is the same as for brickwork as are the other aspects of construction considered in the previous section.

Coursed masonry may be corbelled out to a total projection not exceeding the thickness of the wall below. Each course may project a distance equal to half the thickness of the flue wall below it, provided the corbel stone is bonded into the wall a distance equal to at least twice its projection.

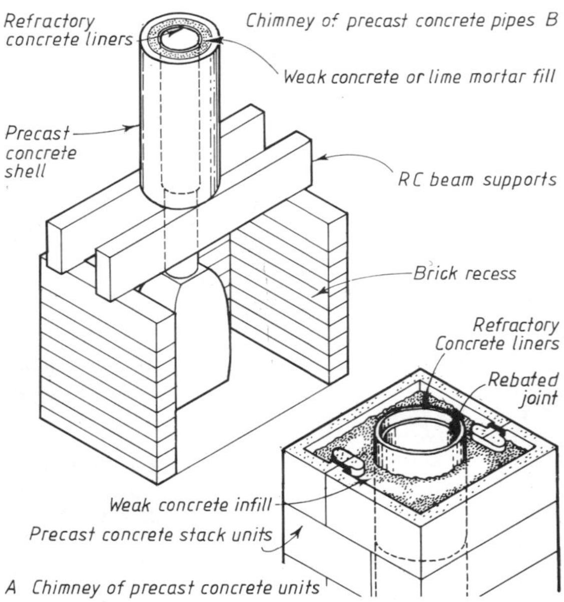

9.12 *Chimney construction*

9.4.3 Concrete chimneys

Concrete chimneys can be constructed in three ways:

(i) With in situ concrete

(ii) With precast concrete units

(iii) By a combination of (i) and (ii).

Concrete for in situ work may be either plain or reinforced and where in contact with the flue gases, that is in the absence of flue liners, should be of an acid-resisting refractory type. Lightweight concrete made with foamed slag or expanded clay aggregates, or no-fines concrete, can also be used, provided protection is given by flue liners. The mix for dense concrete should not be too rich in order to reduce shrinkage and to resist the effects of heat satisfactorily crushed brick, slag, clinker or crushed limestone should be used as aggregate.

The concrete should be at least 100 mm thick and unless increased to at least 150 mm where penetrating the roof should be rendered to provide adequate protection against damp penetration.

Up to a height of seven times its least horizontal dimension the effect of wind pressure on a plain, dense concrete chimney need not be considered. Oversailing projections should form an angle of not more than 30 degrees with the vertical unless the projection is reinforced. The height of in situ lightweight or no-fines concrete chimneys should be limited to four times their least horizontal dimension and all oversailing or projecting parts should be formed with dense concrete, reinforced as necessary. The open-textured internal surface of such chimneys should always be lined and the external surfaces should be rendered. With cast in situ chimneys of all types liners are invariably used as they form permanent shuttering. Damp-proof courses are not generally required in chimneys of dense concrete, but for cast in situ lightweight concrete damp-proof precautions should be taken as for masonry construction.

A variety of precast units of dense or lightweight concrete are available for forming chimneys. There are three approaches to this form of construction — one by the use of solid or hollow precast blocks bonded to form the walls and withes of the chimney as normal masonry; another by the use of precast blocks incorporating an integral liner and interlocking joints and another by forming the internal and external surfaces of the chimney with precast units and filling the intervening cavity with lightweight concrete as shown in figure 9.12 *A* and *B*. Dense vibrated concrete blocks will generally withstand damp penetration without rendering the external surfaces, and such constructions automatically provide a sufficiently smooth surface to the flue. As with in situ cast flues of lightweight concrete flue liners are essential with lightweight blocks, and these are incorporated in the manufacture.

9.4.4 Insulated metal chimneys

These are factory made chimneys consisting of a circular external metal casing and either a stainless steel or ceramic liner with a space between filled with thermal insulating material. The chimney is made up of sections in a range of lengths and diameters which have rebated joints secured by some simple mechanical fixing, such as a twist action lock. These are accompanied by a range of fittings including tee junctions, elbows, flashing and support elements. The chimney may be supported at the foot, by loadbearing components designed to bear the weight of the whole system, at the floors through which it may pass (figure 9.13) or at roof level and must be easily accessible throughout its length for maintenance. In certain locations these chimneys must be cased in non-combustible material.[1] Where the chimney passes through a roof space it should be enclosed by a wire cage 1 m high and 100 mm greater in diameter than the chimney to prevent contact of stored materials with the outer casing.

In areas of moderate exposure a chimney projecting up to 1.8 m above the roof line is likely to be stable. When the projection is greater lateral support should be provided. Where it will be exposed to high winds support may be required even when the projection is less than this.

Flue diameters for these chimneys will vary, as with other flues, with the size of appliance the chimney serves and its setting and in respect to this reference can be made to table 9.1 (see also Part 2, page 274).

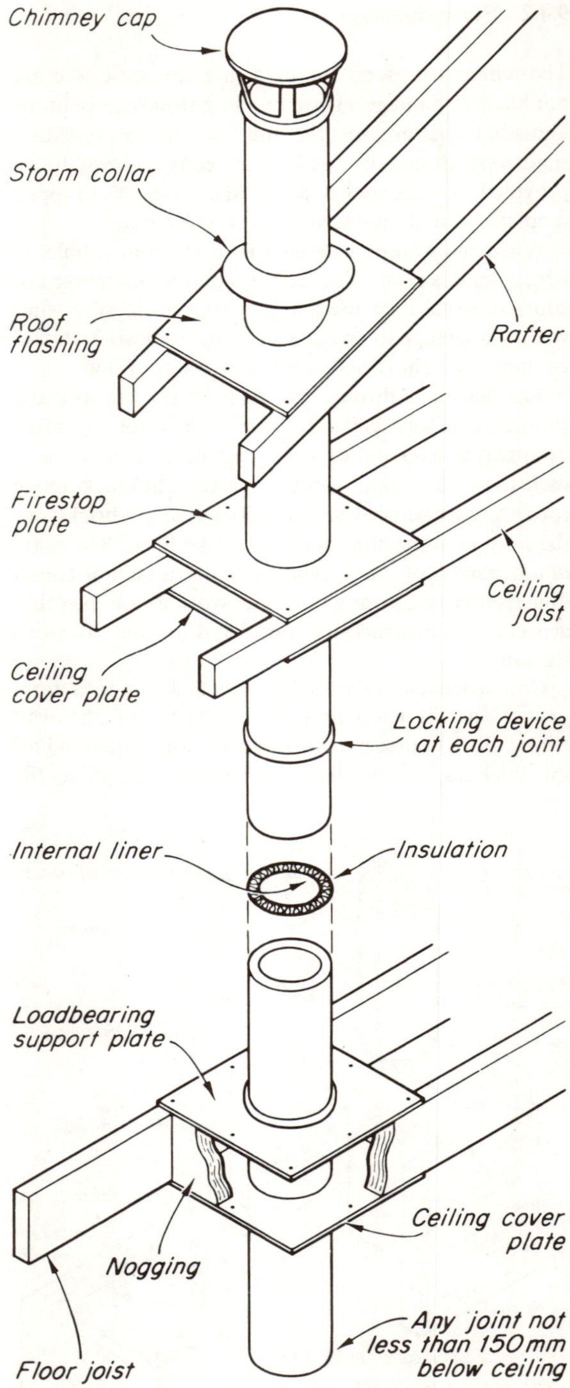

9.13 *Insulated metal chimneys*

[1] See AD J1/2/3, 131.

9.4.5 Metal flue pipes

Metal flue pipes have poor thermal insulation value and are not suitable for external use unless insulated. They should be used only for flues within the room containing the appliance or to connect the latter to a chimney. Metal flues may be made of various types of steel or of cast iron.[1] The pipes should be frequently supported, usually at every joint or at intervals not exceeding sixteen times the internal diameter. The joints should be airtight and allowance should be made for the expansion and contraction of the pipes at the joints and at the supports.

The Building Regulations prohibit flue pipes from passing through any internal wall or partition or a floor or roof space. Where they pass through an external wall or the roof of a habitable area as in figure 9.14 all combustible material in the structure must (i) be separated from the pipe by an air space at least three times its external diameter or (ii) be separated from the pipe by at least 200 mm of solid non-combustible material. Alternatively the pipe may be enclosed by 25 mm of non-combustible insulating material and a sleeve of metal or other non-combustible rigid material from which the combustible material must be separated either by an air space as shown in figure 9.14 or by at least 115 mm of solid non-combustible material if the pipe passes through solid construction. In each case the insulation sleeve must extend beyond the combustible material as indicated in figure 9.14.

A flue pipe must not be nearer than one and a half times its external diameter to the surface of adjacent combustible material, and at any distance less than three times its external diameter the effect of radiant heat must be reduced by a non-combustible shield placed at least 12·5 mm in front of the combustible surface. The width of the shield should be at least four times the external diameter of the pipe.

Flue pipes connecting an appliance to a chimney should be kept as short as possible and should not make an angle of more than 45 degrees with the vertical unless the connection is from a back outlet appliance when the horizontal section should not exceed 150 mm. Necessary bends in the connection should not exceed two.

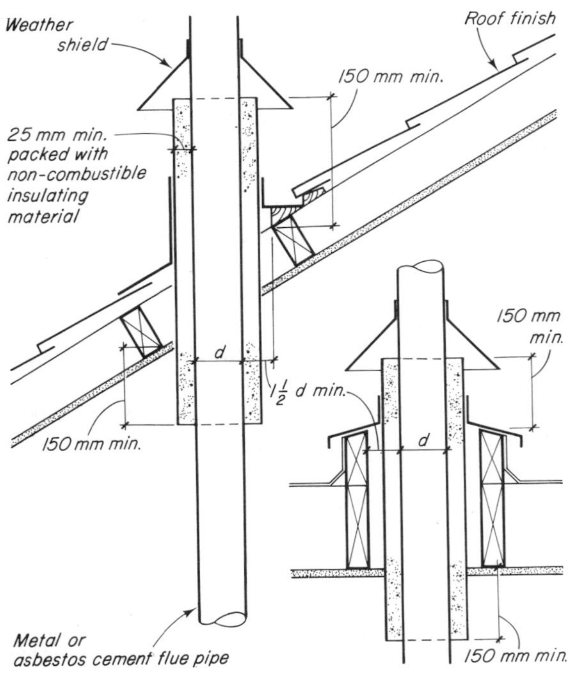

9.14 *Metal flue pipes*

9.4.6 Chimneys for domestic boilers

Reference has already been made to the need to avoid loss of heat from flue gases. This is particularly important in the case of flues to any form of slow combustion boiler, heater or cooker. During long periods of burning at low temperature the gases from these appliances are likely to cool below their dewpoint and excessive condensation may occur which may contain sulphur compounds, ammonia, tar and soot. As the moisture containing these can be absorbed by mortar, brickwork or stone this may result in sulphate attack on the mortar joints. This is often greatest on the side of the stack most exposed to rain and, since sulphate attack is accompanied by expansion of the mortar, the stack will often tilt over (figure 9.15 *A*). The tar and soot content result in stained joints and plaster finishes.

For this reason, therefore, adequate insulation of the flue should be ensured by surrounding the lining with insulating material and the size of the flue should not exceed 175 mm diameter (see Table 9.1) in order to reduce the area through which the flue gases can lose heat. These measures will limit the risk of condensation

[1] See AD J1/2/3, 1.20 for details of approved materials.

to periods when the appliance is first lit and the flue surfaces are cold or when it is burning at low temperature and the gases are likely to cool below their dew-point. Such condensate must be prevented from coming into contact with the fabric of the chimney by the use of acid resistant liners which are impervious to liquids and vapour, such as those referred to on page 216 which are required by the Building Regulations.

Provision should be made for the trapping and removal of possible condensate either by a removable vessel or by a container fitted with a drainage pipe. The construction of such flues is shown in figure 9.15.[1]

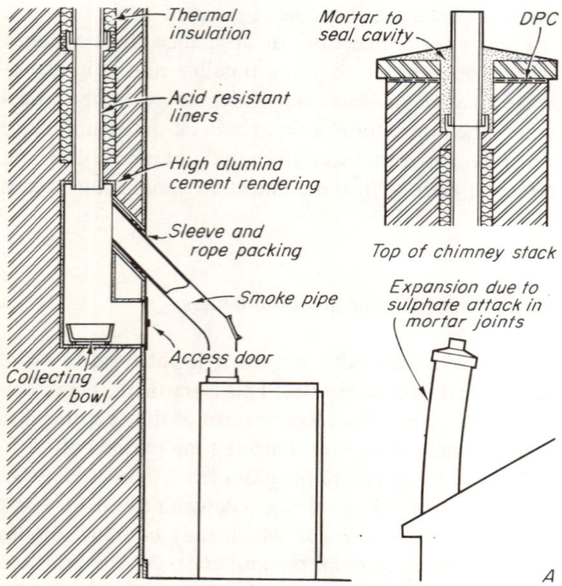

9.15 *Flues for domestic boilers*

9.4.7 Chimneys for oil-fired boilers

Oil-fired domestic boilers are now in common use and reference has been made under *Size and shape of flue* on page 213 to the relative areas and heights of flues for solid-fuel and oil-fired boilers. The draught requirements of a solid-fuel fired boiler operating on natural draught and that of an oil-fired unit are different. The former requires sufficient height of flue to provide draught for drawing combustion air through the fire-bed whilst in the latter air is generally provided by a fan incorporated in the burning equipment. For this reason the chimney height of an oil-fired unit can be less than that of a solid fuel installation. As with solid-fuel installations sharp bends and offsets should be avoided and the use of linings is necessary in order to protect the flue from condensation which may occur when the boiler is operating efficiently and the temperature of the flue gas is low. Liners, in addition, combat the effects of high temperatures which can occur as a result of bad installation or poor operation.

9.5 CONSTRUCTION OF CHIMNEYS FOR GAS-FIRED APPLIANCES

Some gas appliances require no flue provided they are used where ventilation is good. BS 5440: Part 2: 1976 gives a table of such appliances with the minimum area of air vent required.[2] Others with larger heat inputs, or used continuously, must be provided with flues.

The principles of the design of gas flues are the same as for the design of those serving solid-fuel appliances.[3]

9.5.1 Flue size and outlet

A procedure for determining the minimum flue size for gas appliances according to their rating and the height of the flue is given in BS 5440: Part 1: 1978. The Building Regulations require a flue to a gas fire to have an area of at least 12000 mm² if it is circular or 16500 mm² if it is rectangular, with a minimum dimension of 90 mm. Flues to other appliances should have an area at least equal to that of the appliance outlet.

The positioning of the flue outlet externally should be governed generally by the same considerations as for other flues and it must be situated so that wind can blow across it at all times and it is well clear of other higher parts of the building such as lift motor rooms and tank enclosures. The outlet must be at least 600 mm from any opening into the building (see Part 2, figure 7.8).

Flues, other than those to gas fires, with outlets having any dimension across the axis less than 175 mm should be provided with a terminal designed to allow

[1] For a full consideration of the design and construction of flues for solid fuel see BS 6461: Part 1: 1984 *Installation of chimneys and flues for domestic appliances burning solid fuel.*

[2] *Code of practice for flues and air supply for gas appliances of rate input not exceeding 60 kW.*

[3] For balanced flues see Part 2, 7.2.3.

free discharge of the flue gases and to prevent the access of birds, rain and snow. Any wall position is less satisfactory than a terminal placed above a roof and should be avoided when possible. However, if such a position is essential a surface-fitting wall terminal should not be used, but the flue should project a short distance from the wall and be fitted with a terminal designed to assist in the extraction of the flue gases.

At certain heights of flue, depending upon the type of appliance and the degree of exposure of the flue to the external air, the flue gases will cool below their dew-point and condensation will occur. Table 7.2 in Part 2 gives some indication of the maximum desirable heights to which gas flues should be taken, in order to avoid such condensation. Where possible flue lengths should be kept within these limits; for greater lengths insulation may be necessary, either applied or in the form of factory-made metal chimneys. When coal gas is burnt the condensate is usually corrosive; the flue materials must, therefore, resist corrosive attack and where heavy condensation is anticipated special precautions must be taken (see Part 2, 7.2.2).

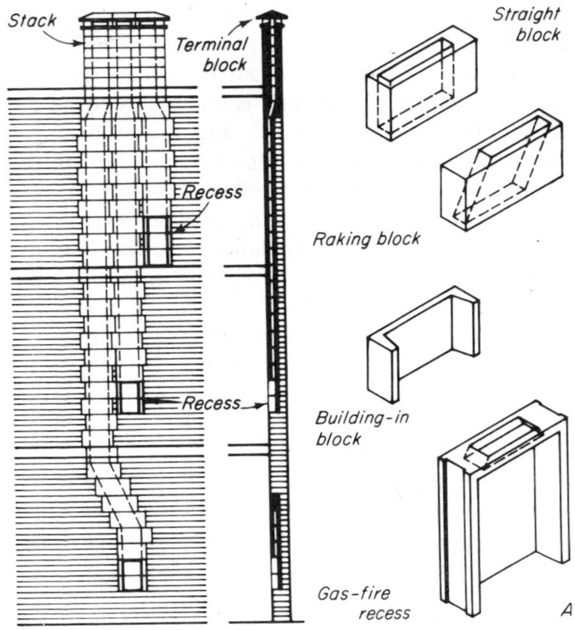

9.16 *Concrete block gas flues*

9.5.2 Chimneys for gas-fired appliances

These may be constructed of masonry or factory-made insulated metal chimneys described on page 218 may be used. The former are normally built of brickwork with a lining or of precast concrete flue blocks.

225 mm × 225 mm brick flues may be used for gas appliances, but are larger than required for the smaller appliances and the large surface area may result in excessive heat loss and condensation. They should be used only if already existing or in new work only when provision must be made for the flue to be capable of serving future appliances burning different fuels. They may be lined with any of the materials described on page 216 and fixed in the same way.

The connection between the outlet of an appliance and a masonry flue may be made by a short length of vitreous enamelled steel or cast iron pipe, so arranged that the appliance can be removed easily. Gas fires can be stood in precast concrete fireplace opening units similar to that shown in figure 9.16 *A*.

Standard precast concrete flue blocks[1] for constructing chimneys to serve gas appliances are designed to bond in with brickwork and provide the smaller flues desirable for these appliances.

Standard fittings are available for all normal requirements and some are shown in figure 9.16. These blocks are set in cement mortar and to assist in preventing the bedding mortar squeezing into the flue space and so reducing the flue area blocks having a modified spigot and socket joint can be used. Blocks having small rectangular flue openings are generally not suitable for flues to water heaters or for flues exceeding about 15·25 m in height. In these cases the larger types of flue blocks or tubular flues should be used. Walls to gas flues should be not less than 25 mm thick with greater thicknesses if forming part of a separating or compartment wall or if passing through a compartment wall or floor.[2] A stack built of small aperture blocks is illustrated in figure 9.16.

9.5.3 Flue pipes

In addition to the materials permitted for flue pipes for solid fuel and oil burning appliances (see page 219) flue

[1] BS 1289 *Precast Concrete Flue Blocks for Domestic Gas Appliances.*

[2] See AD J1/2/3, 2.26.

pipes for gas may be made of sheet metal and asbestos cement.

Gas flues and appliances do not present a serious fire hazard. The flue temperature is reasonably low since flames do not reach the flue and the products of combustion are diluted with cool air on leaving the appliance. Nevertheless, if the flue is in the form of a flue pipe its proximity to combustible materials is limited by the Building Regulations, the requirements of which in this respect are explained in Part 2, 7.2.2 and figure 7.8. Pipes should be supported as indicated on page 219.

For further consideration of the use of flue pipes for gas appliances see Part 2, Section 7.2.2.[1]

[1] For methods of discharging the flue gases to the atmosphere other than those described here reference should be made to Part 2, Chapter 7.

10 Stairs

A stair is a number of steps leading from one level to another, the function of which is to provide means for movement between different levels. This function in building is twofold: firstly, that of normal everyday access from floor to floor, and secondly, that of escape from upper floors in the event of fire.

Functional requirements

In order satisfactorily to perform this twofold function the stair must fulfil certain requirements, which are the provision of adequate:

strength and stability
fire resistance
sound insulation.

Strength and stability

Stairs, like the floors they link, must carry loads; not only the weight of people using them but also the weight of any furniture or equipment being carried up or down them. Timber stairs for small domestic buildings are usually constructed on the basis of accepted sizes for the various parts which are known to meet the loading requirements for that class of building. In other buildings, the stair must be designed on the basis of superimposed loads related to the class of building of which the stair is part.

Although the strength of balustrades in domestic stairs is not calculated in terms of horizontal pressure, this is sometimes desirable in other stairs, particularly where very large crowds of people are likely to use the stair. BS 6399: Part 1: 1984, gives a guide to the loads to be assumed for this purpose and these, together with the superimposed loads for stairs, are given in Part 2 chapter 8.

Fire resistance

Apart from the function of the stair itself as a means of escape in the event of fire during which its structural integrity must be maintained, the staircase links the floors throughout a building and can act as a path by which fire can spread from floor to floor. The requirements relating to stairs and staircases in respect of fire protection are discussed in the chapter on *Fire protection* in Part 2.

Sound insulation

As a stair links together the various floors in a building it may transmit noise for considerable distances, particularly impact noise when the walking surfaces are finished with hard material. In some circumstances the only way to prevent this is to make a complete structural break between the stair and the structure of the building as described in Part 2 chapter 8.

Definition of terms

Many terms are used in connection with the design and construction of stairs. Some of the main ones are defined here in order to make the subsequent descriptions more easily understood.

Step A step is a short horizontal surface for the foot to facilitate ascent from one level to another. It commonly consists of a horizontal element called a *tread* and a vertical element called a *riser*. The external junction of the tread and riser, or the front edge of the tread if it projects beyond the face of the riser, is called a *nosing* (figure 10.7). The riser, of course, is not essential to the step and ladders and many staircases are designed with no risers. Particular names are given to steps according to their shape on plan: a *flier* is the normal parallel step, uniform in width and rectangular on plan: a *tapered step* is one of which the nosing is not parallel to that of the step above it (figure 10.17 *A*).

Stair This has already been defined as a number of steps leading from one level to another.

Flight A series of steps between floors or landings.

Landing A platform between two flights. A landing serves as a rest between flights and also as a means to turn a stair. A *half-space landing* extends across the width of two flights and on it a complete half-turn is made; a *quarter-space landing* is one on which a quarter-turn only is made from the end of one flight to the beginning of the next (figure 10.15).

Staircase This term is applied to a stair together with that part of the building which encloses it, although it is also commonly used in reference only to the complete assembly of flights, landings and balustrades in a single stair.

String or stringer An inclined member which, if fixed to a wall may act simply as a housing for the steps as in a timber stair. If it is not fixed to a wall, it then acts as an inclined beam supporting the steps (figure 10.6).

Balustrade This provides protection on the open side or sides of a stair; it may be either solid or open. An open balustrade consists of vertical bars called *balusters* supporting a handrail.

Rise The rise of a step is the vertical distance between the upper surfaces of two consecutive treads and the rise of a flight is the total height between the floors or landings it connects.

Going (or run) The going of a step is the horizontal distance between the nosings or risers of two consecutive steps, and of a flight, the horizontal distance between the top and bottom nosings.

Line of nosings This is an imaginary inclined line touching the nosings of a flight.

Pitch or slope The angle made between the line of nosings and the line of the floor or landing.

Headroom The vertical distance between the line of nosings and any obstruction over the stair, usually the soffit of an upper flight or the lower edge of a floor or landing.

Walking line The average position taken up by a person ascending or descending the stair and generally taken to be 457 mm from the centre of the handrail.

Design of stairs

Apart from economic factors a number of others, related to comfort and safety in use, must be considered in the design of a stair. These are concerned with ease of ascent and descent, and with protection and support at the sides.

The dimensions of a stair will depend on the volume of traffic it must carry and also on the nature of furniture and equipment which is likely to be carried on it; in this respect the widths of the flights and landings are important, particularly at the turns. In addition, the dimensions of the treads and risers should be proportioned to give easy ascent and descent.

These factors, which will be considered here for staircases generally, are particularly important in the case of escape stairs and reference to the requirements in respect of these is made in the chapter on *Fire Protection* in Part 2. Reference should also be made to the Building Regulations, the requirements of which in respect of these factors vary according to the type of building in which a stair may be situated.

Width In most cases it will be necessary to allow sufficient width for two persons to pass which requires a minimum of 1015 mm to 1065 mm. For domestic stairs, 865 mm from wall face to the handrail is accepted as a reasonable width with 800 mm as a minimum. Stairs which may be used by disabled persons should not be less than 1010 mm nor more than 1800 mm wide. Although landings are usually made the same width as the flights, in some circumstances it may be desirable to make them considerably wider in order to allow for the passage of large pieces of furniture or equipment accommodated in certain types of building.

Slope or pitch A safe stair allows persons using it to move naturally and to ensure this natural movement a stair should be designed with the following in mind:

(a) The slope should be neither too steep nor too gradual. Excessive steepness calls for excessive effort and upsets the balance; if the pitch is too small, the heel tends to strike the nosings of the steps on descent, and the stair takes up a great deal of space.

(b) The treads should be wide enough for the foot to be placed on the step when descending without the leg touching the step above.

(c) The nosing should project sufficiently to prevent the heel striking the face of a solid riser above.

(d) All steps in a stair must be uniform to permit a regular movement and the dimensions of the rise and going of the steps should be properly related.

The pitch should not exceed 42 (38 for common stairs to flats, etc) nor be less than 25 degrees; for stairs in regular use, a maximum of 35 degrees should be taken. For most stairs, where a minimum nose projection of 19 mm is provided, a minimum going of 254 mm and a maximum of 305 mm should be adopted, although in domestic stairs a minimum of 230 mm is acceptable. A rise from 140 mm to 178 mm is usually satisfactory.

Comfort in use of a stair depends largely upon the relative dimensions of the rise and going of the steps and rules for determining the proportion are based to some extent upon the assumptions that about twice as much effort is required to ascend as to walk horizontally and that the pace of an average person measures about 585 mm. This, and the fact that a 305 mm going with a rise of 140 mm or 150 mm is generally accepted as comfortable, results in the rule that the going plus twice the rise should lie between 585 mm and 610mm. The dimensions for minimum going and maximum rise permitted by the Building Regulations are respectively smaller and greater than given here, except for the going of a stair in an institutional or an assembly building. For the 'going plus twice the rise' relationship a wider range than that given above is permitted.[1]

Steps to stairs which may be used by disabled persons for access or escape should have a going of at least 280 mm and a rise not greater than 150 mm.

Flights Long flights of stairs without landings as points of rest can be dangerous, especially when used by children, invalids or elderly people, or as a means of escape. Twelve steps in a flight is usually taken as a comfortable maximum, but much of course depends upon the pitch of the stair and with low pitches a greater number would be satisfactory. The Regulations require at least one change in direction of the stair between flights of at least 30 degrees if there are more than thirty six risers in consecutive flights. They also impose a maximum of sixteen risers in a flight in staircases in public buildings and shops.

Flights in stairs used by disabled persons should not rise more than 1200 mm.

Headroom Sufficient headroom must always be provided and the clear distance measured vertically between the line of nosings and the edge of a landing above or the soffit of a flight above should be not less than 2·00 m. This is satisfactory for low pitches but for pitches around 35 degrees, 2·15 m is preferable.

Handrails and balustrades The function of a handrail is to assist people using the stair, whether it is enclosed by walls or open, and the function of a balustrade or other form of guarding is to prevent falls from the open side of a stair or landing. Handrails should be of such a size and shape that they are easy to grip and should be placed at a natural height: for domestic stairs with pitches about 35 degrees this is 840 mm measured vertically from the line of nosings to the top of the handrail. For lower pitches this height should be somewhat greater and for horizontal handrails it should be 900 mm.[2] Handrails should be provided on at least one side if the flight is less than 1 m wide and on both sides if wider. They are not required to extend over the bottom two steps.

Balustrading or other form of guarding is required at the sides of a flight or landing where there is a drop of more than 600 mm. It must be of sufficient height and, if open, must have sufficient vertical or raking members to stop people who may slip on the stair from falling through.[3] For domestic stairs the heights given above for handrails are sufficient for balustrades guarding the open side of flights and landings, but for others somewhat greater heights are required by the Buildings Regulations.[4]

As mentioned earlier, guardings must be sufficiently strong and rigid to withstand side pressure, particularly in public buildings where large numbers of people are likely to use a stair at the same time (see Part 2, 8.1).

Types of stair

There are several ways of classifying stairs according to plan form or construction. The following is that based on plan form:

Straight flight This has no landings and is a useful form of stair when the total rise is not too great, otherwise the absence of landings makes it tiring to ascend (figure 10.6). The space required is rather long but only needs to be the width of the stair. Where floor heights are great and it is essential to use a straight stair with no turns, it may be necessary to introduce a landing in the length of the stair. This is in order to keep the number of risers in the flights within the limits referred to above.

Dog-leg This is most common in timber construction. In this stair there are two flights which return on each

[1] See AD K1, Table 1 and Cl.1.7.

[2] AD K1 sets a maximum height of 1000 mm for handrails. It should be noted, however, that AD M requires handrails to be not more than 850 mm above the line of nosings if the stair is to be used by disabled persons.

[3] Reference should be made to AD K1, 1.27 which restricts the size of openings in balustrades and other forms of guarding in buildings where young children are likely to be present.

[4] See AD K1, Table 3.

other about a single newel (see below), so that the handrail of the lower flight stops short against the underside of the upper flight (figure 10.15). It is useful when the going is restricted and the space available is only sufficient to take the combined width of the two flights.

Open well In this form the stair has two flights returning on each other but with a space called the *well* between the two flights (figure 10.15). This improves the appearance of the stair and permits the lower handrail to extend the full length of the flight. With adequate width of well a third flight may be introduced between the other two with quarter-space landings between the flights as shown in figure 10.15 (*C*).

The incorporation of landings to permit flights to turn gives greater flexibility in planning and shortens the flights of a stair. It does, however, require a greater floor area than that for a straight flight.

The following classification is based on the construction adopted:

Newel stair Vertical posts, called newels or newel posts, are used in timber stairs at the end of the flights to support the strings and connect them to the floors, and to support the handrails and any necessary bearers (figure 10.6). In concrete, steel or stone, a circular stair may have a central newel from which the treads radiate (figure 10.1).

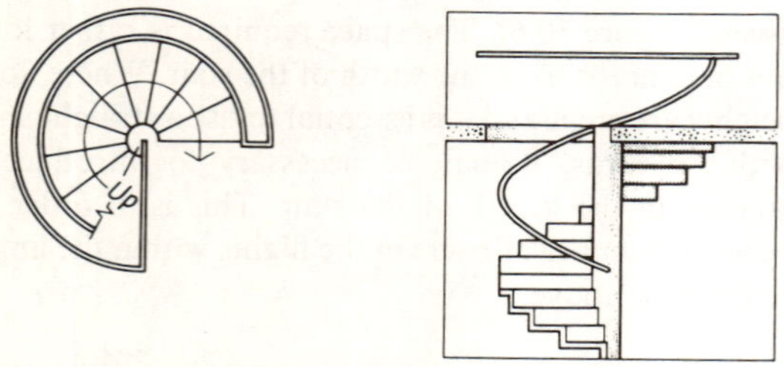

10.1 *Circular newel stair*

Geometrical stair In this form there are no newels, the strings and handrails being continuous from floor to floor round a well. In some forms of concrete stairs and in stairs in which the steps cantilever out from a wall, there is no string as such and the handrail only is continuous (figure 10.2 *A*). The geometrical stair may be rectangular, circular or elliptical on plan, the steps in the latter being tapered steps, and it may be constructed in timber, concrete, steel or stone. Most present-day stairs other than domestic come within the category of geometrical stairs.

The stair is often quite detached from its surrounding wall so that it becomes a design element in space (figure 10.2 *B*). These free-standing stairs can be placed in two broad groups: ramp and open-riser or 'ladder' types:

(i) *Ramp stair* This can be visualised as a ramp or as the floor flowing from one level to another, a conception which is emphasised when close or up-stand strings are employed to shield the saw-tooth line of the steps. The greatest effect is obtained when a single flight, unbroken by landings, can be employed.

(ii) *Open-riser or 'ladder' stair* This consists of treads only carried on strings, and in its simplest form is like a ladder spanning between floor slabs. In some forms there may be only one string with the treads cantilevering on one or both sides.

Apart from the visual effect upon the stair itself as an element, the absence of risers enhances the actual size of the space within which the stair is situated. In domestic work where, for reasons of economy, circulation areas are reduced to a minimum, this is an advantage, but regard must be paid to AD K1, 1.11 which restricts the size of openings in risers as in balustrades (see footnote to page 225).

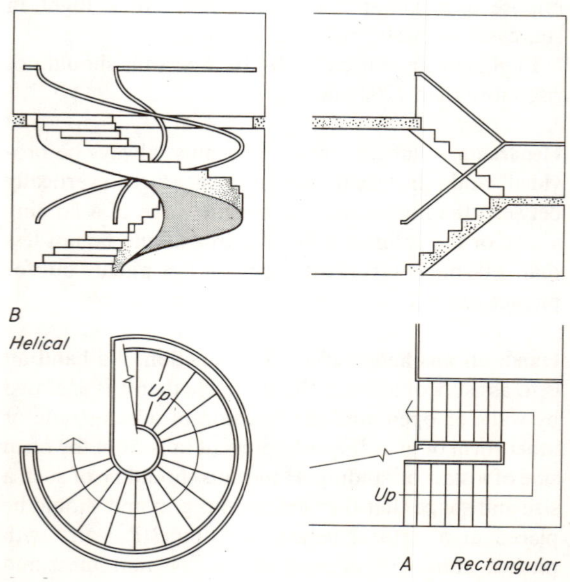

10.2 *Geometrical stairs*

10.1 CONSTRUCTION OF STAIRS

10.1.1 Brick stairs

Brick is used for external steps and stairs and occasionally for internal use.

The steps must be formed of good hard, square bricks to withstand wear and are bedded in cement mortar on concrete as shown in figure 10.3. If the steps are not built on a natural slope of ground the deep hardcore filling must be carefully and well consolidated to avoid settlement. The bonding of the bricks will depend on the dimension of tread and riser and the bricks will normally be laid on edge to expose the face sides and ends.

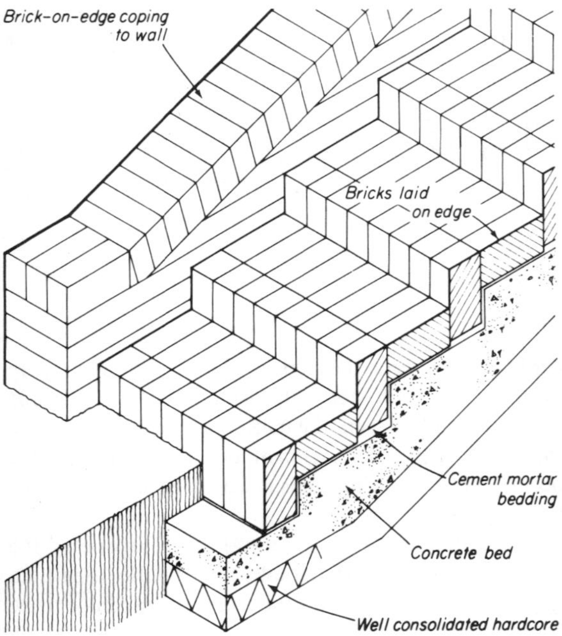

10.3 *Brick stair*

10.1.2 Stone stairs

These are not very much used today, mainly because of the cost, and will only be mentioned briefly here. Stone stairs may be in the form of steps simply supported on end walls or as cantilever flights and landings, or in the form of a circular newel or turret stair. Simply supported or cantilevered steps can be either rectangular blocks giving a stepped soffit as in figure 10.4, or

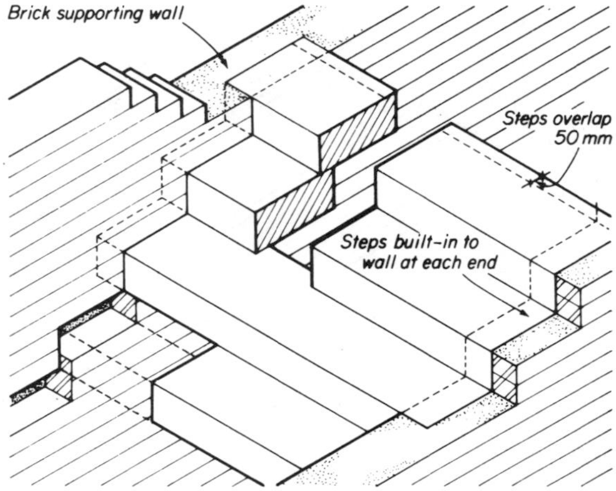

10.4 *Stone stair*

spandrel steps (see page 228) splayed on the underside to give a smooth soffit as shown for precast concrete stairs in figure 10.5 *B*. Cantilever stone steps should not usually exceed 1·50 m to 1·80 m in projection, the safe maximum depending upon the type of stone used; if landings are large, these are made up of a number of slabs with joggled joints (see figure 5.35). The newel stair is similar to the spiral newel stair in precast concrete. However, because of the transverse weakness of stone, the steps are not designed to cantilever out from the central newel where the stone is thinnest, but the outer ends are built into the enclosing wall so that each is a step simply supported on the wall and the newel.

10.1.3 Concrete stairs

Concrete stairs are widely used in all types of buildings. They have a degree of fire resistance, are strong, and make possible a wide variety of forms. They may be cast in situ or be precast as whole flights or in separate parts. In situ cast stairs and the larger types of precast stairs are described in Part 2. The simpler forms of precast stairs will be discussed here.

Reinforced concrete rectangular steps may, of course, be supported at the ends on walls to form a stair in the same way as stone (figure 10.4) but since they can be reinforced to resist bending stresses they may be thinner in cross-section and can easily be cantilevered as in the open-riser stair shown in figure 10.5 *A*. The thickness of each step will depend upon the projection,

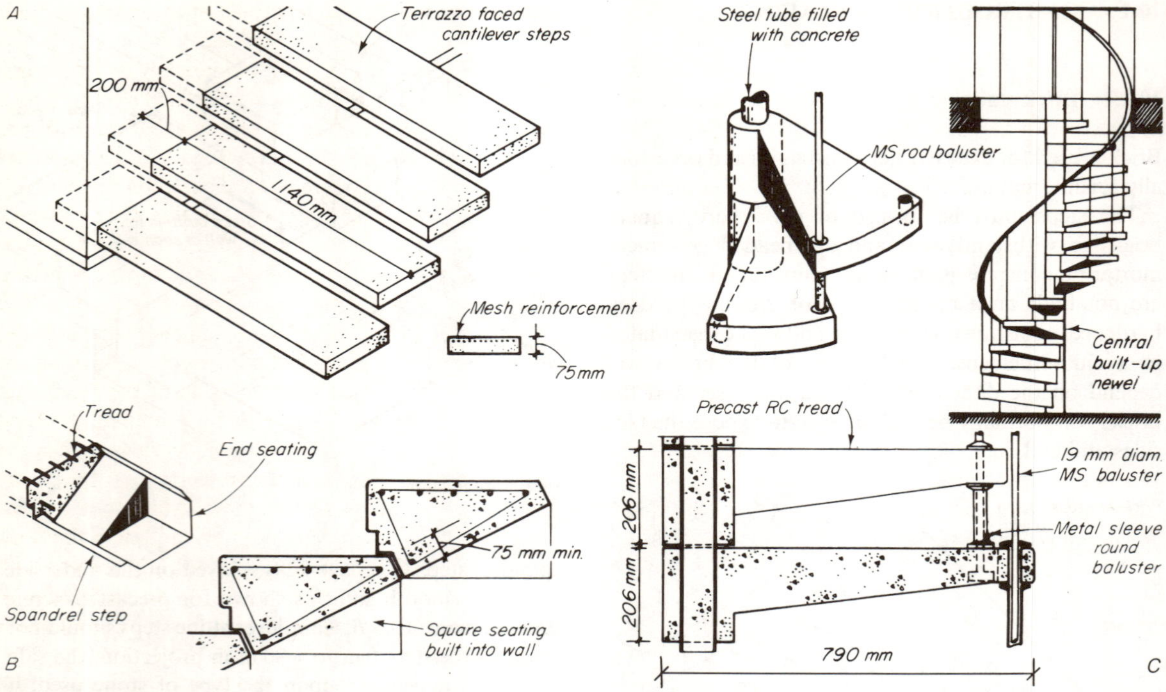

10.5 *Concrete stairs*

that is the width of the stair, and upon the loading on the stair. The thickness shown would be suitable for the normal domestic loading of 1·50 kN/m². The steps should be built into the wall not less than 200 mm to 230 mm and the wall, if of masonry construction of any form, should be built in cement mortar for at least 305 mm above and below the line of the cantilever stair. A closed-riser stair could be formed by taking advantage of the adaptability of concrete and casting each step as an L-section, the vertical arm forming the riser to each tread as in the example in Part 2. This has the advantage of increasing the stiffness of each step by the increased structural depth given by the riser.

Such a closed-riser stair results in a 'stepped' soffit similar to that shown in figure 10.4. When a smooth soffit is required solid steps of triangular section must be used with rectangular end blocks cast on to form a square seating on the wall as shown in figure 10.5 *B*. These are called *spandrel* steps. The 'point' of the section should not be too thin in order to avoid spalling or breaking off and the minimum thickness at this point, known as the *waist*, should be 75mm. In order to avoid gaps between the steps, and to facilitate lining up on the soffit, each step is birdsmouthed over the one below. As will be seen, the main reinforcement is at the top of the steps where, as cantilevers, the tension zone occurs. In end bearing steps it would be situated at the bottom.

The spiral stair with a central newel is one of the oldest types of stair. It can take up a smaller area than any other type but, as all the steps are tapered, to be comfortable in use it usually requires more space than a rectangular stair. The smaller form of newel stair is often constructed in precast concrete on the lines shown in (*C*). Each step has a section of the newel cast on through which is formed a hole and the stair is built up by threading the steps over a steel rod or tube, with a bedding of mortar between each. The steel core is built into the floors at each end or bolted to them through steel flanges welded on its ends. The steps may be positioned relative to each other by means of the balusters as shown or by means of stubs and mortices formed on the newel sections. The former is simpler and cheaper relative to casting the steps.

10.1.4 Timber stairs

Early timber stairs consisted of rectangular baulks of timber spanning between supporting walls similar to stone stairs of the same type. Later the steps were made of planks of timber forming treads and risers. This economised in timber and produced a lighter structure making possible the self-supporting string stair still used today, consisting of steps of this type framed between and supported by inclined strings spanning from floor to floor.

Timber stairs are commonly used in domestic buildings with either closed or open risers, but by the use of laminated timber strings and treads, it is possible to construct timber stairs of considerable span and width suitable for larger buildings, in situations where combustible materials may be used.

In this volume newel and ladder type stairs suitable for domestic buildings will be described.

Straight flight stair

This may be constructed between walls which give it continuous support or it may be open on one or both sides.

Figure 10.6 *A* shows diagrammatically a closed-riser straight flight domestic stair constructed against a wall with an open balustrade on one side. The sizes of the members are not usually calculated for domestic stairs of normal rise and width but are based on those which have been found satisfactory for this type of work (figure 10.7). Treads are 32 mm thick and risers 25 mm. These are nominal sizes and when the timber is planed will finish about 27 mm and 21 mm respectively. The method of fixing together treads and risers shown in (*A*) is common. The top edge of the riser may, however, be simply butted against the underside of the tread, but the joint should then be covered by a small mould fixed to the tread so that any gap formed by shrinkage of the riser will be concealed. The projection of the nosing should not be much greater than 25 mm to avoid the danger of the toe catching on it on mounting the stair. The nosing profile may be square, slightly splayed with rounded top edge or half-round. As an alternative to solid timber 13 mm plywood may be used for the risers as shown in (*B*).

The ends of the treads and risers are housed into

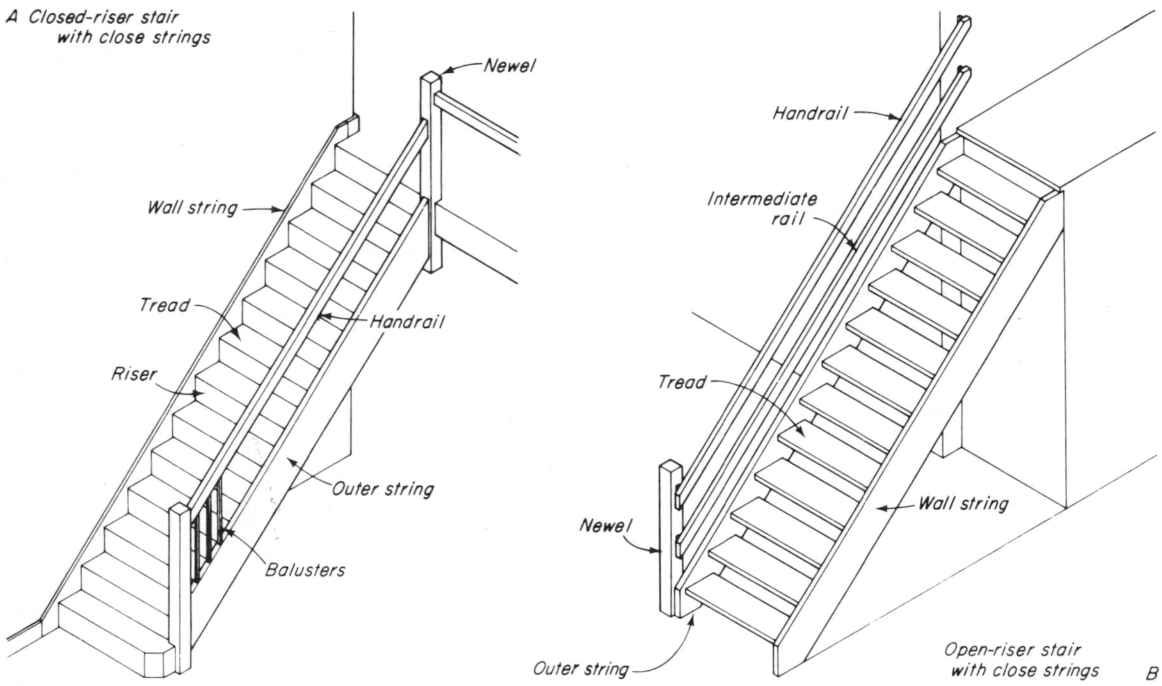

10.6 *Straight flight timber stairs*

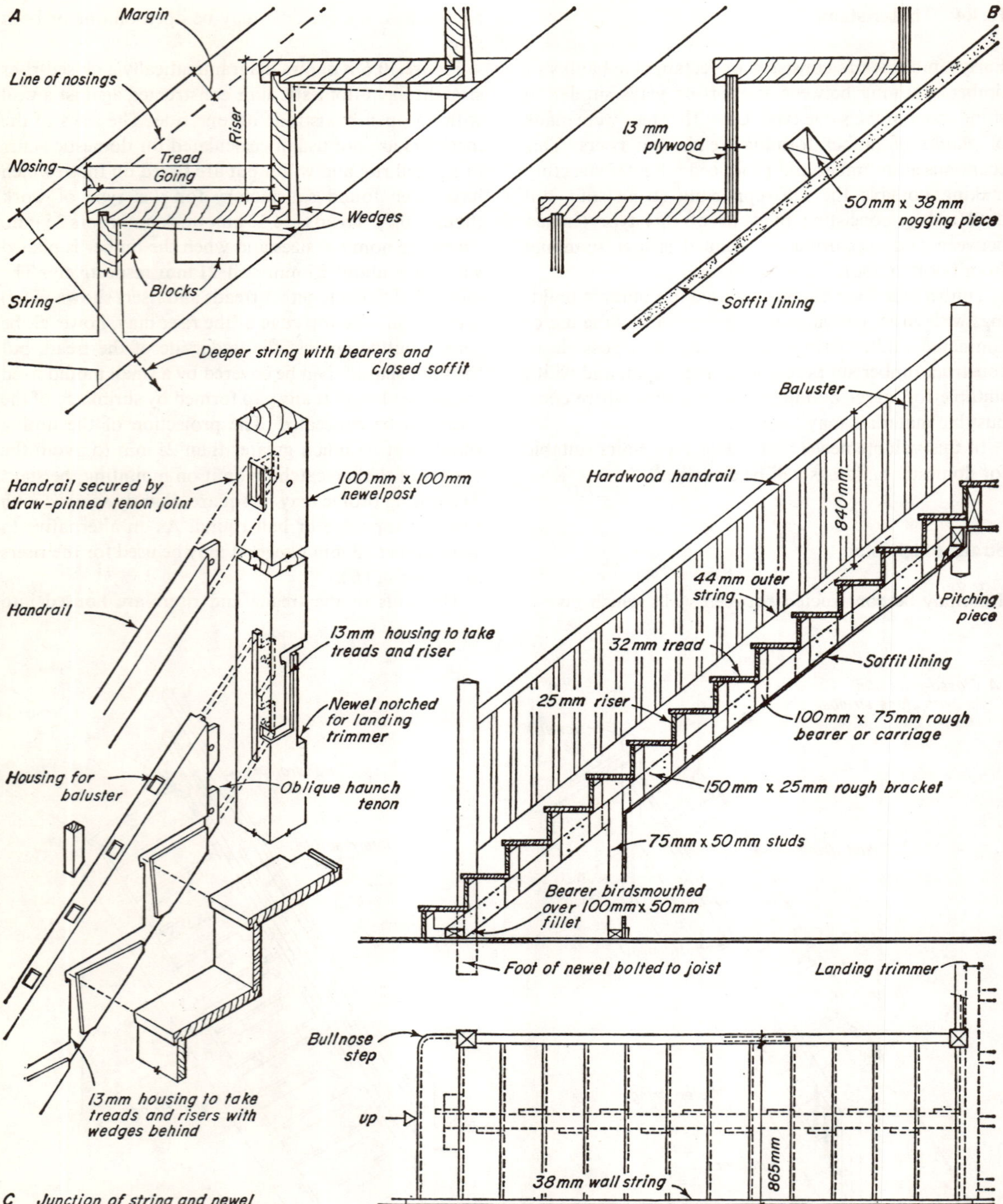

10.7 *Construction of timber stairs*

grooves or housings, about 13 mm deep, formed in the strings (figures 10.7 *C*, 10.8). The housings are wider than the thickness of tread and riser and are tapered so that hardwood wedges, after covering with glue, may be driven behind the treads and risers forcing them tight against the outer faces of the housings as shown. Triangular blocks of wood are glued at the junctions of the treads with the risers and strings to give increased rigidity to the whole staircase.

The wall string should be about 38 mm thick securely plugged to the wall. The upper edge is rebated if it is to take plaster and should be moulded to match the skirtings with which it will join at each floor. The projection of the string beyond the plaster face should be the same as the skirting thickness so that the two flow neatly into each other (figure 10.8). The thickness of the string should, therefore, allow for this.

The outer string should be 44 mm to 50 mm thick. It must be thicker than the wall string as it acts as an inclined beam whereas the former serves as a plate supported by the wall.

For stair widths of 914 mm and over, it is desirable to introduce intermediate support in the form of 100 mm × 75 mm or 100 mm × 50 mm *rough bearers* or *carriages* under the steps; a single central bearer is sufficient for a width of 914 mm, as in figures 10.7, 10.8, with an additional one for each 380 mm increase in the width. The bearers are securely spiked at the feet to the floor if the ends coincide with a joist or, to spread the load, they are birdsmouthed over a short fillet fixed to the flooring (figure 10.7). At the top they are birdsmouthed to the landing trimmer or, if this does not project low enough relative to the stair, to an additional cross-bearer called a *pitching piece* as shown. It is sufficient to arrange the rough bearers to touch the bottom edge of the steps; notching is not necessary, direct support to the treads being provided by short pieces of 25 mm timber, called *rough brackets*, which are nailed on alternate sides of the bearer with their upper edges tight against the treads (figure 10.7 and 10.17).

For domestic stairs the minimum depth of a close string necessary for framing in the steps, including a margin of 38 mm above the nosings (see figure 10.7 *A*), is structurally sufficient if the outer strings are not less than 44 mm thick.

If the stair is less than 914 mm wide so that rough bearers are not required a closed soffit of plasterboard or other lining material may be fixed direct to such strings and to 50 mm × 38 mm nogging pieces fixed at 455 mm centres up the flight (figure 10.7 *B*). The provision of a rough bearer, however, necessitates deeper strings if the soffit is to be lined in order to provide edge fixing for the soffit lining (figure 10.8). Deep strings may be cut out of one piece of timber or may be formed of two pieces tongued together. No more than the minimum depth of string is required if the soffit is hidden and is not to be lined, whether or not rough bearers are provided.

The outer string is framed into 100 mm × 100 mm newels at top and bottom of the flight. The strength of newel stairs such as this depends largely on the rigidity of the joint between the string and newel and the normal method of joining the two is shown in figure 10.7 *C*. This consists of a *draw-pinned joint* consisting of two oblique haunch tenons on the end of the string fitted into mortices formed in the newel, the whole being secured by a slightly tapering hardwood dowel at each tenon. It will be seen that the newel, like the strings, is housed to take the treads and risers and is, in addition, notched to fit over the landing trimmer to which it is nailed or, preferably, bolted. The junction with the lower newel is similar but the joint is reversed as shown for the upper string in figure 10.16. The foot of the lower newel should be taken through the floor and bolted to a convenient joist, using a packing piece if necessary, to give a firm and secure connection. The upper newel extends a short way below the string. This is termed a newel 'drop' (figures 10.7, 10.9).

Ends of handrails should be housed slightly into the newels and fixed by draw-pinned tenon joints and the ends of balusters should be housed or tenoned into handrail and string (figure 10.7). When flights are not too long the omission of balusters and the provision of intermediate rails fixed between the newels below the handrail avoids the cost of preparing and fixing a large number of balusters (figure 10.6 *B*). In domestic and certain other types of building the spacing of the rails will be determined by the restrictions laid down by the Building Regulations on the size of openings in balustrades (see footnote to page 225).

When a flight of stairs is constructed between two walls a handrail must be fixed to the wall on one side. The handrail is usually of simple section, often circular (called a *mopstick* handrail), supported at 915 mm to 1·20 m centres by metal handrail brackets screwed to the underside and fixed to plugs set in the wall. The projection of the brackets should be such as to leave a clearance of not less than 38 mm between handrail and wall (figure 10.8).

The trimmer to the upper floor landing is faced with an *apron lining* tongued and grooved at the top to a

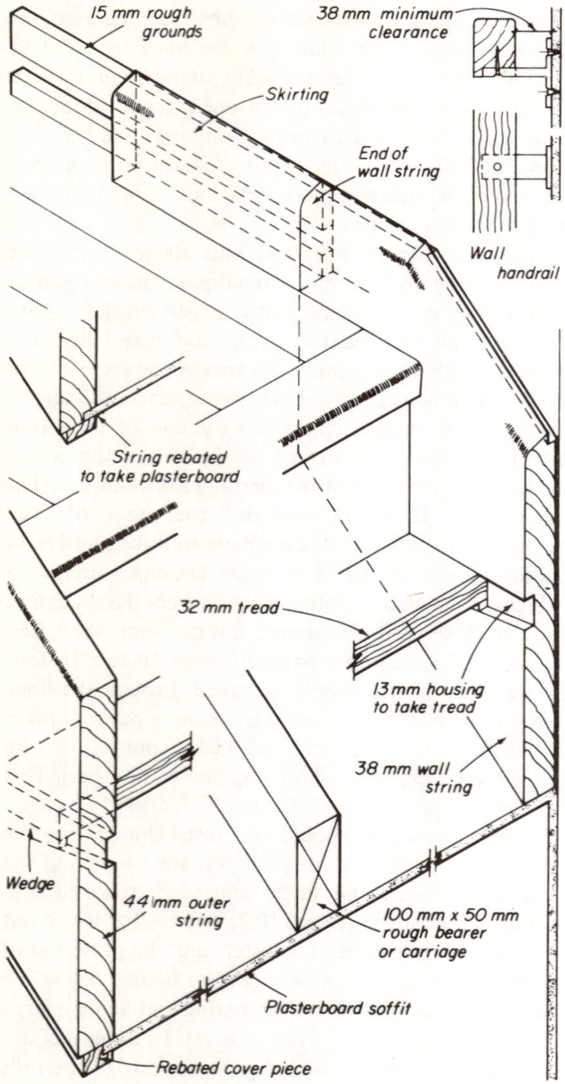

10.8 *Construction of timber stairs*

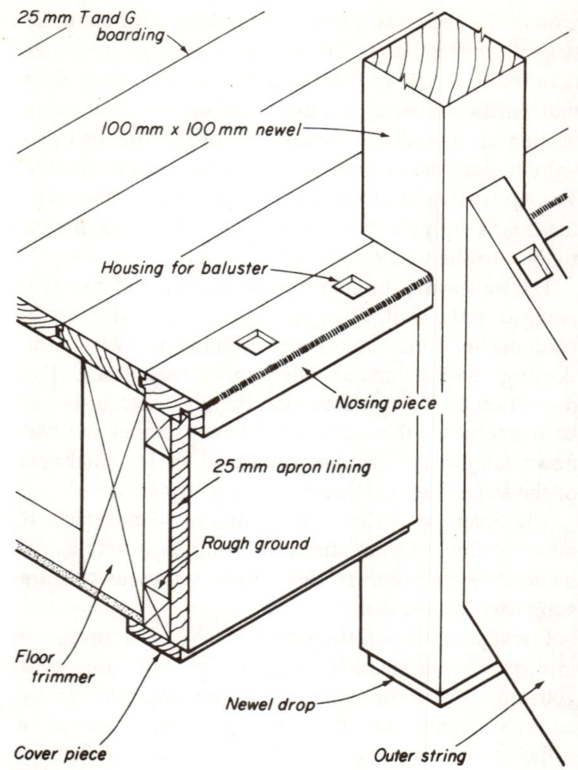

10.9 *Apron lining to landing*

nosing piece, preferably the same thickness as the stair treads, into which any landing balusters are housed and the floorboards tongued and grooved (figure 10.9). A similar nosing finishes the top step of the flight. The use of grounds or packing pieces behind the apron lining permits any balusters to be kept central with the newel. The apron piece may be of 6·4 mm plywood instead of solid timber as shown. A cover piece at the junction with the ceiling plaster masks the joint and the lower ground.

The outer string may be formed as a cut string as shown in figure 10.10. In this form the string is cut to the profile of the steps and the treads and risers cannot be wedged to the string. The tread projects to the face of the string and the nosing is carried round as a planted moulding stop returned at the end; the ends of the risers are mitred to the string. An alternative method is to project both treads and risers beyond the string to give a continuous stepped line running up the stair. This is most satisfactory with close-grained hardwoods which do not need mouldings to cover the end grain. Rough bearers are required immediately inside the strings since the effective structural depth of the latter is reduced by cutting for the steps.

For architectural reasons the newel at the bottom of a staircase is usually set back one or sometimes two risers. The entry to the stair is less abrupt and, particularly where one side of the stair abuts a wall, may be made slightly from the side as mounting commences. A specially shaped end to the bottom step or steps must be formed as shown in the examples illustrated. That shown in figure 10.7 is common and is called a *bullnose step*. An extension of this into a semi-circular end, called a *curtail step*, may be used where space on plan permits. Both are constructed on similar lines as shown in figure 10.11. The curved form of the riser necessitates special treatment, its thickness being reduced to a veneer of about 1·6 mm to 2 mm (depending on the sharpness of the curve) to permit it to be bent round a built-up block. The dovetail keyed end and the folding wedges enable the veneer to be drawn tight against the block which is built up from three to four pieces of timber laid cross-grained like the plies in plywood in order to minimise shrinkage and pulling away from the riser veneer. Prior to applying to the block the veneer is steamed or wetted with boiling water to prevent it cracking when bent. The veneer and the face of the block, together with the wedges, are coated with glue and after the wedges are driven the block is screwed from the back to the riser. The face screws necessary at the end of the curtail step because of the shape of the block are hidden by the newel into which it is set. For the sake of clarity the housings for tread and second riser have been omitted from the newel to the bullnose step.

A splayed rather than curved end can be more simply formed, the pieces of the normal riser of which it is constructed being mitred and tongued to each other and to the riser end (figure 10.6 *A*).

Open-riser or ladder stair

These may be constructed with close or cut strings. When close strings are used as shown in figure 10.6 *B*, although the treads are sunk into the strings, the connection between the ends of the treads and the strings is not so good as in a closed-riser stair since there are no wedges or side blocks connecting the two; the strings should, therefore, be tied together by 10 mm or 13 mm diameter metal rods with sunk and pelleted ends placed under every fourth tread (figure 10.12 *A*). Screw fixing between string and tread is not very strong as the screws enter the end grain of the tread. Glued dowels are better than screws for this purpose. Cut strings are tied together by the treads which rest upon the string and are screwed and pelleted to it (*B*). It should be borne in mind that the effective depth of a cut string is that of the waist at the narrowest point. A

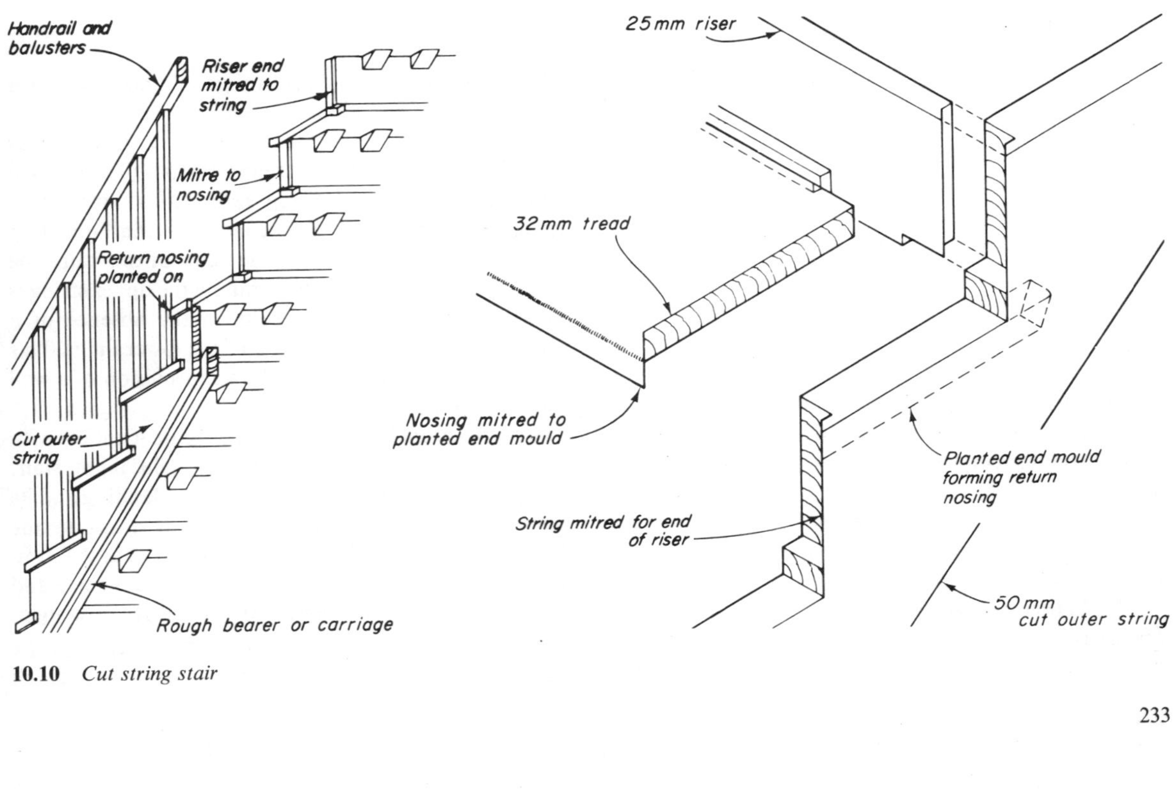

10.10 *Cut string stair*

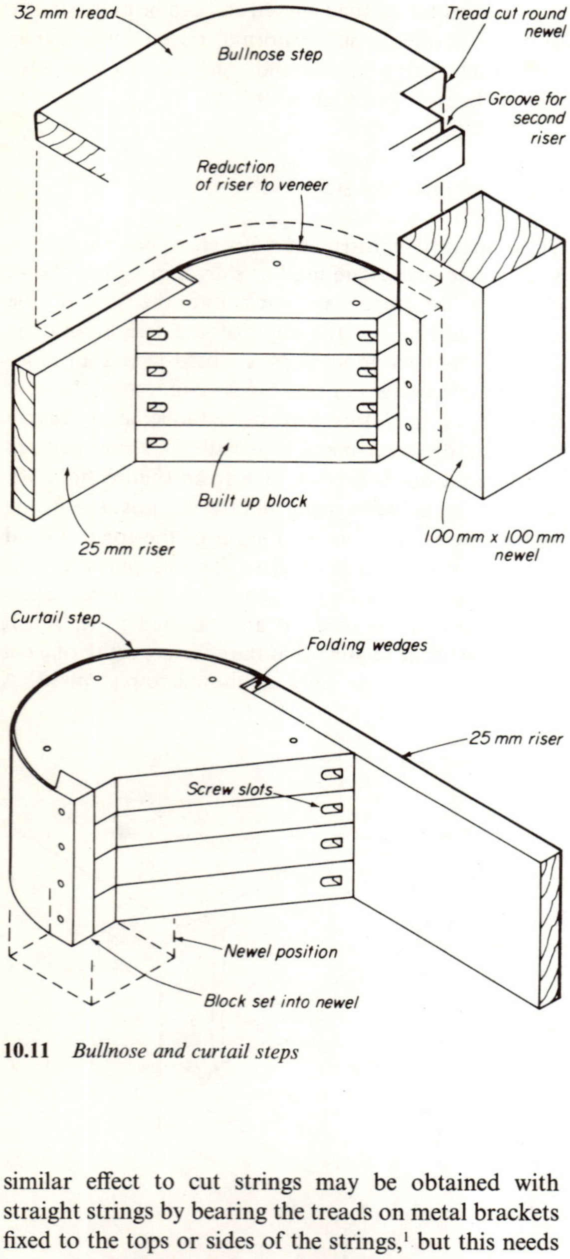

10.11 *Bullnose and curtail steps*

similar effect to cut strings may be obtained with straight strings by bearing the treads on metal brackets fixed to the tops or sides of the strings,[1] but this needs to be carefully detailed to give a satisfactory appearance.

With open-riser stairs no support is given to the tread by a riser so the treads should be at least 38 mm to 44 mm thick.

The Building Regulations restriction on the size of openings in risers, referred to on page 226, precludes a full open riser in domestic buildings and a detail such as that shown in figure 10.12 *E* must be used. The treads, which should overlap at least 16 mm, are generally of hardwood as the application of carpeting is not wholly satisfactory in appearance unless the expensive method of sinking a panel of carpet into the tread is adopted, and even then hardwood is desirable because the nosing is exposed to wear.

It will be seen that there are no newels in the stair illustrated in figure 10.13 and this necessitates a direct fixing of the strings to the floors. In this particular example MS plates to which the strings are bolted are cast in the concrete floors. As the handrail receives no support from newels the balusters must be such as to provide this, necessitating the use of metal balusters as in this example and in figure 10.14 *A* or of substantial timber balusters securely screwed or bolted to the outer face of the string (figure 10.14 *B*). The size of the openings in the risers in figure 10.13 and in the balustrades shown in figures 10.13 and 10.14 would, it should be noted, be too great to meet the requirements of the Building Regulations for domestic and certain other types of buildings.

Alternative methods of securing the top of an open-riser stair are shown in figure 10.12, by fixing the strings directly to a wall face (*C*) or by fixing to the upper floor by means of a head piece dowelled to the tops of the strings as in (*D*).

Dog-leg stair

This consists of two short straight flights, the lower rising to a half-space landing, the upper returning and rising from the landing in the opposite direction as shown diagrammatically in figure 10.15 *A*. The use of a single newel at the landing into which both outer strings are framed produces on elevation the V-junction of strings which gives rise to the name of the stair. The interception of the lower handrail by the upper string is visually unpleasant and results in the absence of a handrail for support at the top of the lower flight. This deficiency can be made good by the provision of a wall handrail at the lower flight.

Constructional details of the flights are identical with those already described for the straight flight stair, except at certain points at the half-space landing and its newel. The latter is usually continued down to the lower floor for the sake of rigidity and fixed at the foot, preferably direct, to a floor joist. It is notched and bolted to the landing trimmer (figure 10.16). The junc-

[1] See illustrations of steel string stairs in Part 2 where similar detailing is used.

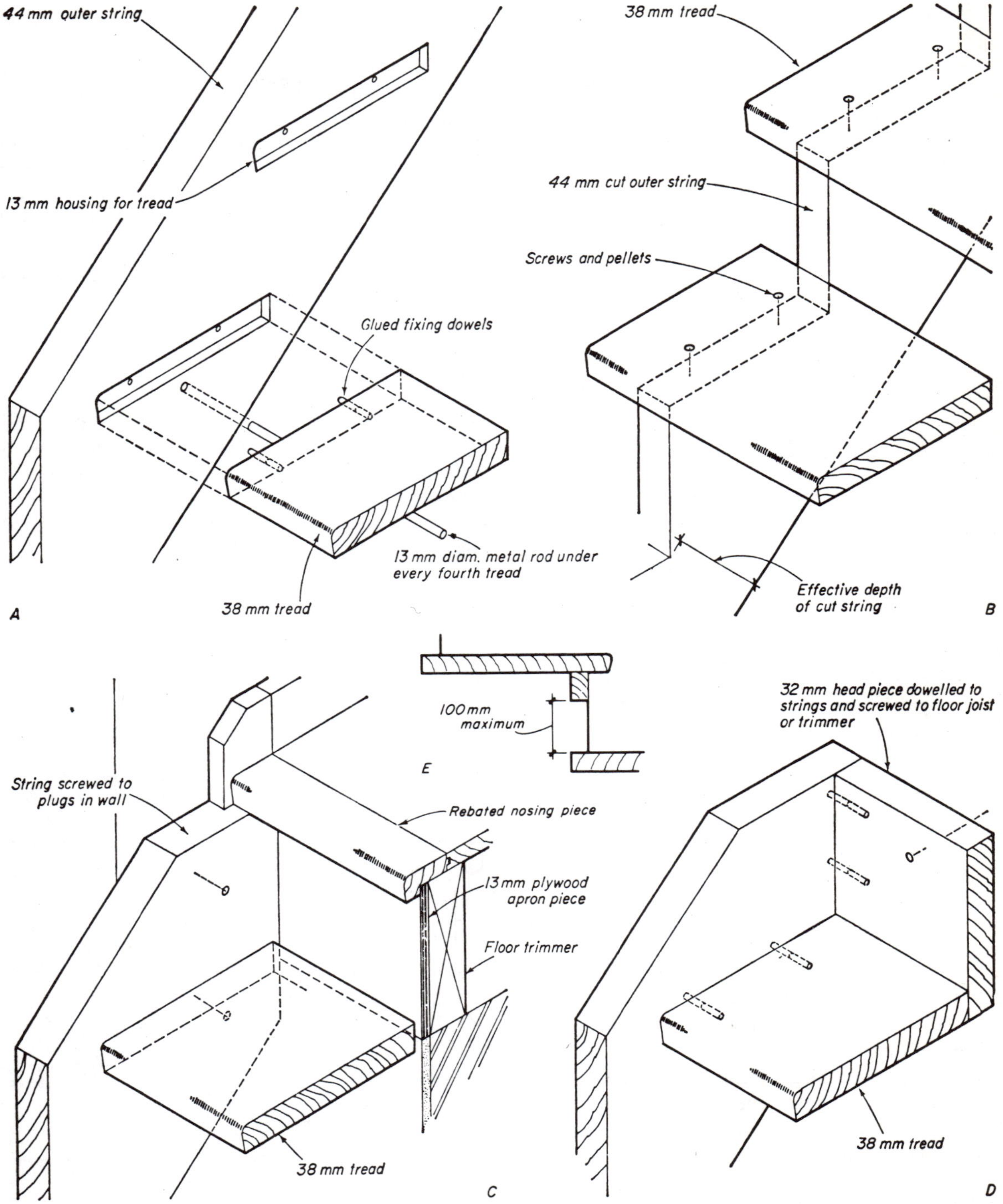

10.12 *Open-riser stair construction*

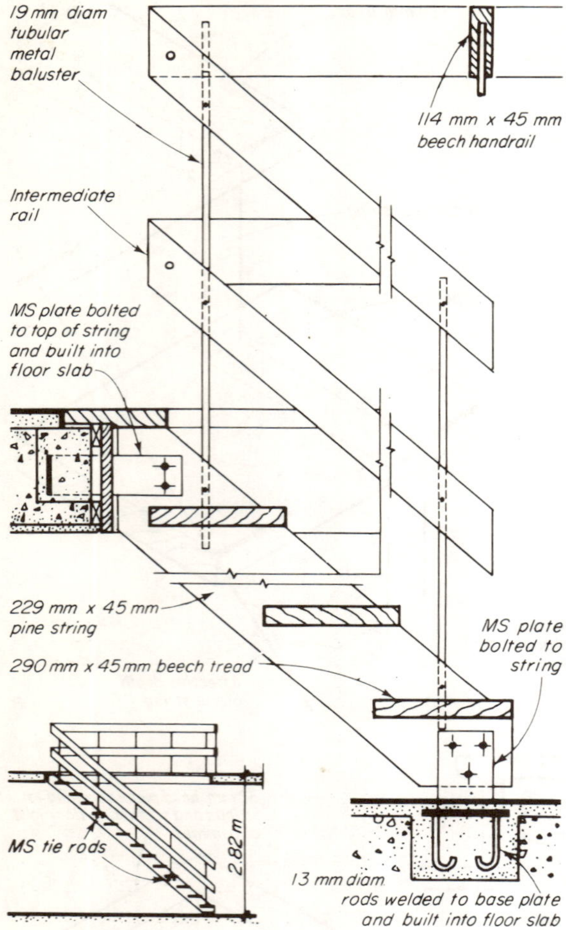

10.13 *Open-riser stair*

tions of both outer strings with the landing newel are by means of draw-pin joints already described. The junction of the lower string in figure 10.16 is identical with that shown in figure 10.7. The strings butt against each other on a horizontal line, about 50 mm wide, outside the face of the newel.

The half-space landing is formed of 100 mm × 50 mm joists supported by the trimmer at one end and by the staircase wall at the other. Although the trimmer to the landing may be deeper than the landing joists in order to provide a bearing for rough bearers to the upper flight, it will not usually be deep enough to provide a bearing for any lower bearers. The upper ends of these must, therefore, be birdsmouthed to a pitching piece, as shown in figure 10.16. The feet of the upper bearers are notched over a small fillet or plate fixed to the face of the trimmer.

In order to obtain a satisfactory junction with the lower handrail, if this is wider than the string, a cover piece on the bottom of the upper string should be provided not less in width than that of the handrail against which the latter may terminate. The spandrel under the lower flight of a closed-riser dog-leg stair, which will normally be quite low, is usually filled in and the area under the stair and landing formed into storage space, with access by a door under the landing trimmer below the top flight.

Open well stair

As already described, this stair in its simplest form, like the dog-leg stair, consists of two straight flights turning on a half-space landing, but with a space or well between the outer strings. This gives a better appearance and a continuous handrail up to the landing newel (figure 10.15 *B*).

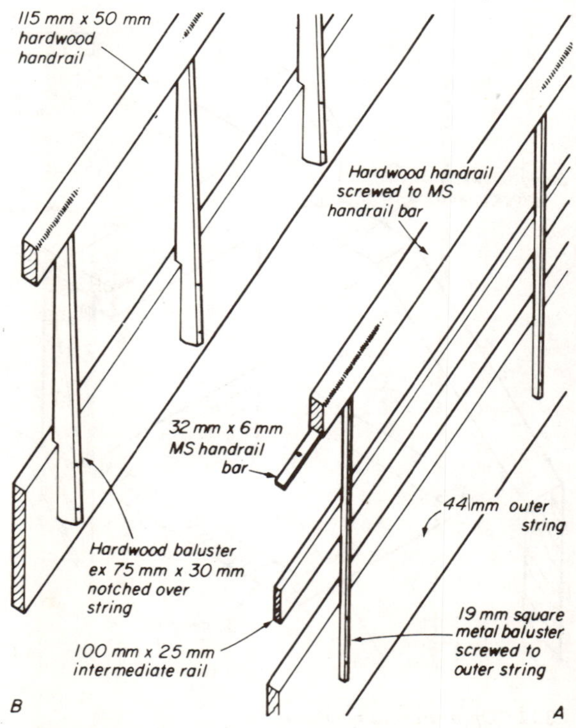

10.14 *Balustrading*

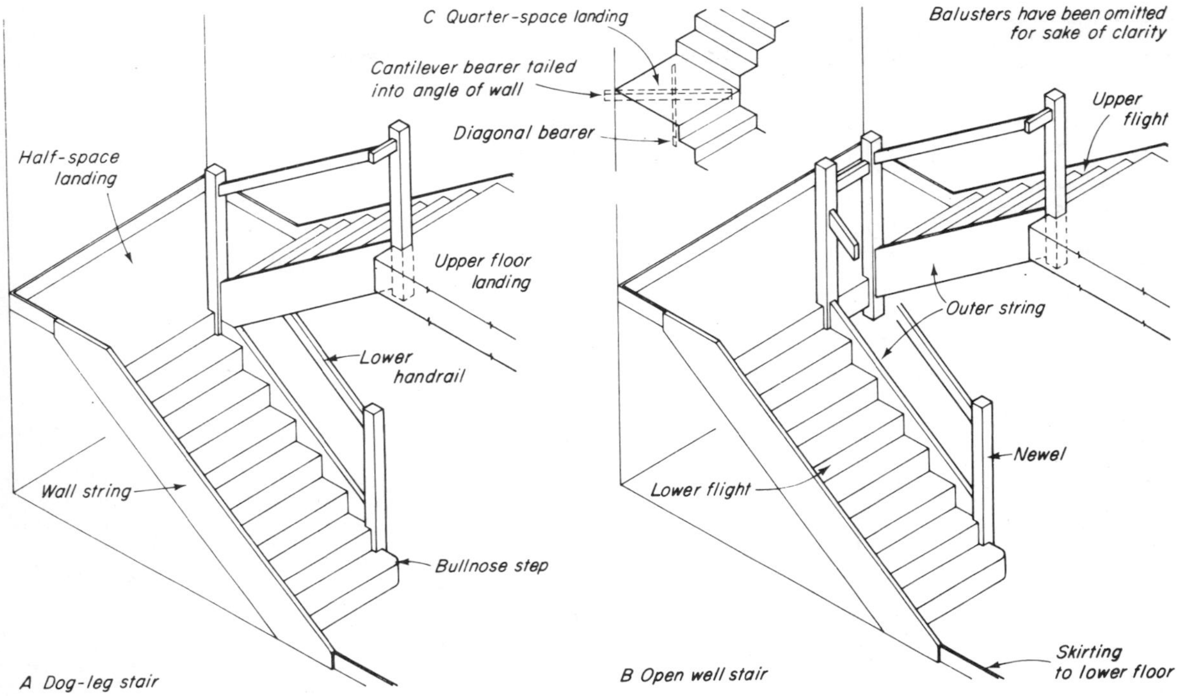

10.15 *Dog-leg and open well stairs*

All the relevant details are similar to those described for the previous stairs. If the landing is half-space the landing newels may terminate just below the landing as on the upper floor. The section of landing exposed between the newels is finished with an apron lining. If a very narrow well is adopted, say 75 mm to 100 mm, a single newel about 230 mm wide is preferable to avoid an extremely small space between a pair of newels and its accompanying problems at handrail and landing level.

If an intermediate flight is incorporated quarter-space landings will be formed. In small domestic stairs a quarter-space landing will usually support itself provided the two surrounding walls are capable of withstanding the lateral thrust which the stair imposes on them and secure fixings are made between strings and bearers and the landing structure. In larger stairs it is necessary to provide support to the trimmers of the landings. This may be done most simply by carrying the landing newels down to the floor below as in the case of the dog-leg stair. If this is not desired, particularly in the case of the upper newel below which headroom will not be unduly restricted, the landing and newel may be supported on cantilever construction. This consists basically of a diagonal bearer under the landing built-in to the staircase walls carrying a diagonal cantilever bearer, one end of which tails in to the angle of the wall, the other of which carries the ends of the landing trimmers at the newel (figure 10.15 *C*). Alternatively, a double-cranked bearer may be formed of laminated timber to span across the width of the staircase and to which the newels are secured.

Tapered steps

Tapered steps are essential in circular stairs and may be used in rectangular stairs in order to save space when this is limited. In the latter case they should be placed only at the bottom of the stair as in figure 10.17 *A*. An even number of tapered steps produces an unpleasant and difficult junction between the centre riser and the internal angle of the wall string, and an odd number, producing a centre 'kite' step as in figure 10.17 *A*, is preferable.

The rise and going should be within the limits given for parallel steps on page 224 including the 'going plus twice the rise' relationship. If the width of the flight is

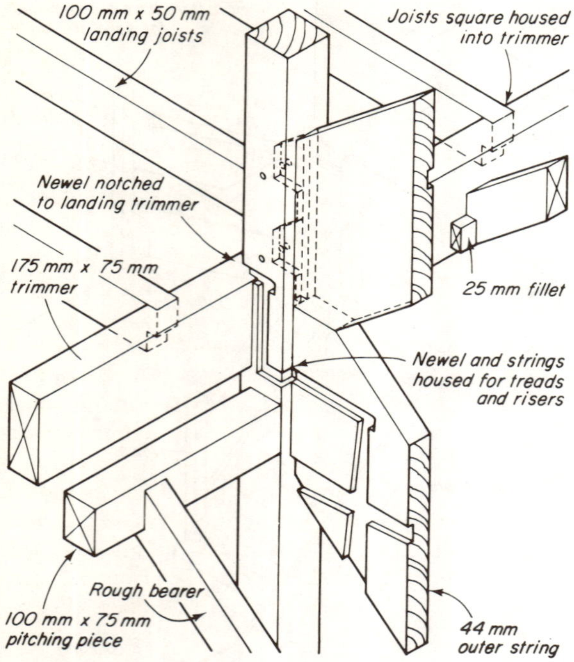

10.16 *Dog-leg stair construction*

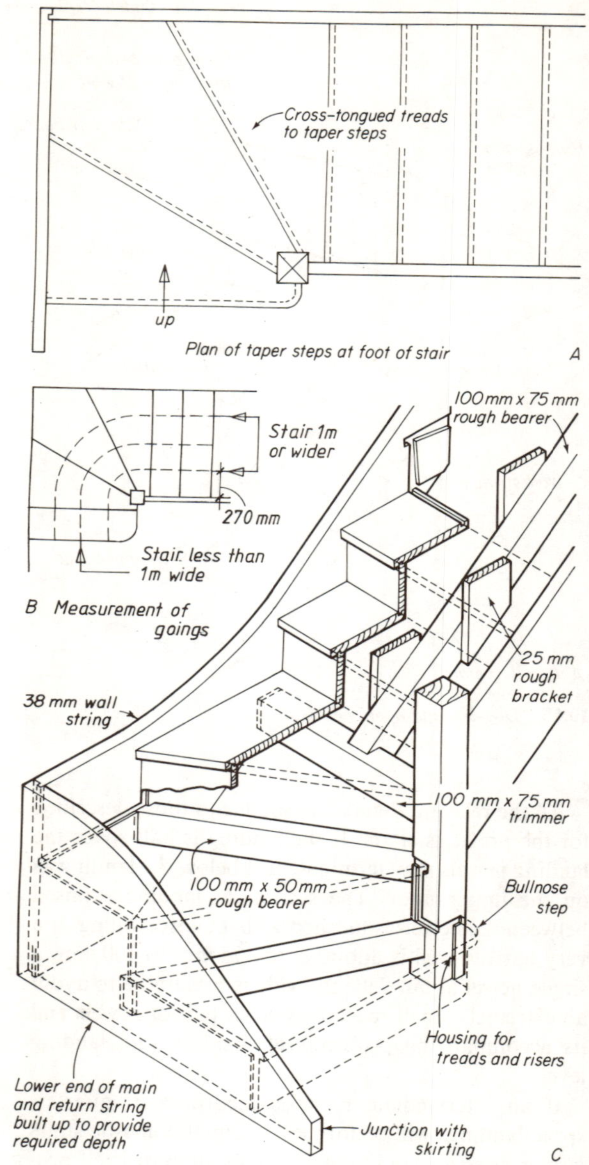

10.17 *Tapered steps*

less than 1 m the Building Regulations require the going to be measured between the centre points of the tread; if 1 m or wider it is to be measured between points at 270 mm from both ends of the tread, in order to establish acceptable maximum and minimum goings. Where the treads are of unequal length as at the turn of a rectangular stair (as in figure 10.17) these points are to be based on the parallel steps above and below the set of tapered steps (figure 10.17 *B*). The Regulations require the going at the narrower end to be at least 50 mm and, as in parallel steps, the goings throughout should be uniform.

The method of constructing tapered steps is shown in (*C*). Cross-tongued treads must be used to give the necessary width and the lower end of the main wall string and the return string are also built up in the same way to accommodate the ends of the steps. The top edges of these strings are shaped or 'eased' to a line giving the required margin at the nosings of the steps.

Unless the stair is very narrow the length of the risers will be greater than 914 mm and some support will be necessary. This is provided by a rough bearer under each riser housed into newel and wall string. If the width of the main flight necessitates a rough bearer this is terminated on a trimmer as shown, over which it is birdsmouthed.

11 Temporary works

Construction work involves a large number of operations concerned with temporary constructions of various types which are a necessary part of the whole building process. These are called *temporary works.* They include (i) means of supporting the soil in excavations (ii) support to parts of the work, such as arches, during their construction (iii) the provision of moulds into which concrete may be cast (iv) the provision of working platforms at different heights by means of scaffolding (v) the provision of temporary support to buildings or parts of buildings by means of shoring.

Some of these are introduced here. They are discussed more fully in Part 2, chapter 11, along with those works not referred to at this point.

11.1 TIMBERING FOR EXCAVATIONS

Excavation of the soil in relation to building work is required for various purposes: for foundations, for basements, for sewers and drains. This will involve the cutting of relatively narrow trenches, the excavation of large areas to a considerable depth or the sinking of shafts or the formation of tunnels. Except for shallow trenches in good firm soil some form of temporary support must be given to the soil as excavating proceeds in order to prevent collapse of the sides. At this point the support provided to shallow trenches, that is those not exceeding 1·20 m in depth, is considered. That to other types of excavation is described in Part 2.

11.1.1 Trenches

The cutting of trenches should be carried out with considerable care, particularly if the trenches are to be left open for any length of time, as there is a danger of the moisture draining or drying out and the sides of the excavation falling in. As a rule in firm soil, if the trench can be filled in reasonably quickly, it may be sunk to a depth of from 1·20 m to 1·80 m without support to the sides. But above 1·80 m in depth any soil should be timbered as vibration or the withdrawal of water may cause the sides to collapse.

For shallow trenches in firm ground, open timbering as shown in figure 11.1 *A* can be employed. This consists of pairs of 225 mm × 38 mm *poling boards*, 900 mm to 1·20 m in length, placed at intervals of 1·80 m and fixed by struts. In ground that is less firm the second method of open timbering shown at (*B*) is used. Here the poling boards are placed along the sides of the trench about 230 mm apart, and horizontal timbers from 150 mm to 225 mm wide by 50 mm to 100 mm thick, called *walings*, are placed against the poling boards on each side of the trench and are strutted apart by stout struts. The walings should be placed in the centre of the polings. In trenches above 1·80 m in depth, or in loose soil, the sides should be *close boarded*, in which case the polings are placed close together and waled and strutted as before (*C*). It is essential to prevent the escape of soil from between the boards, an occurrence liable to take place after heavy rains, as any lessening of the resistance behind the boards will cause the timbering to collapse with little warning. Where the trenches are liable to remain open for any length of time (with deep trenches this must always be the case), the walings and struts must be of ample dimensions as the pressures will be considerable.

Struts are usually square in section and should not be less than 100 mm × 100 mm. A rule-of-thumb method of estimating the size of square struts which are not intermediately supported over spans from 1·20 m to about 4 m is to make the breadth of face one-twelfth of the span. It is common practice now to use adjustable tubular steel struts in place of timber struts (*E*).

Shallow trenches in very soft ground are sometimes *sheeted*, that is, the sides are lined with 225 mm × 38 mm boards laid horizontally, as shown at (*D*). The ground is excavated 225 mm at a time and the sides lined with a pair of boards, these being temporarily strutted. When the full depth is reached poling boards are placed in pairs, one on each side of the trench against the sheeting and strutted. The temporary struts are then removed.

11.2 CENTERING

The need for temporary support for arches during their construction is referred to on page 90. The framework of timber used for this purpose is known as *centering*, the shape and form of construction being dependent

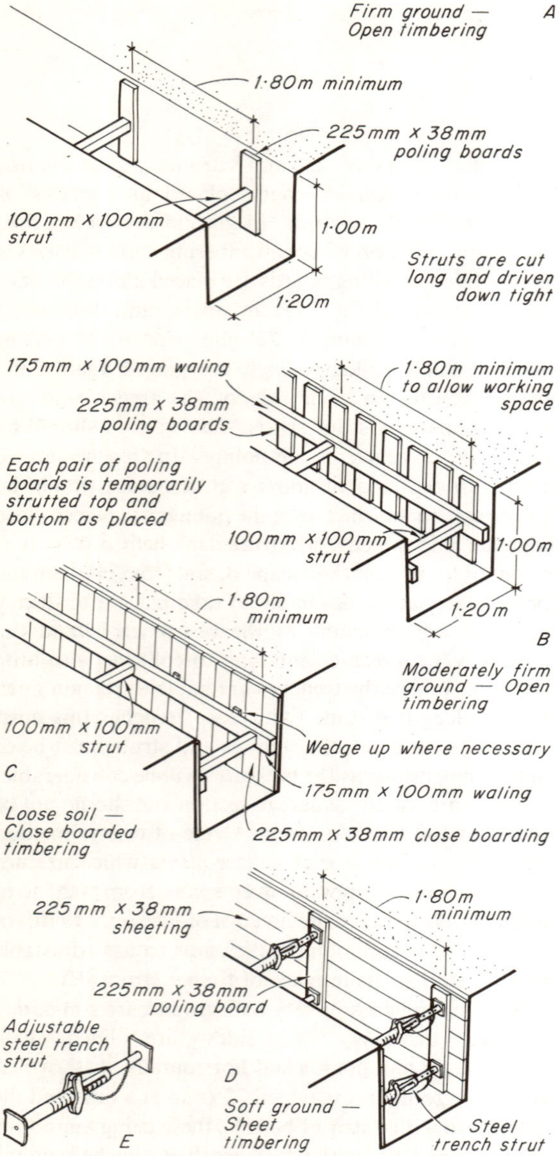

11.1 *Shallow trench timbering*

upon the arch to be supported. It ranges from a single timber beam with a curved top face on timber props (called a *turning piece*) for camber and segmental arches to a fully framed construction of ribs and struts for wide and deep arches.

The centering must be rigid enough to bear its temporary load and capable of vertical adjustment to permit 'easing the centre', that is slightly lowering it, before the mortar has quite set. The 'striking' of the centre, that is its removal, takes place only well after the mortar in the arch joints has thoroughly set.

A 102·5 mm thick arch can be supported on a simple turning piece referred to above, similar to that shown in figure 11.2 *A*, and about 100 mm wide. Wider soffits require a centre formed with *ribs* and *laggings* which provide support to the soffit as shown in (*B*).

The two ribs are tied together at each end by *seating bearers* or *bearing pieces* on the underside and the laggings, which are narrow battens, are nailed on their tops. The laggings will be 16 mm to 25 mm or more in thickness depending upon the distance apart of the ribs and will be spaced apart 19 mm to 25 mm for axed and rough arches and close together for gauged arches. At each end the seating bearers rest upon a pair of folding wedges which permit 'easing' of the centres as well as its initial positioning. The wedges are supported by a bearer on a pair of props (*C*) or, for arches not wider than 215 mm, directly by a 50 mm plank 175 mm to 200 mm wide (*B*). The props bear on a timber cill piece on the wall at the base of the opening.

The props are strutted apart across the opening by timbers cut slightly longer than the distance between the props and wedged tight. This type of centre is suitable for camber and segmental arches, the curvature and thus the depth of the ribs being made in accordance with the curvature of the arch.

Centres for semi-circular arches necessitate the use of built-up ribs as shown in (*C*). The segments of the ribs are secured at the joint by an upper tie and the feet of each rib are secured by a pair of ties to prevent spread. The ribs are secured to each other by the laggings and seating bearers. For larger spans requiring more than two segments to a rib stiffening at the joints is necessary to prevent deformation of the curve. This is accomplished by radial struts from the centre of the lower ties to the junctions of the segments as shown in (*D*). A large centre requires stiffening laterally by horizontal and diagonal cross braces between the ribs.

11.3 FORMWORK

Concrete must be given form by casting it in a mould and the term covering all types of mould for in situ concrete is *formwork*.

For the large-scale works the formwork will be a major construction in its own right and some of the considerations involved are discussed in Part 2.

At this point the relatively simple construction

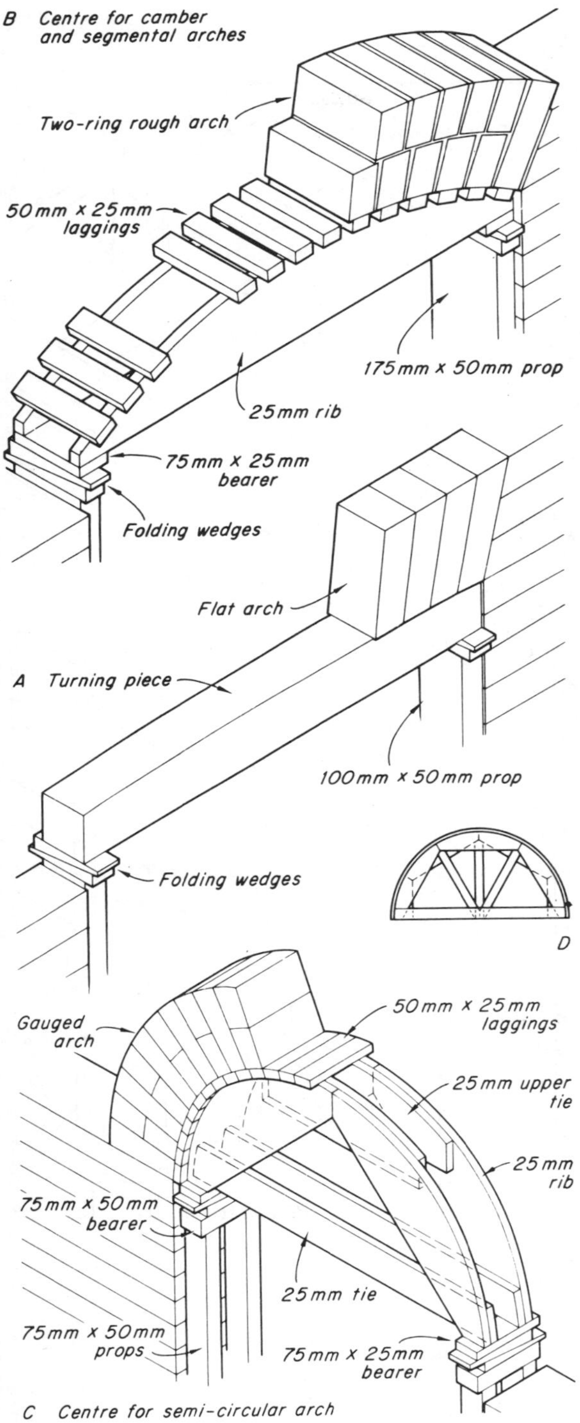

11.2 *Centering*

required for a concrete lintel and for a short span concrete slab will be described.

11.3.1 Formwork for a concrete lintel

As indicated on page 86 in situ cast lintels require formwork to be erected at the head of the opening. A typical method of constructing this is shown in figure 11.3 *A*. The soffit of 32 mm or 38 mm T and G boards spans the full width of the opening and rests on 75 mm × 25 mm bearers at each end. These are supported in the same manner as arch centering by props and folding wedges which serve the same purpose. The sides are 25 mm or 32 mm thick nailed to the soffit and overlapping the wall face beyond the bearings. For additional security timber cleats are fixed at the ends between the bearers and sides as shown.

When the opening is wide the soffit boards should be battened together at intervals and since the lintel and, therefore, the sides of the form will be deeper than shown in (*A*), the bearers should be extended to take struts as shown in (*B*), which provide resistance to the thrust exerted on the sides by the wet concrete.

11.3.2 Formwork for a concrete roof slab

In situ cast concrete floor and roof slabs for small-scale work are normally cast on a formwork of timber, often in this context called shuttering (see Part 2, chapter 11).

This consists of a *decking* of 25 mm boards or 19 mm plywood on which the concrete is placed, supported by 100 to 150 mm × 50 mm timber joists as shown in figure 11.3 *C* (overleaf). The joists are supported by lateral members called *ledgers*, the size of which will vary with the spacing of the vertical struts or *props* which support them.

The props are usually about 75 or 100 mm × 100 mm and are braced in both directions in order to avoid movement. They bear on folding wedges resting on a 150 mm × 25 mm sole piece, the purpose of the wedges, as in the lintel form, being to permit final adjustment in the height of the decking and to facilitate the ‘striking’ or removal of the formwork.

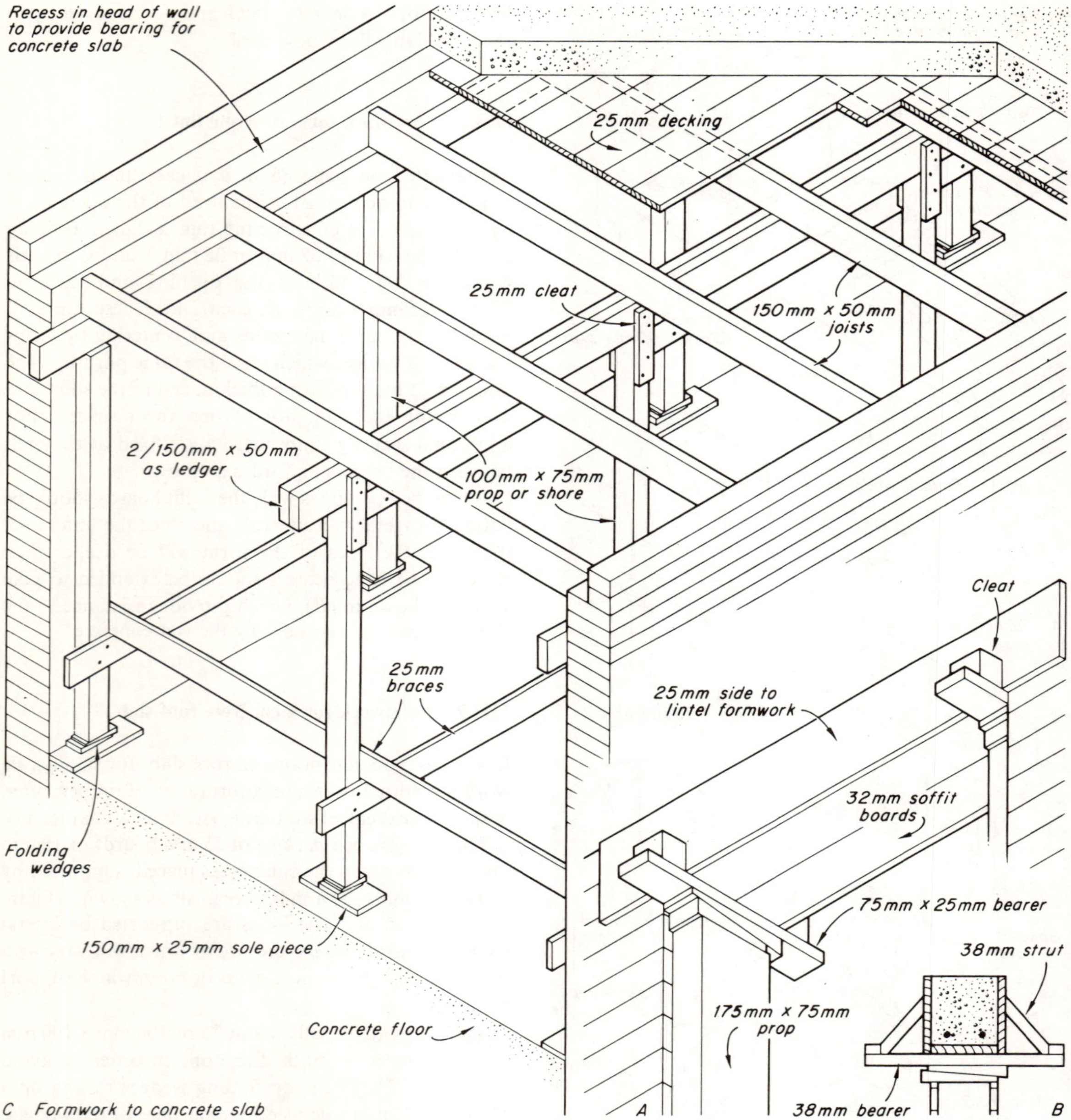

C Formwork to concrete slab

11.3 *Formwork*

SI Units

Quantities in this volume are given in SI units which have been adopted by the construction industry in the United Kingdom. Twenty-five other countries (not including the USA or Canada) have also adopted the SI system although several of them retain the old metric system as an alternative. There are six SI basic units. Other units derived from these basic units are rationally related to them and to each other. The international adoption of the SI will remove the necessity for conversions between national systems. The introduction of metric units gives an opportunity for the adoption of modular sizes.

Multiples and sub-multiples of SI units likely to be used in the construction industry are as follows:

Multiplication factor	*Prefix*	*Symbol*
1 000 000	10^6 mega	M
1 000	10^3 kilo	k
100	10^2 hecto	h
10	10^1 deca	da
0·1	10^{-1} deci	d
0·01	10^{-2} centi	c
0·001	10^{-3} milli	m
0·000 001	10^{-6} micro	μ

Further information concerning metrication is contained in BS PD 6031 *A Guide for the use of the Metric System in the Construction Industry*, and BS 5555: 1976 *SI units and recommendations for the use of their multiples and of certain other units.*

Quantity	*Unit*	*Symbol*	*Imperial unit × Conversion factor = SI value*		
LENGTH	kilometre	km	1 mile	=	1·609 km
	metre	m	1 yard	=	0·914 m
			1 foot	=	0.305 m
	millimetre	mm	1 inch	=	25·4 mm
AREA	square kilometre	km^2	1 $mile^2$	=	2·590 km^2
	hectare	ha	1 acre	=	0·405 ha
	square metre	m^2	1 $yard^2$	=	0·836 m^2
			1 $foot^2$	=	0·093 m^2
	square millimetre	mm^2	1 $inch^2$	=	645·16 mm^2
VOLUME	cubic metre	m^3	1 $yard^3$	=	0·765 m^3
			1 $foot^3$	=	0·028 m^3
	cubic millimetre	mm^3	1 $inch^3$	=	1 638·7 mm^3
CAPACITY	litre	l	1 UK gallon	=	4·546 litres
MASS	kilogramme	kg	1 lb	=	0·454 kg
	gramme	g	1 oz	=	28·350 g
			1 lb/ft (run)	=	1·488 kg/m
			1 lb/ft^2	=	4·882 kg/m^2

A guide to SI metric system

continued overleaf

continued

Quantity	*Unit*	*Symbol*	*Imperial unit × Conversion factor = SI value*		
DENSITY	kilogramme per cubic metre	kg/m^3	1 lb/ft^3	=	16·019 kg/m^3
FORCE	newton	N	1 lbf	=	4·448 N
			1 tonf	=	9 964·02 N
				=	9·964 kN
PRESSURE, STRESS	newton per square metre	N/m^2	1 lbf/in^2	=	6 894·8 N/m^2
	meganewton per square metre	MN/m^2† or N/mm^2	1 $tonf/ft^2$	=	107·3 kN/m^2
			1 $tonf/in^2$	=	15·444 MN/m^2
			1 lb/ft run	=	14·593 N/m
			1 lb/ft^2	=	47·880 N/m^2
			1 ton/ft run	=	32 682 kN/m
	*bar (0·1 MN/m^2)	bar			
	*hectobar (10 MN/m^2)	h bar			
	*millibar (100 MN/m^2)	m bar			
VELOCITY	metre per second	m/s	1 mile/h	=	0·447 m/s
FREQUENCY	cycle per second	Hz	1 cycle/sec	=	1 Hz
ENERGY, HEAT	joule	J	1 Btu	=	1 055·06 J
POWER, HEAT FLOW RATE	watts	W	1 Btu/h	=	0·293 W
	newtons metres per second	Nm/s	1 hp	=	746 W
	joules per second	J/s	1 ft/lbf	=	1·356 J
THERMAL CONDUCTIVITY (k)	watts per metre degree Celsius	W/m deg C	1 Btu in/ft^2h deg F	=	0·144 W/m deg C
THERMAL TRANSMITTANCE (U)	watts per square metre degree Celsius	W/m^2 deg C	1 Btu/ft^2h deg F	=	5·678 W/m^2 deg C
TEMPERATURE	degree Celsius (difference)	°C	1 °F	=	$\frac{5}{9}$ °C
	degree Celsius (level)	°C	°F	=	$\frac{9}{5}$ °C + 32

* Alternative units, allied to the SI, which will be encountered in certain industries
† BSI preferred symbol

CI/SfB

CI/SfB

The following information from the *Construction Indexing Manual 1976* is reproduced by courtesy of RIBA Publications Ltd.

Used sensibly and in appropriate detail, as explained in the manual, the CI/SfB system of classification facilitates filing and retrieval of information. It is useful in technical libraries, in specifications and on working drawings. *The National Building Specification* is based on the system, and BRE Digest 172 describes its use for working drawings.

The CI/SfB system comprises tables 0 to 4, tables 1 and 2/3 being the codes in most common use. For libraries, classifications are built up from:

Table 0	Table 1	Tables 2/3	Table 4
-a number code	-a number code in brackets	-upper and lower case letter codes	-upper case letter code in brackets
eg 6	eg (6)	eg Fg	eg (F)

An example for clay brickwork in walls is: (21) Fg2, which for trade literature, would be shown in a reference box as:

CI/SfB 1976 reference by SfB Agency
(21) Fg2

The lower space is intended for UDC (Universal decimal classification) codes — see BS 1000A 1961. Advice in classification can be obtained from the SfB Agency UK Ltd at 66 Portland Place, London W1N 4AD.

In the following summaries of the five tables, chapter references are made to the six related volumes and chapters of *Mitchell's Building Series* in which aspects of the classifications are dealt with. The following abbreviations are used:

Environment and Services	*ES*
Materials	*M*
Structure and Fabric, Part 1	*SF (1)*
Structure and Fabric, Part 2	*SF (2)*
Components	*C*
Finishes	*F*

Table 0 **Physical Environment**
(main headings only)

Scope: End results of the construction process

0	Planning areas
1	Utilities, civil engineering facilities
2	Industrial facilities
3	Administrative, commercial, protective service facilities
4	Health, welfare facilities
5	Recreational facilities
6	Religious facilities
7	Educational, scientific, information facilities
8	Residential facilities
9	Common facilities, other facilities

Table 1 **Elements**

Scope: Parts with particular functions which combine to make the facilities in table 0

(0-) **Sites, projects**
Building plus external works
Building systems *C* 11

(1-) **Ground, substructure**
(11) Ground *SF (1)* 4, 8, 11; *SF (2)* 2, 3, 11
(12) Vacant
(13) Floor beds *SF (1)* 4, 8; *SF (2)* 3
(14), (15) Vacant
(16) Retaining walls, foundations *SF (1)* 4; *SF (2)* 3, 4
(17) Pile foundations *SF (1)* 4; *SF (2)* 3, 11
(18) Other substructure elements
(19) Parts, accessories, cost summary, etc

(2-) **Structure, primary elements, carcass**
(21) Walls, external walls *SF (1)* 1, 5; *SF (2)* 4, 5, 10
(22) Internal walls, partitions *SF (1)* 5; *SF (2)* 4, 10; *C* 9
(23) Floors, galleries *SF (1)* 8; *SF (2)* 6, 10
(24) Stairs, ramps *SF (1)* 10; *SF (2)* 8, 10
(25), (26) Vacant
(27) Roofs *SF (1)* 1, 7; *SF (2)* 9, 10
(28) Building frames, other primary elements *SF (1)* 1, 6; *SF (2)* 5, 10
Chimneys *SF (1)* 9
(29) Parts, accessories, cost summary, etc

(3-) **Secondary elements, completion of structure**
(31) Secondary elements to external walls, including windows, doors *SF (1)* 5; *SF (2)* 10; *C* 3, 4, 5, 7
(32) Secondary elements to internal walls, partitions including borrowed lights and doors *SF (2)* 10; *C* 3, 7
(33) Secondary elements to floors *SF (2)* 10
(34) Secondary elements to stairs including balustrades *C* 8
(35) Suspended ceilings *C* 10
(36) Vacant
(37) Secondary elements to roofs, including roof lights, dormers *SF (1)* 7; *SF (2)* 10; *C* 6
(38) Other secondary elements
(39) Parts, accessories, cost summary, etc.

(4-) **Finishes to structure**
(41) Wall finishes, external *SF (2)* 4, 10; *F* 3, 4, 5
(42) Wall finishes, internal *F* 2, 4, 5
(43) Floor finishes *F* 1
(44) Stair finishes *F* 1
(45) Ceiling finishes *F* 2
(46) Vacant
(47) Roof finishes *SF (2)* 10; *F* 7
(48) Other finishes
(49) Parts, accessories, cost summary, etc

(5-) **Services** (mainly piped and ducted)
(51) Vacant
(52) Waste disposal, drainage *ES* 13/*ES* 11, 12
(53) Liquids supply *ES* 9, 10; *SF (1)* 9; *SF (2)* 6, 10
(54) Gases supply
(55) Space cooling
(56) Space heating *ES* 7; *SF (1)* 9; *SF (2)* 6, 10
(57) Air conditioning, ventilation *ES* 7; *SF (2)* 10
(58) Other piped, ducted services
(59) Parts, accessories, cost summary, etc
Chimney, shafts, flues, ducts independent *SF (2)* 7

(6-) **Services** (mainly electrical)
(61) Electrical supply
(62) Power *ES* 14
(63) Lighting *ES* 8
(64) Communications *ES* 14
(65) Vacant
(66) Transport *ES* 15
(67) Vacant
(68) Security, control, other services
(69) Parts, accessories, cost summary, etc

(7-) **Fittings** with subdivisions (71) to (79)
(74) Sanitary, hygiene fittings *ES* 10

(8-) **Loose furniture, equipment** with subdivisions (81) to (89)
Used where the distinction between loose and fixed fittings, furniture and equipment is important.

(9-) **External elements, other elements**
(90) External works, with subdivisions (90.1) to (90.8)
(98) Other elements
(99) Parts, accessories etc. common to two or more main element divisions (1-) to (7-)
Cost summary

Note: The SfB Agency UK do not use table 1 in classifying manufacturers' literature

Table 2 **Constructions, Forms**

Scope: Parts of particular forms which combine to make the elements in table 1. Each is characterised by the main product of which it is made.

A Constructions, forms — used in specification applications for Preliminaries and General conditions
B Vacant — used in specification applications for demolition, underpinning and shoring work
C Excavation and loose fill work
D Vacant
E Cast *in situ* work *M* 8; *SF (1)* 4, 7, 8; *SF (2)* 3, 4, 5, 6, 8, 9

Blocks
F Blockwork, brickwork
Blocks, bricks *M* 6, 12; *SF (1)* 5, 9 *SF (2)* 4, 6, 7
G Large block, panel work
Large blocks, panels *SF (2)* 4

Sections
H Section work
Sections *M*9; *SF (1)* 5, 6, 7, 8; *SF (2)* 5, 6
I Pipework
Pipes *SF (1)* 9; *SF (2)* 7
J Wire work, mesh work
Wires, meshes
K Quilt work
Quilts
L Flexible sheet work (proofing)
Flexible sheets (proofing) *M*9, 11
M Malleable sheet work
Malleable sheets *M* 9
N Rigid sheet overlap work
Rigid sheets for overlappings *SF (2)* 4; *F* 7
P Thick coating work *M* 10, 11; *SF (2)* 4; *F* 1, 2, 3, 7
Q Vacant
R Rigid sheet work
Rigid sheets *M* 3, 12, 13; *SF (2)* 4; *C* 5
S Rigid tile work
Rigid tiles *M* 4, 12, 13; *F* 1, 4
T Flexible sheet and tile work
Flexible sheets eg carpets, veneers, papers, tiles cut from them *M* 3, 9; *F* 1, 6
U Vacant
V Film coating and impregnation work *F* 6; *M* 2
W Planting work
Plants
X Work with components
Components *SF (1)* 5, 6, 7, 8, 10; *SF (2)* 4; *C* 2, 3, 4, 5, 6, 7, 8
Y Formless work
Products
Z Joints, where described separately

Table 3 **Materials**

Scope: Materials which combine to form the products in table 2

a **Materials**
b, c, d Vacant

Formed materials e to o
e **Natural stone** *M* 4; *SF (1)* 5, 10; *SF (2)* 4
e1 Granite, basalt, other igneous
e2 Marble
e3 Limestone (other than marble)

e4 Sandstone, gritstone
e5 Slate
e9 Other natural stone

f **Precast with binder** *M* 8; *SF (1)* 5, 7, 8, 9, 10; *SF (2)* 4 to 9; *F* 1
f1 Sandlime concrete (precast)
Glass fibre reinforced calcium silicate (gres)
f2 All-in aggregate concrete (precast) *M* 8
Heavy concrete (precast) *M* 8
Glass fibre reinforced cement (gre) *M* 10
f3 Terrazzo (precast) *F* 1
Granolithic (precast)
Cast/artificial/reconstructed stone
f4 Lightweight cellular concrete (precast) *M* 8
f5 Lightweight aggregate concrete (precast) *M* 8
f6 Asbestos based materials (preformed) *M* 10
f7 Gypsum (preformed) *C* 2
Glass fibre reinforced gypsum *M* 10
f8 Magnesia materials (preformed)
f9 Other materials precast with binder

g **Clay (Dried, Fired)** *M* 5; *SF (1)* 5, 9, 10; *SF (2)* 4, 6, 7
g1 Dried clay eg pisé de terre
g2 Fired clay, vitrified clay, ceramics
Unglazed fired clay eg terra cotta
g3 Glazed fired clay eg vitreous china
g6 Refractory materials eg fireclay
g9 Other dried or fired clays

h **Metal** *M* 9; *SF (1)* 6, 7, *SF (2)* 4, 5, 7
h1 Cast iron
Wrought iron, malleable iron
h2 Steel, mild steel
h3 Steel alloys eg stainless steel
h4 Aluminium, aluminium alloys
h5 Copper
h6 Copper alloys
h7 Zinc
h8 Lead, white metal
h9 Chromium, nickel, gold, other metals, metal alloys

i **Wood** including wood laminates *M* 2, 3; *SF (1)* 5 to 8, 10; *SF (2)* 4, 9; *C* 2
i1 timber (unwrot)
i2 Softwood (in general, and wrot)
i3 Hardwood (in general, and wrot)
i4 Wood laminates eg plywood
i5 Wood veneers
i9 Other wood materials, except wood fibre boards, chipboards and wood-wool cement

j **Vegetable and animal materials** — including fibres and particles and materials made from these
j1 Wood fibres eg building board *M* 3
j2 Paper *M* 9, 13
j3 Vegetable fibres other than wood eg flaxboard *M* 3
j5 Bark, cork
j6 Animal fibres eg hair
j7 Wood particles eg chipboard *M* 3
j8 Wood-wool cement *M* 3
j9 Other vegetable and animal materials

k, l Vacant

m **Inorganic fibres**
m1 Mineral wool fibres *M* 10; *SF (2)* 4, 7
Glass wool fibres *M* 10, 12
Ceramic wool fibres
m2 Asbestos wool fibres *M* 10
m9 Other inorganic fibrous materials eg carbon fibres *M* 10

n **Rubber, plastics, etc**
n1 Asphalt (preformed) *M* 11; *F* 1
n2 Impregnated fibre and felt eg bituminous felt *M* 11; *F* 7
n4 Linoleum *F* 1

Synthetic resins n5, n6
n5 Rubbers (elastomers) *M* 13
n6 Plastics, including synthetic fibres *M* 13
Thermoplastics
Thermosets
n7 Cellular plastics
n8 Reinforced plastics eg grp, plastics laminates

o **Glass** *M* 12; *SF (1)* 5; *C* 5
o1 Clear, transparent, plain glass
o2 Translucent glass
o3 Opaque, opal glass
o4 Wired glass
o5 Multiple glazing
o6 Heat absorbing/rejecting glass
X-ray absorbing/rejecting glass
Solar control glass
o7 Mirrored glass, 'one-way' glass
Anti-glare glass
o8 Safety glass, toughened glass
Laminated glass, security glass, alarm glass
o9 Other glass, including, cellular glass

Formless materials p to s

p **Aggregates, loose fills** *M* 8
p1 Natural fills, aggregates
p2 Artificial aggregates in general
p3 Artificial granular aggregates (light) eg foamed blast furnace slag
p4 Ash eg pulverised fuel ash
p5 Shavings
p6 Powder
p7 Fibres
p9 Other aggregates, loose fills

q **Lime and cement binders, mortars, concretes**
q1 Lime (calcined limestones), hydrated lime, lime putty, *M* 7
Lime-sand mix (coarse stuff)
q2 Cement, hydraulic cement eg Portland cement *M* 7
q3 Lime-cement binders *M* 15
q4 Lime-cement-aggregate mixes
Mortars (ie with fine aggregates) *M* 15; *SF (2)* 4
Concretes (ie with fine and/or coarse aggregates) *M* 8
q5 Terrazzo mixes and in general *F* 1
Granolithic mixes and in general *F* 1
q6 Lightweight, cellular, concrete mixes and in general *M* 8
q9 Other lime-cement-aggregate mixes eg asbestos cement mixes *M* 10

r **Clay, gypsum, magnesia and plastics binders, mortars**
r1 Clay mortar mixes, refractory mortar
r2 Gypsum, gypsum plaster mixes
r3 Magnesia, magnesia mixes *F* 1
r4 Plastics binders
Plastics mortar mixes
r9 Other binders and mortar mixes

s **Bituminous materials** *M* 11; *SF (2)* 4
s1 Bitumen including natural and petroleum bitumens, tar, pitch, asphalt, lake asphalt
s4 Mastic asphalt (fine or no aggregate), pitch mastic
s5 Clay-bitumen mixes, stone bitumen mixes (coarse aggregate)
Rolled asphalt, macadams
s9 Other bituminous materials

Functional materials t to w

t **Fixing and jointing materials**
t1 Welding materials *M* 9; *SF (2)* 5
t2 Soldering materials *M* 9
t3 Adhesives, bonding materials *M* 14
t4 Joint fillers eg mastics, gaskets *M* 16 *SF (1)* 2
t6 Fasteners, 'builders ironmongery'
Anchoring devices eg plugs
Attachment devices eg connectors *SF (1)* 6, 7
Fixing devices eg bolts, *SF (1)* 5
t7 'Architectural ironmongery' *C* 7
t9 Other fixing and jointing agents

u **Protective and Process/property modifying materials**
u1 Anti-corrosive materials, treatments *F* 6
Metallic coatings applied by eg electro-plating *M* 9
Non-metallic coatings applied by eg chemical conversion
u2 Modifying agents, admixtures eg curing agents *M* 8
Workability aids *M* 8
u3 Materials resisting specials forms of attack such as fungus, insects, condensation *M* 2
u4 Flame retardants if described separately *M* 1

u5 Polishes, seals, surface hardeners *F* 1; *M* 8
u6 Water repellants, if described separately
u9 Other protective and process/property modifying agents eg ultra-violet absorbers

v **Paints** *F* 6
v1 Stopping, fillers, knotting, paint preparation materials including primers
v2 Pigments, dyes, stains
v3 Binders, media eg drying oils
v4 Varnishes, lacquers eg resins
Enamels, glazes
v5 Oil paints, oil-resin paints
Synthetic resin paints
Complete systems including primers
v6 Emulsion paints, where described separately
Synthetic resin-based emulsions
Complete systems including primers
v8 Water paints eg cement paints
v9 Other paints eg metallic paints, paints with aggregates

w **Ancillary materials**
w1 Rust removing agents
w3 Fuels
w4 Water
w5 Acids, alkalis
w6 Fertilisers
w7 Cleaning materials *F* 1
Abrasives
w8 Explosives
w9 Other ancillary materials eg fungicides

x **Vacant**

y **Composite materials**
Composite materials generally *M* 11
See p. 63 *Construction Indexing Manual*

z **Substances**
z1 By state eg fluids
z2 By chemical composition eg organic
z3 By origin eg naturally occurring or manufactured materials
z9 Other substances

Table 4 **Activities, Requirements**
(main headings only)

Scope: Table 4 identifies objects which assist or affect construction but are not incorporated in it, and factors such as activities, requirements, properties, and processes.

Activities, aids
(A) Administration and management activities, aids *C* 11; *M* Introduction, *SF (1)* 2; *SF (2)* 1, 2, 3
(B) Construction plant, tools *SF (1) 2; SF (2)* 2, 11
(C) Vacant
(D) Construction operations *SF (1)* 2, 11; *SF (2)* 2, 11

Requirements, properties, building science, construction technology
Factors describing buildings, elements, materials, etc
(E) Composition, etc *SF (1)* 1, 2; *SF (2)* 1, 2
(F) Shape, size, etc *SF (1)* 2
(G) Appearance, etc *M* 1; *F* 6

Factors relating to surroundings, occupancy
(H) Context, environment

Performance factors
(J) Mechanics *M* 9; *SF (1)* 3, 4; *SF (2)* 3, 4
(K) Fire, explosion *M* 1; *SF (2)* 10
(L) Matter
(M) Heat, cold *ES* 1
(N) Light, dark *ES* 1
(O) Sound, quiet *ES* 1
(Q) Electricity, magnetism, radiation *ES* 14
(R) Energy, other physical factor *ES* 7
(T) Application

Other factors

(U) Users, resouces
(V) Working factors
(W) Operation, maintenance factors
(X) Change, movement, stability factors
(Y) Economic, commercial factors *M* Introduction; *SF (1)* 2; *SF (2)* 3, 4, 5, 6, 9

(Z) Peripheral subjects, form of presentation, time, place — may be used for subjects taken from the UDC (*Universal decimal classification*), see BS1000A 1961

Subdivision: All table 4 codes are subdivided mainly by numbers

Index

n indicates footnotes